Beginner's Guide to SolidWorks 2014 – Level II

Sheet Metal, Top Down Design, Weldments, Surfacing and Molds

Alejandro Reyes, MSME
Certified SolidWorks Professional

Publications

SDC Publications
P.O. Box 1334
Mission, KS 66222
913-262-2664
www.SDCpublications.com
Publisher: Stephen Schroff

ISBN-13: 978-1-58503-842-8
ISBN-10: 1-58503-842-3

Printed and bound in the United States of America.

Acknowledgements

Beginner's Guide to SolidWorks 2014 – Level II is dedicated to the love of my life, my wife Patricia and my kids Liz, Ale and Hector, who have given me the time, support and patience to write this book. To you, all my love.

Also, I wish to thank the hundreds of students, users, professors and engineers whose great ideas and words of encouragement helped me to improve this book.

About the Author

Alejandro Reyes holds a BSME from the Instituto Tecnológico de Ciudad Juárez, Mexico in electro-mechanical engineering and a Master's Degree from the University of Texas at El Paso in mechanical design, with strong focus in Materials Science and Finite Element Analysis.

Alejandro spent more than 8 years as a SolidWorks Value Added Reseller. During this time he was a Certified SolidWorks Instructor and Support Technician, CosmosWorks Support Technician, and a Certified SolidWorks Professional, credentials that he still maintains. Alejandro has over 19 years of experience using CAD/CAM/FEA software and is currently the President of MechaniCAD Inc.

His professional interests include finding alternatives and improvements to existing products, FEA analyses and new technologies. On a personal level, he enjoys spending time with his family and friends.

Notes:

Table of Contents

List of commands introduced in each chapter. Note that many commands are used extensively in following chapters after they have been learned.

Multi Body Parts
Multi Body
Local Operations
Hide/Show Body
Merge
Feature Scope
Delete Body
Body Pattern
Combine Bodies
 Add
 Subtract
 Common
Bodies to keep
Offset from Surface
Sketch Picture

Contour Selection
Regions available
Contour Selection
Shared Sketch
Start From: Condition

Part Editing
What's Wrong
Parent/Child relations
Sketch Editing
Dangling Relations
Delete Absorbed
 Features
Sketch Relations
Over Defined Sketch
Not Solved Sketch
SketchXpert
View Sketch Relations

Equations
Rename Dimensions
Pattern Seed Only
Add Equations
Edit Equations
Delete Equations
Link Values

Top Down Design
New Part
Edit In Context
Assembly Transparency
Internal Part
Externalize Part
Edit Assembly
External References
 In Context
 Out of Context
 Locked
 Broken
List External References

Sheet Metal and Top Down Design
Base Flange
Sheet Metal Thickness
Bend Radius
Bend Allowance
Bend Deduction
K-Factor
Auto-Relief
 Rectangular
 Obround
 Tear
Flat Pattern
Forming Tools
Modify Sketch
Link to Thickness
Normal Cut
3D Content Central
Vent feature
Miter Flange
Unfold/Fold Bend
Edge Flange
Build Library Features
Library Parts
Mate Reference
Break Corners
Jog bend
Flat Pattern Drawing
Bend Notes

Convert to Sheet Metal
Closed Corners
Selection Filters
Sketch Pattern
Feature Driven Pattern
Hem Feature
Creating Forming Tools
Component Pattern
Collision Detection
Flexible Sub Assemblies
Assembly Features

3D Sketch
3D Sketch
3D Sketch Relations
Derived Sketch
Projected Curve

Weldments
3D Sketch review
Cut list
Weldment feature
Structural Member
Corner Treatment
 End Miter
 End Butt
Locate Profile
Rotate Profile
Trim/Extend
Gusset
End Cap
Weld Beads
Weldment Cut List
Weldment Drawings
Cut List Table
Weld Table
Weld Symbols
Save Bodies to
 Assembly
Structural Member
 Libraries

Surfacing	Mold Tools
Revolved Surface	Draft Analysis
Lofted Surface	Direction of Pull
Extruded Surface	Positive Draft
Direction of Extrusion	Negative Draft
Extrude with Draft	Draft
Trim Surface	Neutral Plane
Mutual Trim	Rollback/Roll Forward
Planar Surface	Scale
Filled Surface	Parting Line
Knit Surface	Parting Surface
Constant Width Fillet	Tooling Split
Thicken	Composite Curve
Body Split	Swept Surface
Face Fillet	Shut Off Surfaces
Extend Surface	Move/Copy body
Extrude From	Delete Face
Mirror Bodies	Face Classification
Swept Surface	Manual Parting Line
Twist Along Path	Selection
Sweep Cut	Select Open Loop
	Rib
	Side Core

Notes:

Introduction

Beginner's Guide to SolidWorks – Level II starts where *Beginner's Guide – Level I* ends, following the same easy to read style, but this time covering advanced topics and techniques.

The purpose of this book is to teach advanced SolidWorks functionality including sheet metal, surfacing, how to create components in the context of an assembly referencing other components (Top-down design), propagate design changes with SolidWorks' parametric capabilities, mold design, weldments, and more while explaining the basic concepts of each trade to allow the reader to understand the how and why of each operation.

The book will use simple examples to allow the reader to better understand commands and environments, as well as to make it easier to explain the purpose of each step, maximizing the learning time by focusing on one task at a time. Keep in mind that this book is focused on learning SolidWorks, and the specific details of some of the topics that will be covered go far beyond the scope of this book; entire books have been written about sheet metal processes, mold design, welded structures, and such. With that in mind, please remember that this book will teach you how to use SolidWorks' tools for those trades, and at no time will attempt to teach the trade itself, and will only explain the general details in order to understand the processes.

At the end of this book, the reader will have acquired enough skills to be competitive when it comes to designing with SolidWorks, and while there are many less frequently used commands and options available that will not be covered, rest assured that those commands that are covered are most of the commands used every day by SolidWorks designers.

IMPORTANT: Files required to complete some of the exercises are included in the accompanying disk, or available for download from www.mechanicad.com/download or www.sdcpublications.com. The files include high resolution images of the exercises and the finished files made throughout the book to help you practice and enhance your skills. We hope you learn many new things and enjoy reading this book as much as we enjoyed making it. We love to hear ideas and comments from our readers. If you have any, please share them with us and our readers; we'll try our best to accommodate any suggestions to improve and expand the content of this book, and while we cannot guarantee that they will make it into the next edition, we can assure you that we reply to all email messages.

alejandro@mechanicad.com

Prerequisites
This book has been written assuming the reader has knowledge of the following topics:
- Familiarity with the Windows operating system;
- General knowledge of mechanical design and drafting;
- Previous experience with basic SolidWorks functionality.
- A general understanding of sheet metal, structural and molding processes will help to better understand these topics.

Note about the screen images: The images in your screen *may* be slightly different from this book. The images in the book were made using Windows 7 Professional and SolidWorks' default installation settings. Unless otherwise noted, the only changes made to SolidWorks' default options were adding a white background and in some instances the preview colors were changed to improve clarity in print and/or electronic format.

IMPORTANT INFORMATION ABOUT FILES: High resolution images of the exercises in this book and the files needed for some exercises are included in the accompanying disc or can be downloaded from our website. The files may be included in a ZIP file. Extract all the files to a local folder in your PC to have them ready when needed.

Multi Body Parts, Sketch Editing and Other Tools

Multi body parts

SolidWorks offers a number of tools to help us model components easier and faster with powerful options like multi body parts and contour selection. Multi body parts means that we can have more than one "solid" body in a part. Up until now we have been modeling single body parts, meaning that each feature either added or removed material to the part, but always resulted in a single solid body. Designing with multi body parts allows us to do things that sometimes are very difficult or even impossible using a single body component, like, among other things, combine bodies, remove one body from another or calculate their common volume. Multi body operations are particularly useful when working with mold design and is the essence of the weldments environment (both are covered later in this book).

Contour Selection

Besides working with single body parts, we have also worked with single and multiple contour open or closed sketches. In this section we'll cover how to use Contour selection to work with sketches that contain multiple entities sharing an endpoint and/or having self-intersecting lines. We'll learn how to take advantage of these situations and make the best of them.

Part Editing

Another topic that is very important to modeling in SolidWorks is editing parts to fix errors or make modifications to components. After all, when we are designing, at some point we will most likely have to go back and make changes to our design, and sometimes those changes can cause trouble in subsequent features. In the Editing section we'll cover different options available when editing parts, as well as fixing and recovering from errors.

Equations

One last topic we'll cover in this section deals with maintaining design intent; we'll learn how to add, edit and delete equations in a part, giving us powerful tools to make our designs more flexible, robust and predictable when making design changes.

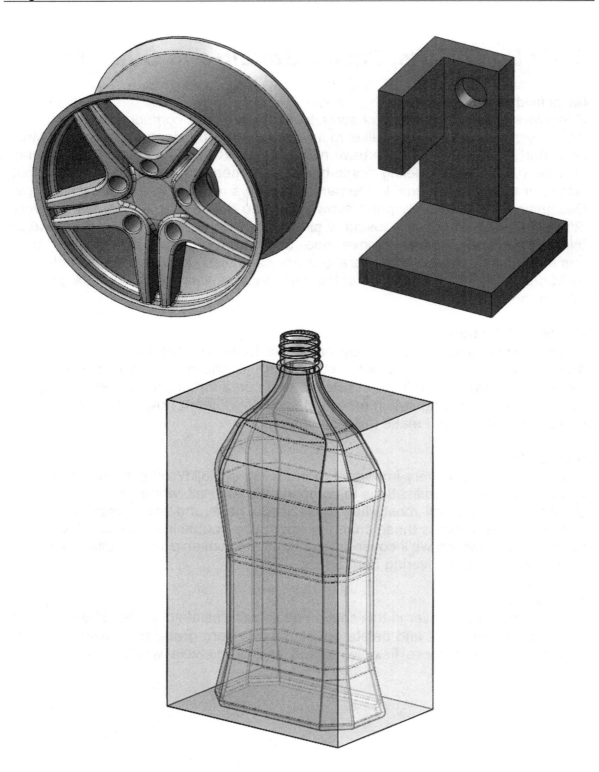

Multi Body Parts

As we stated earlier, multi body parts allow us to do certain operations that would be difficult to accomplish with a single body part. First, we'll talk about how to make multi body parts and after that how to use them

1. – Multi body parts are made by adding material that is not connected to the current solid body, in other words, that is purposely <u>not</u> merged to it. A different way to create a multi body part is by splitting an existing part into multiple pieces (bodies). To show how it works, make a new part, and add a sketch to the *"Front Plane"* as shown. Do not worry about dimensions in this example as we are just illustrating the concept of how multi bodies work.

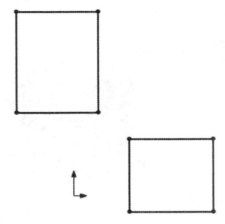

2. – Select the "**Boss Extrude**" command. Notice that we are not given a warning about having two separate bodies; it just works. Extrude any size that looks similar to the following image (dimensions are not important at this time) and click **OK** to finish. (Making two or more extruded/revolved/swept/lofted features would work just the same.)

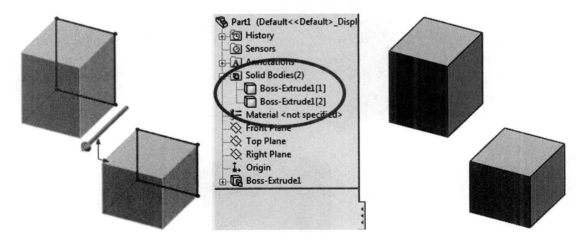

5

The first thing we notice is a new folder in the FeatureManager called "*Solid Bodies (2)*." This folder is automatically added when SolidWorks detects multiple disjointed bodies in a part and lists the number of bodies found in the part (in this case 2). If we expand the folder, we can see the two bodies in our part listed under it. The important thing to know and remember is that **multi body parts are not to be confused or used in place of an assembly**; parts and assemblies have significant differences and each serves a different purpose. A multi body part is used mostly as a means to an end.

3. – The next step is to add a new feature. When working in a multi body part we can make 'local' operations, for example, a shell feature affecting only one body. Applied features like fillets, shells, chamfers, etc. can only be applied to a single body at a time. Select the "Shell" command from the Features toolbar, and shell the bottom body removing the indicated faces.

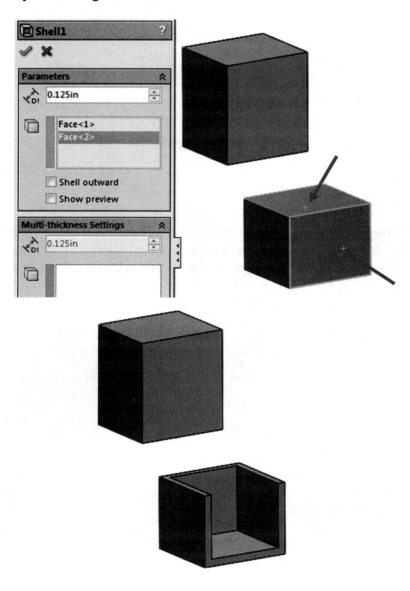

4. – Select the **Fillet** command and round two corners of the top body as shown.

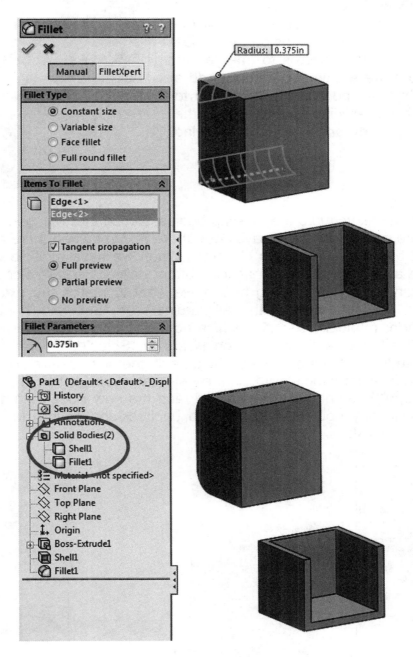

To help the reader understand the concept of local operations better, "local" means that we can add applied features (features that don't require a sketch like Shell, Fillet, Chamfer, Draft, etc.) to one body at a time. Notice the name of a body changes to the last feature applied to it.

5. – If we add another boss extrude feature that intersects with existing bodies, by default it will automatically merge with them. At the time of making the extrusion we can optionally select which bodies to "merge" (or fuse) with to make a new single body. Add a new Boss-Extrude and explore the option to merge it with the existing bodies. Select the front face of a body and make the following sketch in it.

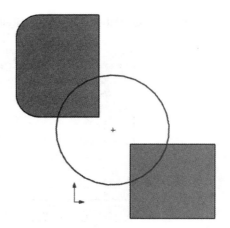

6. – Extrude the sketch *into* the existing bodies. Notice the "Merge result" option in the Extrude command. It's always been there (except when there are no solid bodies) and by default is always checked. When we have multiple bodies in a part we see an additional option at the bottom called "Feature Scope." This is where we can select the existing bodies we want to merge the extrusion with. The options are to merge with "All bodies," or "Selected bodies," either automatically or manually selecting which bodies to affect with the new feature. By default the "Feature Scope" option is set to "Selected bodies" and "Auto-select." These two options mean that, by default, the new feature will merge with any solid body it intersects with. If the "Merge result" option is un-checked, "Feature Scope" is automatically removed, and the new Boss-Extrude will does not merge with any bodies. For this step uncheck the option "Merge Result" and click OK to finish.

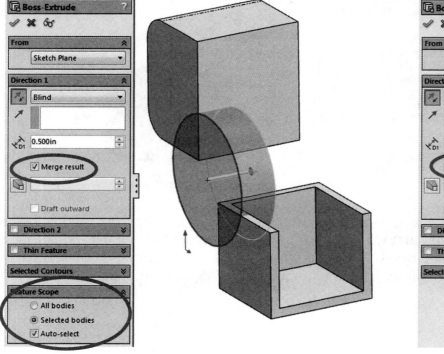

7. – After adding this extrusion we end up with three bodies in our part. See how the different bodies' edges intersect each other. If the "Merge result" option had been checked, we would not see these edges overlapping as they would have merged into a single body.

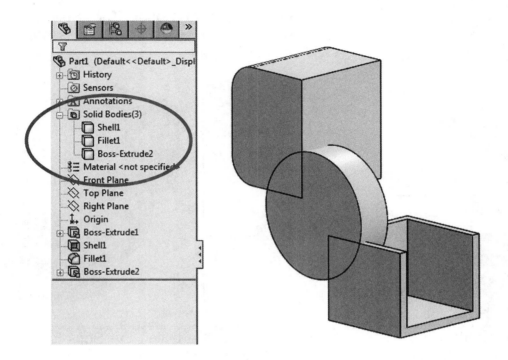

Edit the last extrusion to explore the effect of different "Merge Result" and "Feature Scope" combinations.

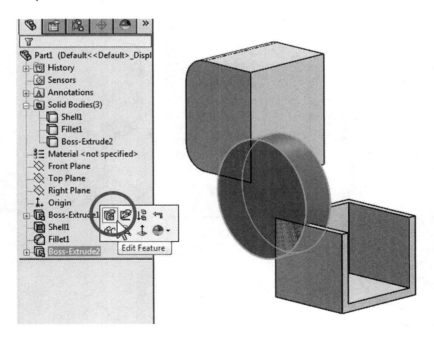

Combination:	Result:
Merge Result: Checked **Feature Scope**: Auto-Select 	Single body. By touching both of the existing bodies "Auto-select" merges with them, fusing them into a single solid body. The "Solid Bodies" folder is no longer visible as there is only one solid body in the part. 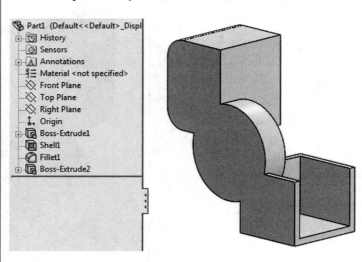
Merge result: Checked **Auto-select**: Unchecked Add shell body to selection box 	Two bodies. We are telling the Boss-Extrude to merge only to the Shell body (or whichever we pick). Notice the edges of the upper body and the Boss-Extrude overlap, as they are different bodies. 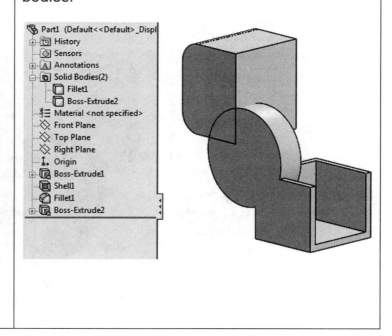

8. –The "Merge result" option works the same way with any feature that adds material to the part, including revolved boss, sweep, loft, etc. Now we'll see how it works when we remove material. Delete the Boss-Extrude feature and keep the sketch. Select the sketch in the FeatureManager and make a Cut-Extrude "Through All." In this case the "Merge Result" is gone and we only have the **"Feature Scope"** option at the bottom with the same selection options: "All bodies" or "Selected Bodies" with/without "Auto-select." Leave the "Auto-select" option on and click **OK** to finish.

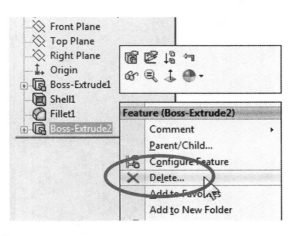

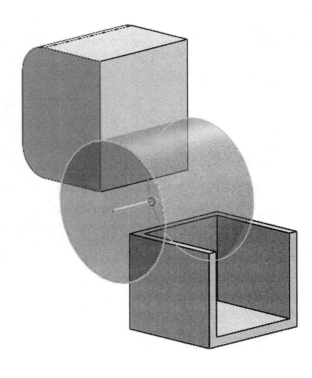

9. – What we end up with is the same two bodies we had before, but now they have a cut through them because the cut automatically selected them.

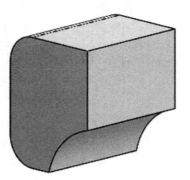

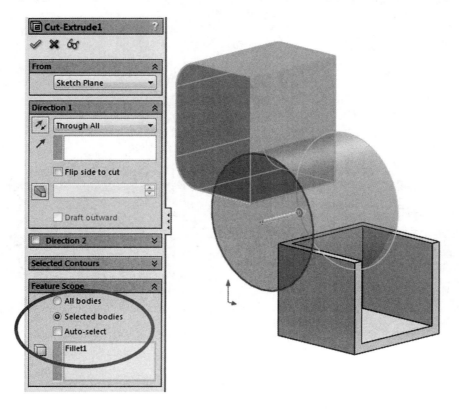

10. – Edit the "*Cut-Extrude1*" definition, turn off the "Auto-select" option in the "**Feature Scope**," and select only the top body. Click **OK** to finish. Now we are only modifying the top solid, even when the cut overlaps the lower body.

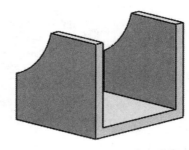

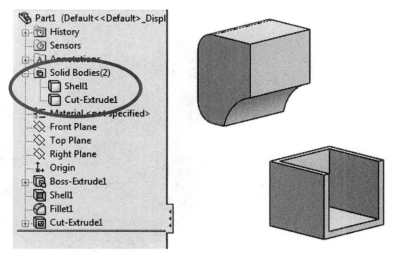

And just as with the features that add material, this technique works the same way with all the features that remove material including revolved cut, sweep cut, loft cut, etc.

11. – Multi bodies is a powerful technique to model parts that would otherwise be difficult to complete. A frequently used technique is called "bridging"; this means to connect two or more bodies by adding material between them to merge them into a single solid body. Reasons to use this technique may include a model where we know what opposite sides/ends of a part look like, but we may not know what the middle (the "*bridge*") should be like. In this example we'll make a car's wheel using multi-body techniques to complete it, including common volume between two bodies and adding volumes. For our example we'll assume that we need to design a car's wheel. We know what the actual tire and hub dimensions should be, but we don't know yet what the spokes will look like; we just know it has to look great ☺.

We'll assume the dimensions for the wheel are as shown in the following sketches. Make a new part; the first feature for the wheel will be the hub or mounting pad. Draw the following sketch in the *"Right Plane"* of the new part and make a 360° Boss-Revolve feature. Our wheel's dimensions will be in millimeters, so be sure to change your model's dimensions accordingly (**Tools, Options, Document Options, Units,** or in the status bar's **Unit System**). Pay attention to the diameter dimensions (doubled about the horizontal centerline).

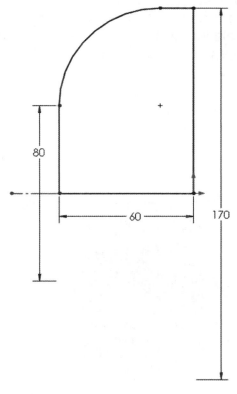

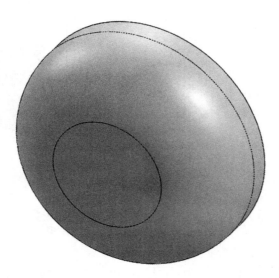

12. – Add a new sketch in the front flat face of the part, and add the following sketch. The arcs are tangent to the circular edges. Make the extrusion 20mm to the front (Direction 1) and "Through All" going to the back (Direction 2.) Turn off the "Merge result" option to have two different bodies.

To make the sketch symmetric about the centerline, draw the centerline first, and while it's selected, turn on the "**Dynamic Mirror**" tool in the menu "**Tools, Sketch Tools, Dynamic Mirror**." After we turn it on the centerline will show an equal sign at both ends letting us know that whatever we draw on one side will be automatically drawn in the other side. At the same time, symmetric geometric relations will be automatically added. To turn it off, select it again in the menu.

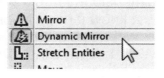

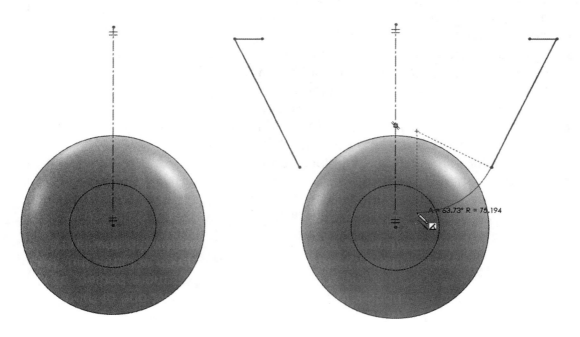

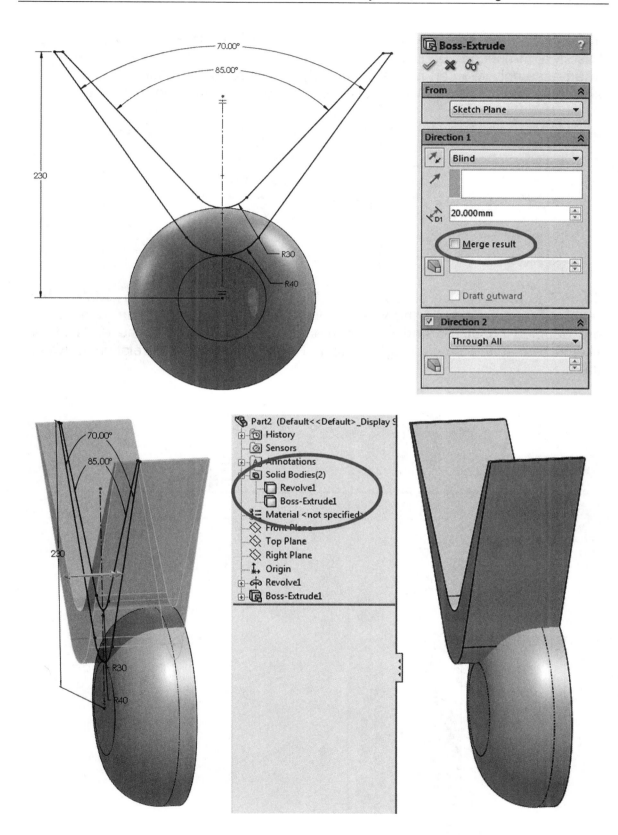

13. – In the next step we'll make the outside rim of the wheel. To better visualize this step, the second body will be hidden from view. Just like we can hide a part in an assembly, we can hide a solid body in a part. In the *"Solid Bodies"* folder, select the *"Boss-Extrude1"* body and select "Hide" from the pop-up toolbar.

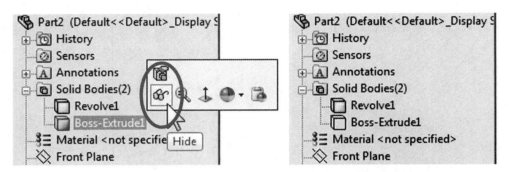

14. – Add the following sketch in the *"Right Plane"*; we'll use it to create a thin revolved feature to build the wheel's rim. The 360mm and 380mm dimensions are doubled about the horizontal centerline. The endpoints at the ends of the sketch are at the same height; add a Horizontal geometric relation between them.

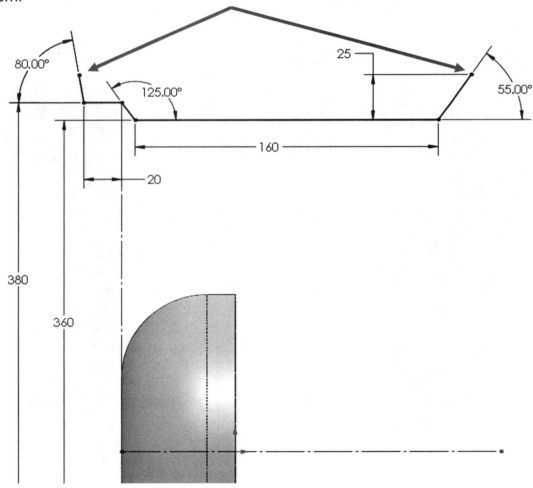

15. – Select the "**Revolved Boss/Base**" command and make a 360 deg. thin revolve. We'll be asked if we want to close the sketch, select "No" to continue. Use the horizontal centerline as the "Axis of Revolution", turn off the "Merge result" checkbox and make the "Thin Feature" 5mm thick going outside.

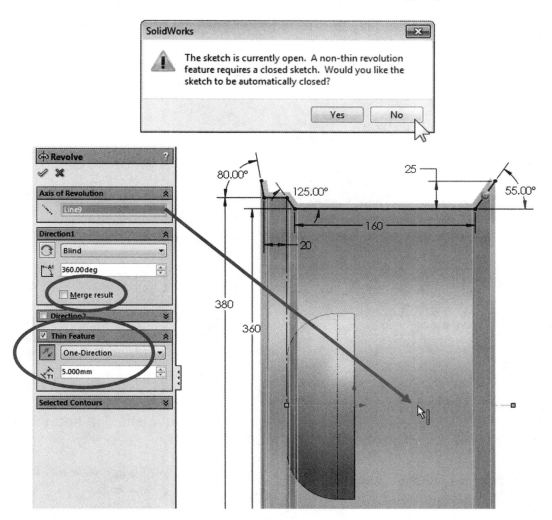

16. – Now we have three different solid bodies in our part. The next feature will be another revolved feature that will be used to build the wheel's spokes. Before making the new solid body, hide the *"Revolve-Thin1"* body and show the *"Boss-Extrude1"* body.

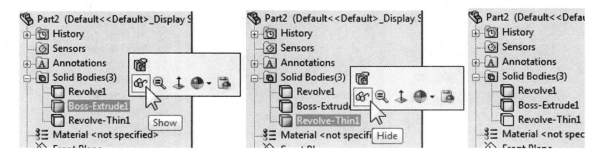

17. – After hiding the revolved thin feature, add a new sketch in the *"Right Plane."* Note: The arc on the left is tangent to the solid body's edge, and the arc on the right is tangent to a vertical centerline. Don't forget the horizontal centerline at the origin to make the revolved feature about it. We can use a **"Three Point Arc"** for this sketch. The part is shown in both shaded and wireframe modes for clarity.

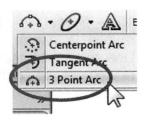

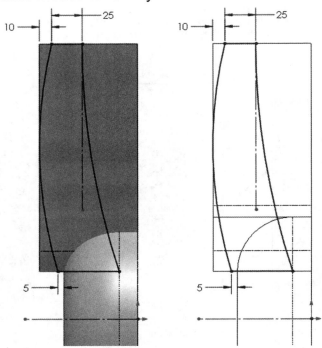

18. – Make a 360 deg revolved boss using the horizontal centerline and turn off the "Merge result" checkbox to create the fourth solid body.

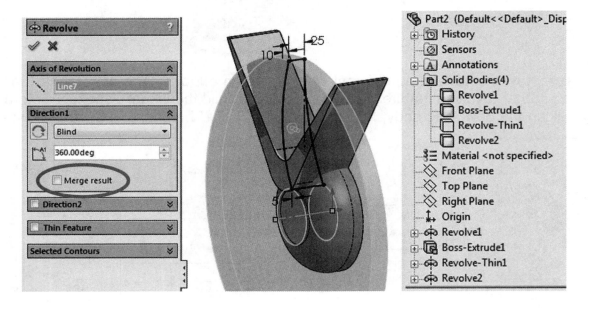

19. – In the next step we will obtain the common volume between two of the bodies to get the spokes of the wheel. Select the *"Boss-Extrude1"* and the *"Revolve2"* bodies in the *"Solid Bodies"* folder holding the "Ctrl" key while selecting, and from the right-mouse-button menu select the command **"Combine"**, or from the menu **"Insert, Features, Combine."**

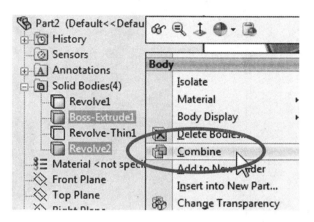

In the "Operation Type" options select "Common" and then press the "Show Preview" button (it will change to "Hide Preview"). The **"Combine"** command is used to Add bodies to form a new one, Subtract one or more bodies from another, or get the Common volume between two or more bodies, as is the case in this step. After clicking OK to calculate the common volume, the "Bodies to Combine" are consumed by the operation and replaced by the new body.

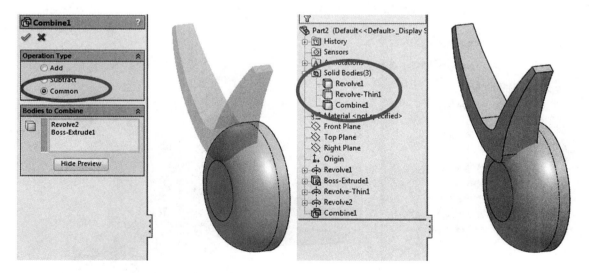

20. – Now that we have the body we are interested in to create the spokes, we need to make a circular pattern using the combined body as a seed feature. Select the "Circular Pattern" command using any circular edge for the "Pattern Axis." For this step, instead of selecting the *"Combine1"* feature to make our pattern, expand the "Bodies to Pattern" selection box and add the combined body to make the pattern. Change the number of instances to 5 equally spaced in 360 degrees. Click OK to continue.

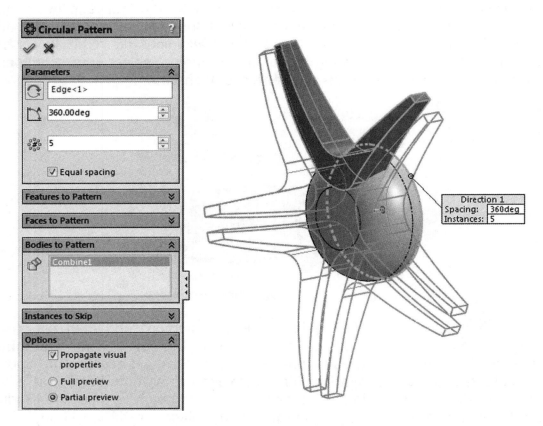

21. – After adding the circular pattern we have 7 different bodies in our wheel.

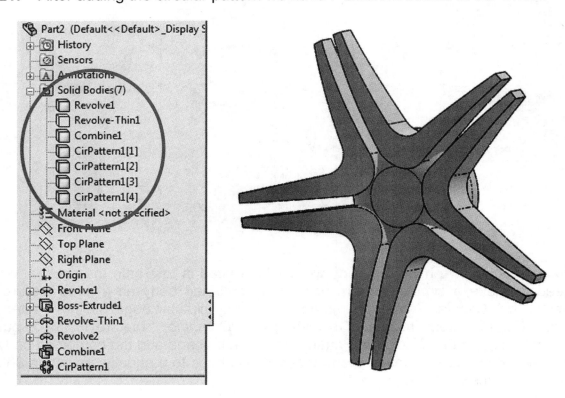

22. – The next step could be done after we combine the bodies into a single body, but we chose to add it at this time to show more multi-body functionality. In this step we'll add the holes for the bolts. Add the following sketch in the *"Right Plane."* The 28mm and 16mm dimensions are doubled about the top centerline; the 125mm is doubled about the lower centerline. Make the sketch big enough to extend well past the left side of the part.

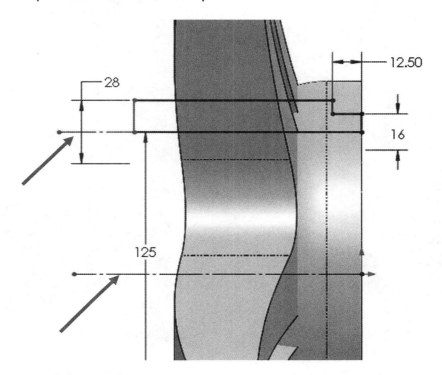

23. – Make a "Revolved Cut" using the top centerline as the axis of revolution, and leave the "Feature Scope" option to "Selected Bodies, Auto-select."

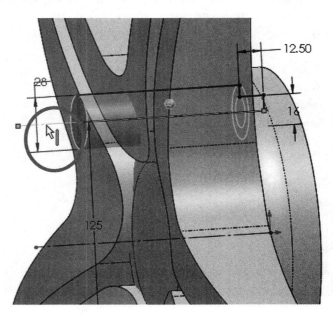

24. – Make a 5 copy circular pattern of the hole we just finished. In the "Feature Scope" use the "Selected Bodies" option, uncheck the "Auto-select" option and select the remaining four bodies made with the circular pattern and the center body (the first revolved feature we did). Click OK to finish the pattern.

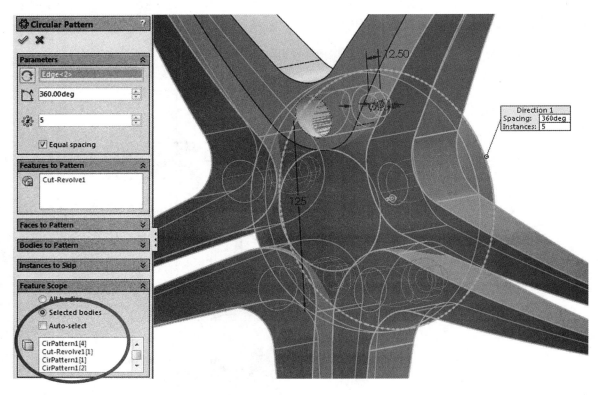

25. – Now that the holes for the bolts are complete we can merge the spokes and the hub into a single solid body. In the "Solid Bodies" folder select the patterned bodies and the hub as indicated, right-mouse-click in any of them and select **"Combine"** from the pop-up menu, or use the menu **"Insert, Features, Combine."** Select "Add" in the "Operation Type" and click OK to finish.

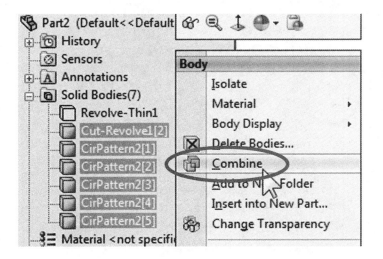
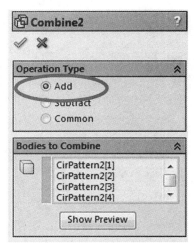

26. – After adding the bodies in the previous step we have two bodies in the part. One is the hub with the spokes and the other (hidden up to this point) is the wheel itself. In the "Solid Bodies" folder show the hidden body. Select it and click in "Show" from the pop-up menu.

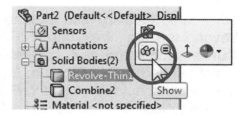

 To hide or show a solid body we can select any feature that modified it in the FeatureManager and use the same "Hide/Show" command from the pop-up toolbar, not just in the "*Solid Bodies*" folder.

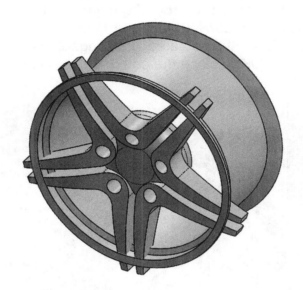

27. – To make the rim better looking add a fillet to round off the sharp edges of the "*Revolve-Thin1*" feature. Add a 5mm to all of the inside and outside sharp edges in the rim. All the edges can be rounded by selecting four faces.

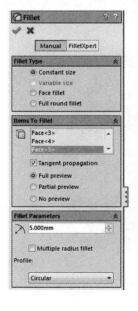

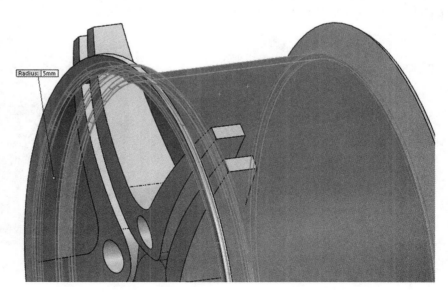

28. – The next step is to trim the excess material from the spokes that protrudes beyond the rim. We'll use a "Revolved Cut" to trim it. Using the "**Intersection Curve**" command we can create a sketch that matches the profile of the rim to cut the spokes. Add a new sketch in the *"Right Plane"* and select the "**Intersection Curve**" command from the "**Convert Entities**" drop down icon, or the menu "**Tools, Sketch Tools, Intersection Curve.**" Select all the faces outside of the rim (or inside, in this case either will give us the same result) using a right-mouse-click

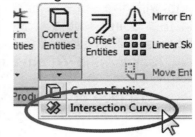

in any face select "**Select Tangency**" from the pop-up menu. This way the selection will propagate through the faces until no tangency is found. (The fillet we added in the previous step created the tangent faces to propagate.) After selecting the faces click OK to close the command and continue.

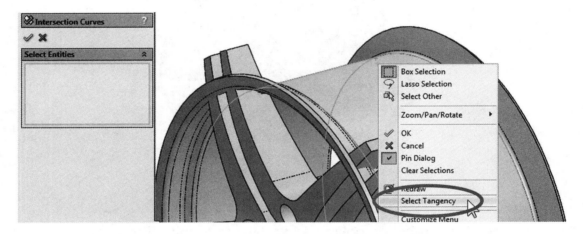

29. – After completing the intersection curve, sketch entities are created where the selected faces cross the sketch plane. Since the faces cross the plane in two different places (top and bottom) we get two profiles. Window-select the bottom profile since we only need the top one for this feature.

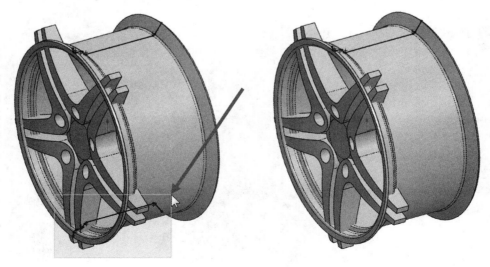

30. – Add a horizontal centerline at the origin and make a "**Revolved Cut**." We'll get the message alerting us that the sketch is open and asking if we want to close the sketch. In this case we want to create a closed profile and select "Yes" when asked. The model is displayed using "Shaded" (No edges) for clarity.

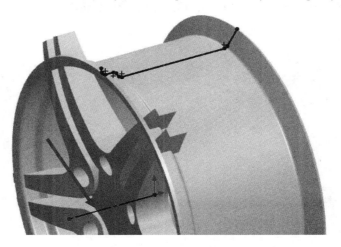

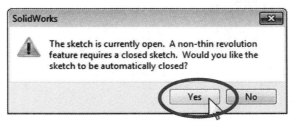

31. – In "Feature Scope", turn off "Auto-select", since we only want to cut the spoke body (*"Combine2"*), add it to the selection list and click OK to continue.

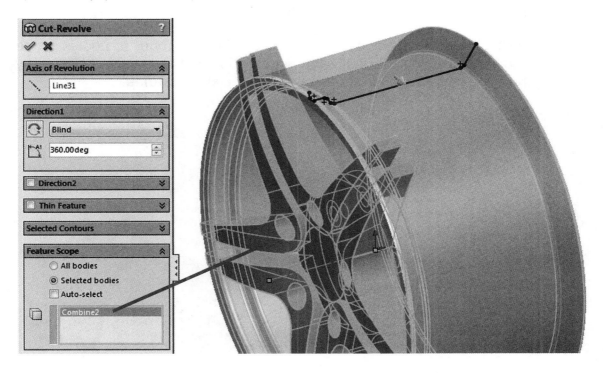

25

32. – After pressing OK to make the revolved cut, the profile we used generates more bodies by trimming the tips of the spokes, and now we are asked which bodies we want to keep. In the pop-up selection box use the "Selected Bodies" option, and pick the body at the center of the spokes to keep it. You will be able to see the preview before selecting it. Click OK to finish.

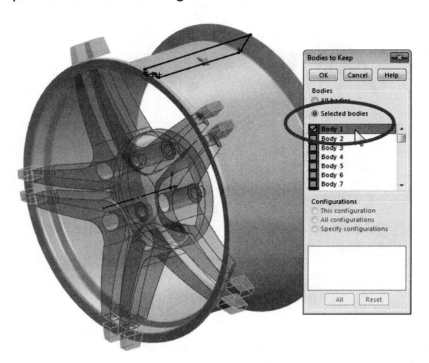

 When a cut operation splits a part into multiple bodies, the largest body is usually the first one listed in the "Bodies to Keep" list.

 If we had kept all the bodies, we'd have a large number of bodies in the "Solid Bodies" folder. In that case we can delete the unwanted bodies by right-mouse-clicking on them and selecting "**Delete Bodies**."

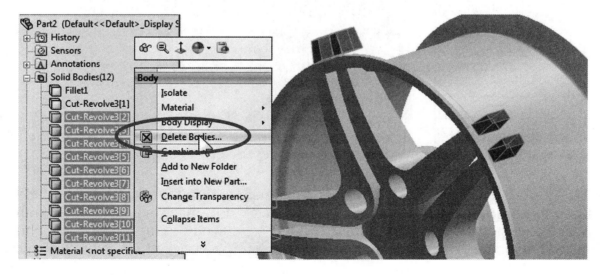

We are asked to confirm which bodies we want to delete. After we click OK to delete them, a new feature is added to the FeatureManager called *"Body-Delete."*

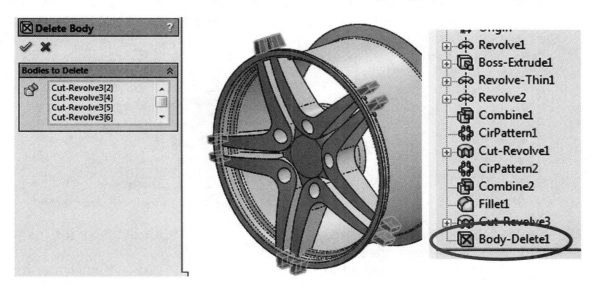

33. – Add a 5mm fillet to the spoke arms selecting the radial edges of the spokes and click OK to finish.

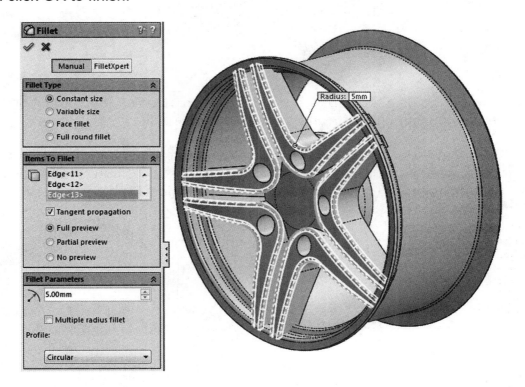

34. – Select the remaining two bodies and use the "**Combine**" command to add them together. In the "Operation Type" select "Add", make sure we have both bodies selected and click OK to finish. After combining the last two bodies, the *"Solid Bodies"* folder goes away and we have a single body left.

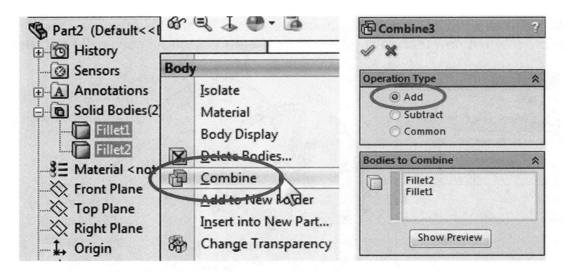

35. – Add a 5mm fillet to the edges in the center. After selecting the edge indicated, we see the expanded selection toolbar. Click in the "Connected to start face" icon to automatically select all the edges needed. Click OK to finish the fillet.

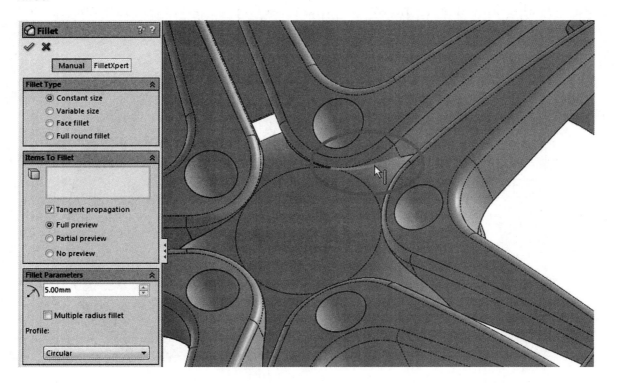

 When we move the mouse over each of the icons in the selection toolbar we can see the edges that would be selected if we clicked on it. This is an easy way to make the selection of multiple edges easier.

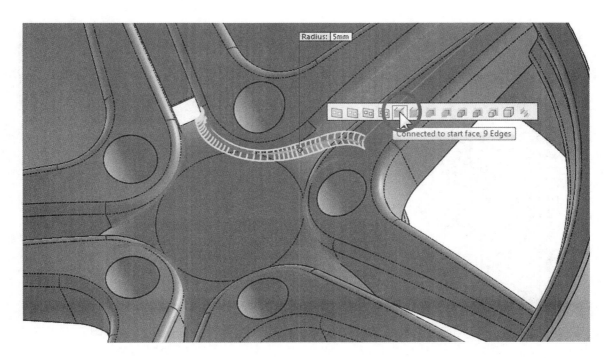

36. – Add a new 2.5mm fillet to the bolt holes and the connections between the spokes and the rim. Use the selection toolbar if needed.

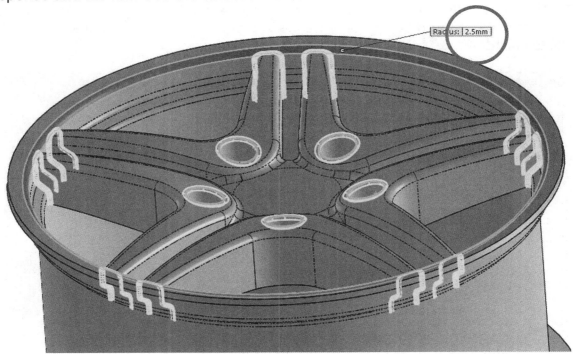

37. – To make the wheel lighter, add an 80 mm diameter hole in the rear of the hub. We want to make the cut deep enough to leave 8mm thick at the bottom. To accomplish this we'll use an end condition called "Offset from Surface."

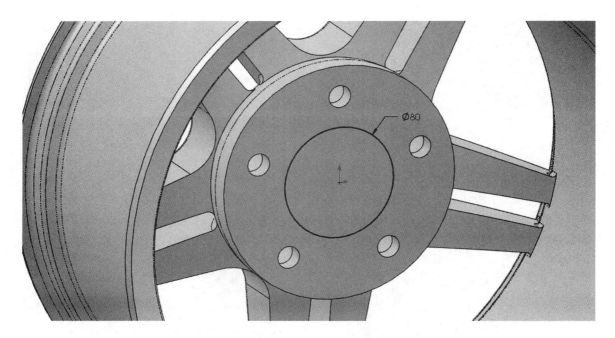

Select the "**Extruded Cut**" command. Use the "Offset from Surface" end condition and select the face on the other side of the cut. Enter 8mm in the distance box. In essence, the cut made will be 8mm away from the selected face. The "Reverse offset" option allows us to measure the distance to one side of the selected face or the other. Rotate the view to select the face if needed and click OK to finish.

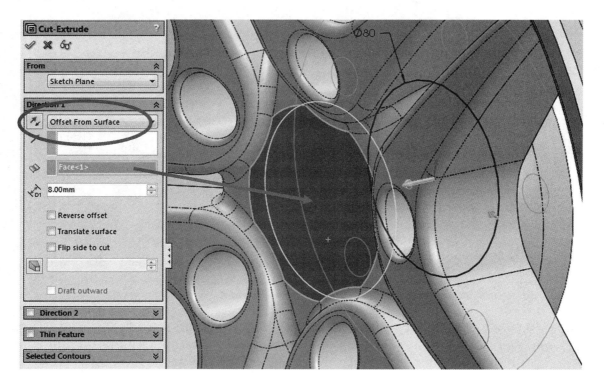

38. – As a finishing touch we'll add a logo to our design. In SolidWorks we can add a picture or image to our models for appearance or presentations. From the included files locate the image file 'logo.png' for use in this step.

Add a new sketch in the front face of the wheel, where the logo will be located. Go to the menu **"Tools, Sketch Tools, Sketch Picture."**

Browse to the location where the logo image was saved and open it. The image is automatically added to the sketch and we are ready to adjust its properties. We can click and drag the corners and the image itself to locate and resize, or enter the image values in the "Properties" box. For this image select the properties shown for size and location.

Another property we can set is the image transparency. We can make the entire image transparent (Full image), use the transparency saved in the image file (From file), or select a color from the image to make it transparent (User defined). Since the image provided has transparency saved with it*, select the "Transparency" option "From file" and note the image updating in the screen.

 After adding the Sketch Picture we can edit it by double clicking on it in the FeatureManager.

Transparency None   From file

File types that support transparency are GIF, PNG and TIFF. Transparency can be added with standard image editing software.

39. – Set the transparency to "From file," click OK to finish the image properties and exit the sketch. In the FeatureManager the image is absorbed by the sketch and it can be hidden by hiding the sketch.

 Feel free to add your favorite automobile logo. Due to copyright restrictions we cannot use registered logos in this book, but we can confirm that there are some that look VERY good!

This is a good example to show how to work with multi body parts and perform local operations. Save the part and close it.

40. – The next area to cover in the multi bodies topic will be to learn more about combining bodies. The other two operations we can do when combining bodies is to subtract one or more bodies from another, or get the common volume between them. As we saw in the previous example, to obtain the common volume between bodies we need two bodies that intersect. Make a new part and draw the following sketch in the *"Right Plane."*

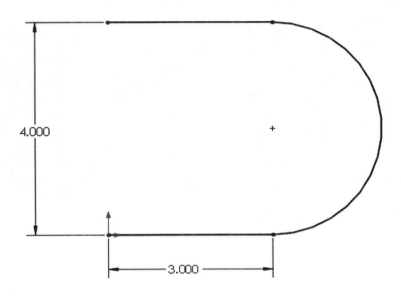

41. – Extrude the sketch as a Thin Feature. The extrusion's depth will be 3″ and the thickness of the part 0.5″ inside.

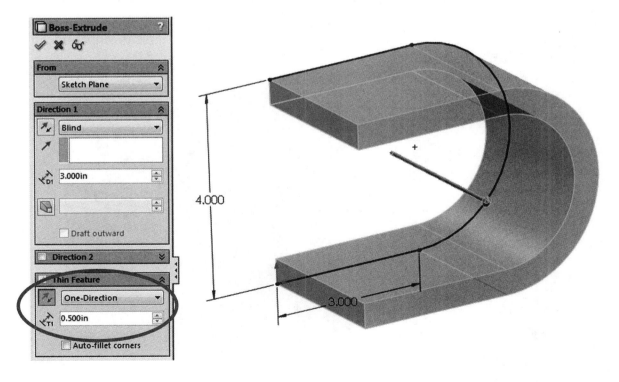

42. – For the next feature, add a sketch in the top face of the first feature. Looking at it from a top view should look like this:

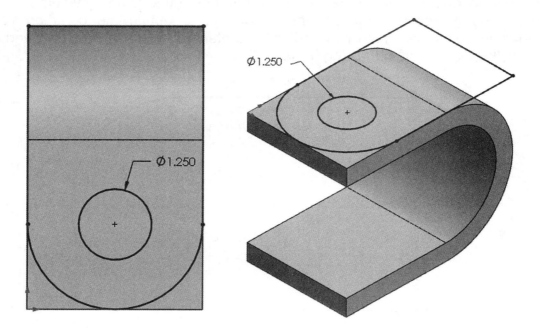

 Notice the only dimension we need to add is the hole diameter; everything else is defined with geometric relations (Coincident and Tangent relations).

43. – Extrude the second sketch downwards with the **"Through All"** end condition, and uncheck the **"Merge result"** checkbox. We want to have two overlapping bodies.

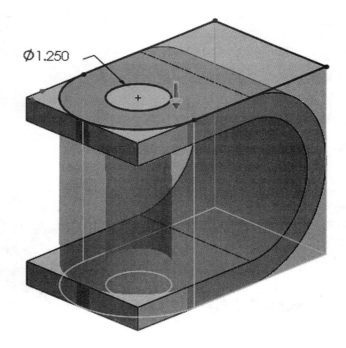

44. – Now that we have two separate bodies, we can combine them to get the common volume. Select the menu "**Insert, Features, Combine**" and select both bodies, or from the *"Solid Bodies"* folder in the FeatureManager pre-select both bodies; right-mouse-click and select "**Combine**" from the pop-up menu.

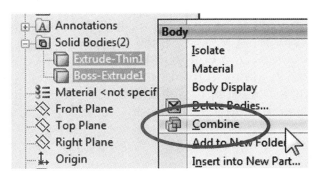

45. – In the Combine operation select the "Common" option under "Operation Type." Click in the "Show Preview" button to see what the resulting body will look like and click OK to finish.

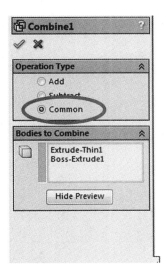

The finished part will look like this.

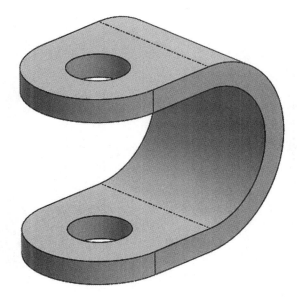

46. – So far we have covered adding bodies and extracting their common volume. The next multi body operation we'll cover is the difference between bodies. One common use for a body difference is to obtain a mold's core and/or cavity (later in the book), but here we'll show how to obtain the volume capacity of an irregularly shaped bottle. Locate the part *'Bottle.sldprt'* from the included files and open it. In order to obtain a volume with the inside capacity of the bottle, we need to make a new solid body that will enclose the bottle up to the fill level. Add a sketch in the *"Front Plane"* and draw a rectangle as shown. Make the sketch and extrusion big enough to cover the entire bottle. Use the "Mid Plane" end condition and uncheck the "Merge result" option before extruding the new body.

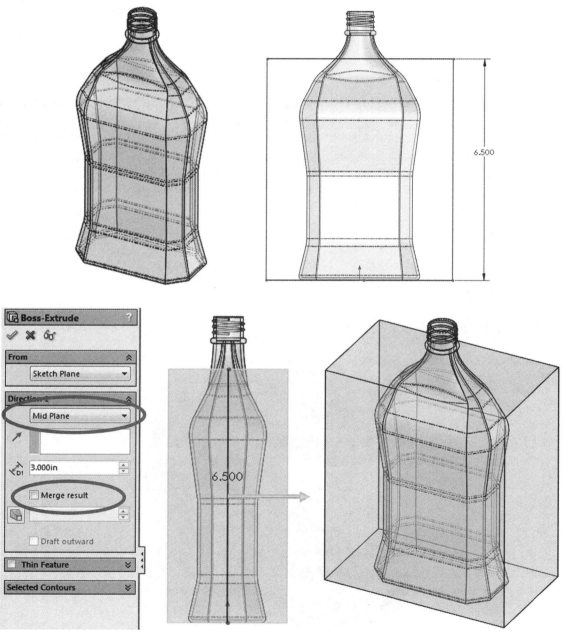

To change a part's transparency use the "**Display Pane**." Activate the transparency at the part level (can also be done to each body individually).

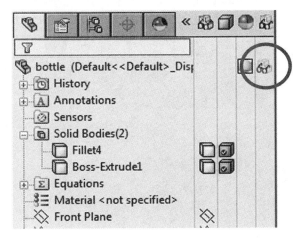

47. – Now that we have the two solid bodies, select the menu "**Insert, Features, Combine**." Under "Operation Type" select the "Subtract" option. In order to get a difference, we need to select the body that we want to remove material from (Main Body) and the body (or bodies) that we want to remove from it. In the "Main Body" selection box, select the body just created, and under "Bodies to Subtract" select the bottle. Click OK when done to continue.

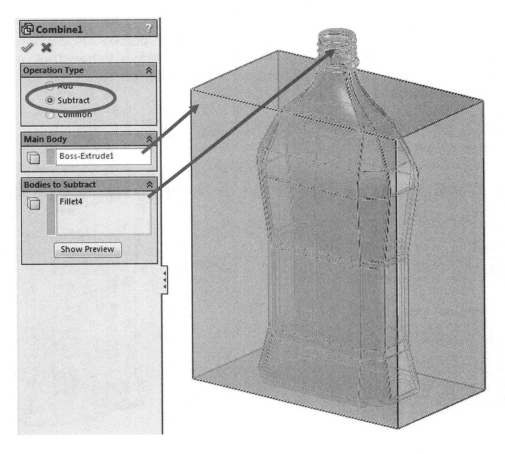

48. – After pressing OK we are immediately presented with the "Bodies to Keep" dialog, just like when we trimmed the ends of the spokes in the wheel. There are two bodies resulting from the operation, but we are interested only in the inside body. Select the inside body to keep it and click OK to finish.

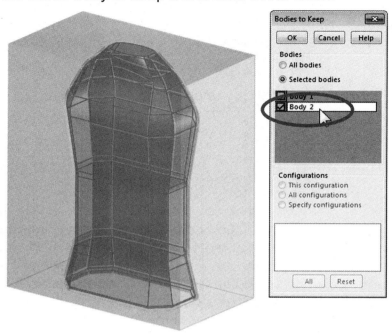

49. – Now we have the actual volume of the bottle up to the fill line. Notice that when we subtract one body from another, the original bodies are consumed and we are left only with the difference. After running a "**Mass Properties**" analysis, we can see that the volume of liquid inside the bottle up to the fill line is 25.1 cubic inches. Save and close the part.

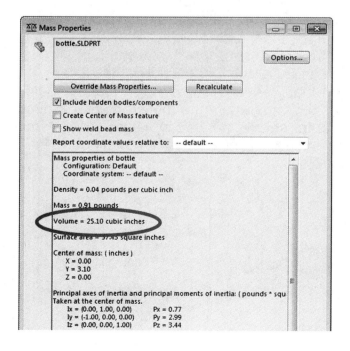

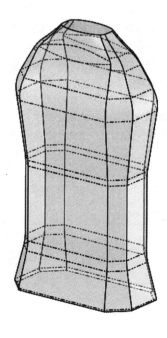

Contour Selection

Contour Selection is a way to work with a sketch that has intersecting entities, endpoints shared by multiple entities and many of the common problems that would otherwise prevent us from using a sketch for a feature. We'll also learn how to reuse a sketch for multiple features and a previously unused Extrude/Cut option.

50. – Make a new part and add a sketch to the *"Front Plane"* as shown. In this example we will not worry about dimensions to simplify the explanation and concentrate on how contour selection works.

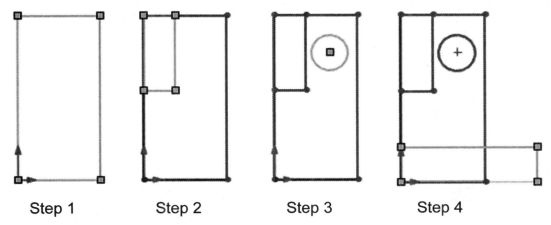

| Step 1 | Step 2 | Step 3 | Step 4 |

 In this sketch we are drawing three rectangles and one circle. Note that the rectangles are overlapping and sharing endpoints.

51. – When we try to create a feature using a sketch with overlapping and/or shared endpoints, the Contour Selection tool is automatically activated and the **"Selected Contours"** selection box is open. Also notice we don't get an Extrusion/Cut preview until we select the region(s) or contour(s) that we want to use in the feature. Select the **"Boss-Extrude"** icon. In the graphics area move the mouse pointer around and see the different regions available for selection. We can select single or multiple regions *and/or* complete closed contours by selecting their perimeter.

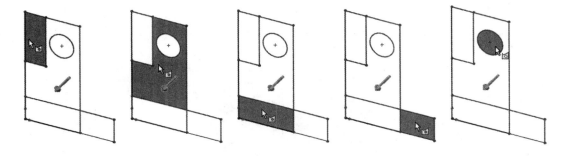

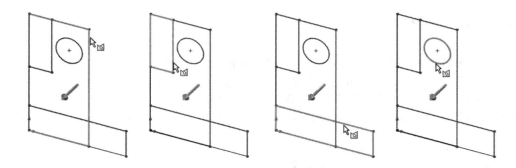

52. – Select the bottom square profile (or both bottom regions) and extrude *approximately* as shown. Just like with any other feature the sketch is automatically hidden after we finish. Do not worry about size at this point.

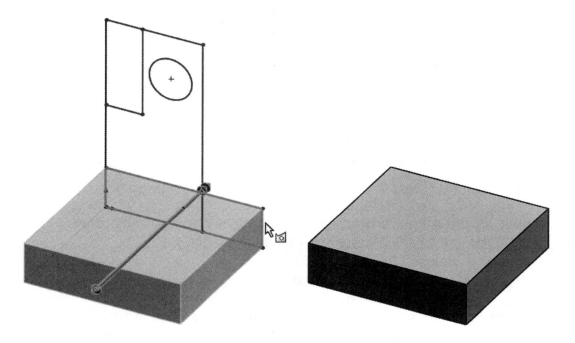

53. – To re-use the sketch for more features, expand the *"Boss-Extrude1"* feature, select the sketch in the FeatureManager and click "**Show**" in the pop-up menu. Note the "**Contours**" icon next to the sketch's name to let us know the feature was made using contours.

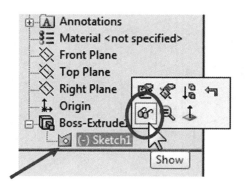

54. – We will now re-use the sketch from the first extrusion to create a new feature. To activate the "**Contour Selection Tool**" we have to make a right-mouse-click in the graphics area (or the sketch itself.) Be aware that by default the "**Contour Select Tool**" option is not available in the pop-up menu; you have to expand the menu at the bottom to make it visible.

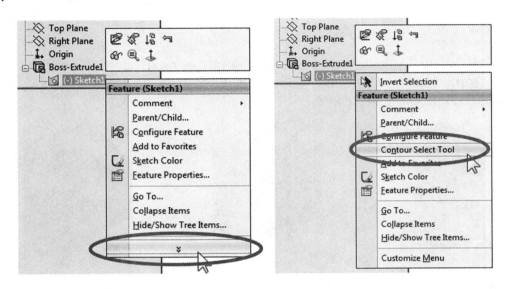

55. – After activating the "**Contour Select Tool**" select the regions indicated and extrude *approximately* as shown. You may have to click in the sketch to enable (activate) selection of regions, and either: hold down the Ctrl key to pre-select all three regions and then extrude, OR select the Extruded Base command *and then* select the regions. Either way works the same.

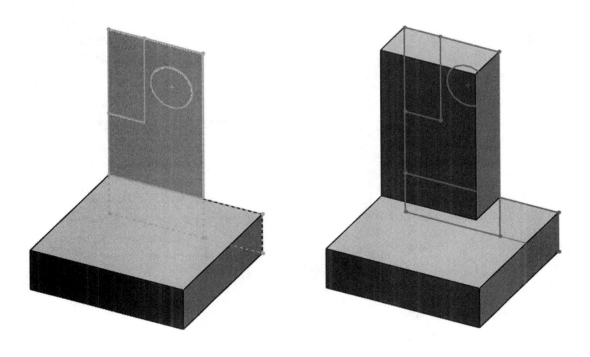

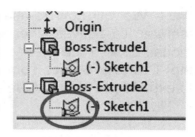

After using the same sketch for two or more features we can see a slightly different icon next to the sketch name with a little hand under it. This means the sketch is "shared" by more than one feature, and the sketch name is the same. Also note that the sketch remains visible after we use the "**Show**" command.

56. – Repeat the previous process to select the next contour and extrude as shown. Remember at this time we are only showing how it works and are not concerned about the dimensions.

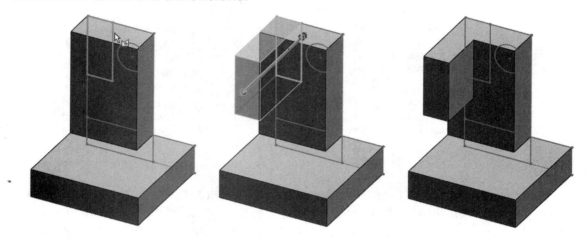

57. – For the final feature select the "**Contour Select Tool**" and select the circle. What we have done up until now is to make features that start at the sketch plane. SolidWorks has a powerful (yet sometimes under used feature) that allows us to start the feature somewhere *other* than the sketch plane. After selecting the circle using "**Contour Select Tool**", click in the "**Extruded Cut**" command; in the "From" start condition's drop-down menu select "Surface/Face/Plane". (We can also use a vertex or an offset distance from the sketch plane.)

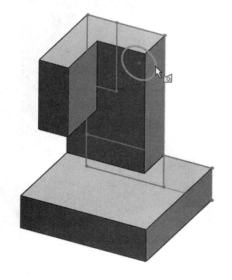

58. – After we select "Surface/Face/Plane" a new selection box is revealed to select where we want the feature to start. Select the front face of the second extrusion as indicated to start the cut feature in this face. Make the feature's depth *about* half way deep, and click OK to finish. By using the "Start Condition" option for feature creation we can easily save time by not having to create auxiliary planes or geometry.

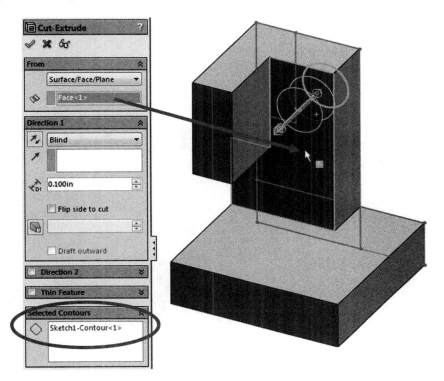

59. – Now that we have made four features from the same multiple contour and self-intersecting sketch, we can hide it. Save and close the file.

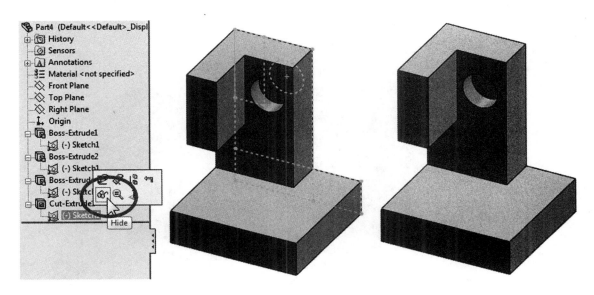

Exercises: Build the following parts using the knowledge acquired in this lesson. Try to use the most efficient method to complete the model. Open the required files from the accompanying disk or download them from our website.

Multi Body Exercise

Open the part *'Multibody Exercise.sldprt'* to build a mesh basket using a multi body part. Make a circular body pattern of the *"Extrude-Thin1"* feature (Blue body) using the centerline in the *"Circular Pattern Axis"* sketch with 7 instances spaced 30 degrees. Add a linear body pattern using the *"Extrude-Thin2"* (Green body) with 9 instances spaced 1".

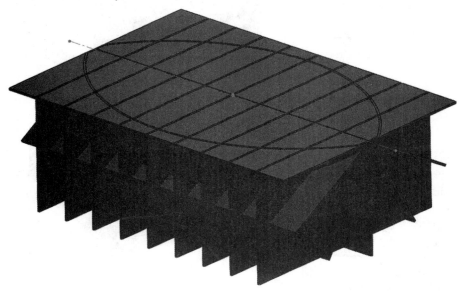

Combine (Add) all the bodies except the *"Revolve-Thin1"* (Red body) to make a single body with them.

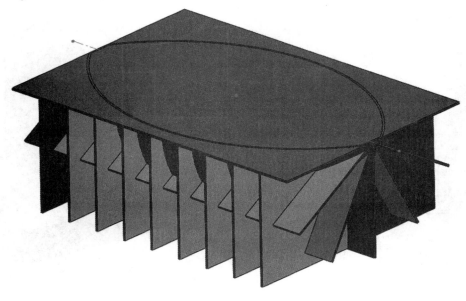

Finally combine the remaining two bodies to get their common volume.

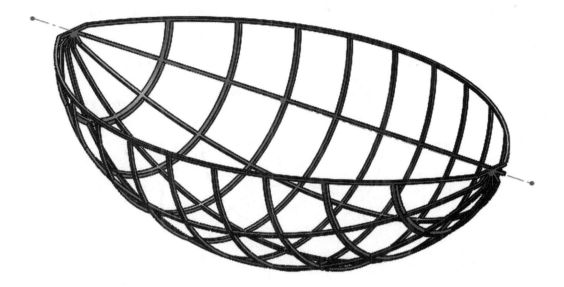

As an optional finishing touch, add a reinforcement in the ends of the basket and fillet the intersections with a 0.050" radius. TIP: Use the expanded selection tool to automatically select all the small edges.

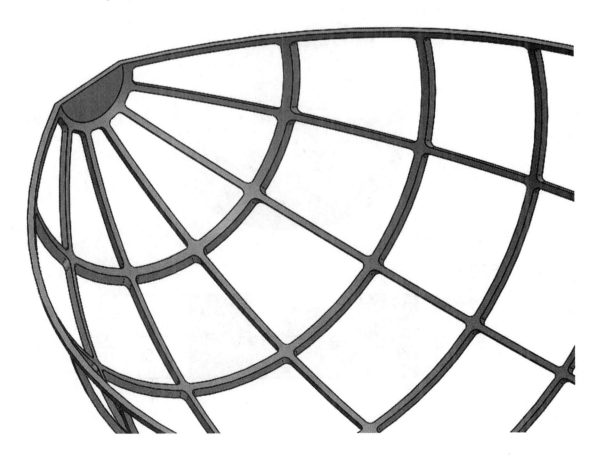

Contour Selection Exercise

Make the following sketch and make all four features off of it using the "Contour Selection" tool.

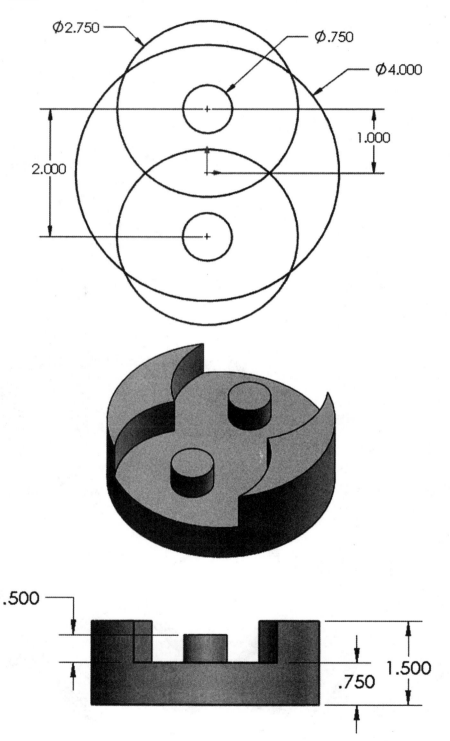

Part Editing

A very important skill to have when modeling in SolidWorks (or any CAD package for that matter) is to be able to change a model and fix errors. Let's face it: the only constant in design is change, and when we make changes to our model, chances are we may cause errors down the road, that is, the Feature-Manager. For example, if we have a part with round edges (fillets) and we change a previous feature and eliminate an edge, the fillet will give us an error because it cannot find it. For the most part those are the types of errors we are talking about.

60. – To practice editing models and fixing errors, open the part *'Repair.sldprt'* from the included files. After we are done modifying and fixing the part, it will look like this:

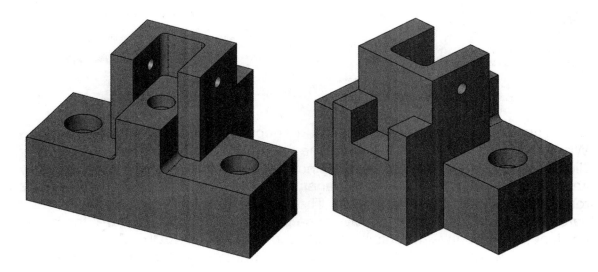

61. – When we open the file we are asked if we want to rebuild it. Select "Rebuild" from the dialog to continue.

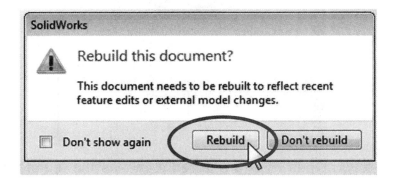

62. – When the model is rebuilt we see a list of errors and no geometry at all.

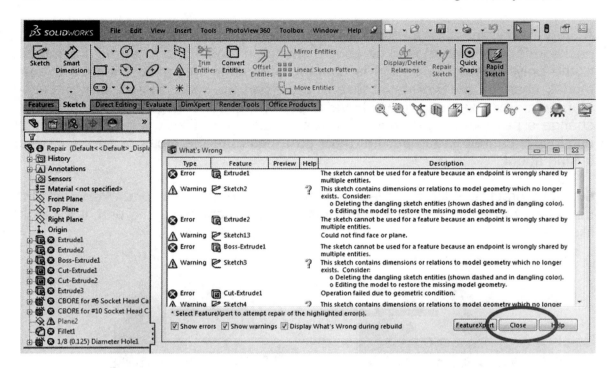

63. – This part has so many errors that no geometry can be generated, and the **"What's Wrong"** dialog box contains the full list of things that need to be fixed. Click the "Close" button for now. Since features are added chronologically starting at the top in the FeatureManager, the logical order to start fixing errors is from the top and work our way down. The reasoning behind this order is that, if a 'parent' feature has an error, it *may* cause problems in a 'child' feature; therefore, the dependency between features is referred to as **"Parent/Child"** relations. For example, if a sketch is added to a face of another feature, or a dimension references another feature's geometry, a **"Parent/Child"** relation is generated. To identify these relations, right-mouse-click in the *"Extrude1"* feature and select **"Parent/Child."**

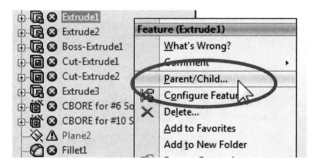

 There are two types or errors: a red X means the feature failed to build; the yellow warning triangle means the feature has an error, but SolidWorks was still able to build it.

64. – Here we can see the features that *"Extrude1"* depends on (Parents), and which features depend on it (Children). The higher we go in the FeatureManager, the more Children a feature may have. Click "Close" to continue.

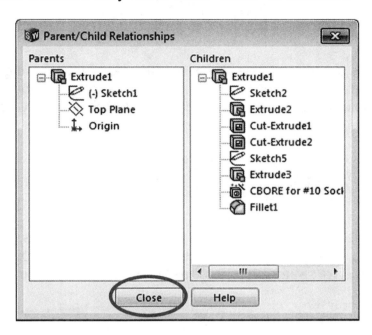

65. – To start fixing errors, we'll check what the error at the *"Extrude1"* feature is. Right mouse click on it and select "**What's Wrong?**"

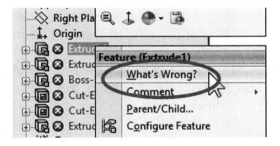

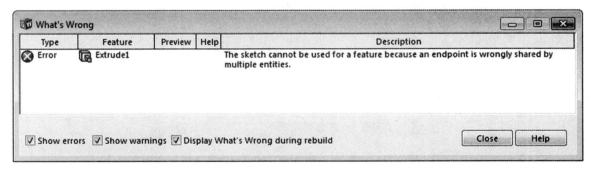

This error means that the sketch has intersecting lines and/or two or more lines are connected to the same endpoint. One common cause of this problem is when sketching we accidentally add overlapping lines. Close the "What's Wrong" dialog and edit the *"Extrude1"* sketch.

66. – Sometimes it's easy to see the geometric elements causing the problem in a sketch and we can correct it, but sometimes it's not that obvious. To help us identify the problem, select the menu "**Tools, Sketch Tools, Check Sketch for Feature**."

 Setting the option to show sketch endpoints is usually a good idea. Go to the menu "**Tools, Options, System Options, Sketch, Display entity points in part/assembly sketches**" to identify overlapping geometry

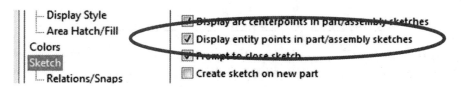

If the sketch has already been used for a feature, the feature type will be pre-selected in the drop-down menu. If the sketch has not been used for a feature we have to select the type of feature that we intend to use it for. In our case "Base Extrude" is pre-selected. Click on "Check" to analyze the sketch. Immediately we see the same error message that we got using the "What's Wrong" command. Click OK to dismiss it and continue.

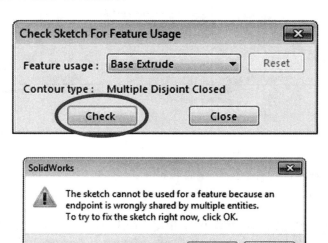

67. – SolidWorks immediately reorients the sketch and places the magnifying glass on top of the geometric element suspected of causing the problem. We are given two areas of concern: the first problem is that we have overlapping entities, and the second, multiple (2 or more) elements sharing the same endpoint.

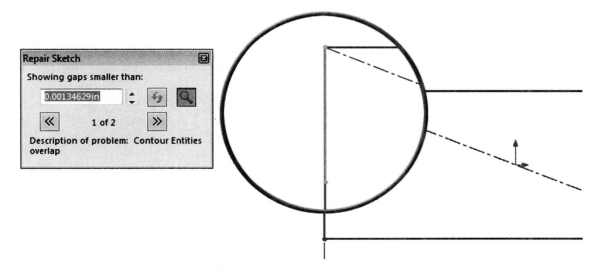

68. – The advantage of using this tool is that we can quickly identify where the problem is and correct it, instead of hunting down small line segments in a sketch. Usually, small line segments would be pre-selected to show them and, if that line is not part of our design, we can delete it. If no line is pre-selected, we know where the problem is; window-select the overlapping line and delete it. Click "**Refresh**" to confirm that we don't have any more problems and close the "Repair Sketch" window.

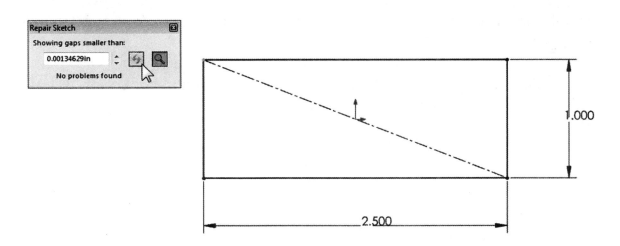

 Window-selecting from left to right selects geometry completely enclosed by the window; window-selecting from right to left selects geometry crossed by the window.

69. – Exit the sketch (or rebuild the model) to continue. We are notified that a subsequent feature has an error and we are asked if we wish to repair it now or continue with the error. Click on "Continue (Ignore Error)" and also close the **"What's Wrong?"** message.

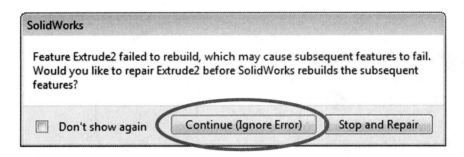

We still have some errors, but now we can see some features and (more importantly) the error from the first feature is gone.

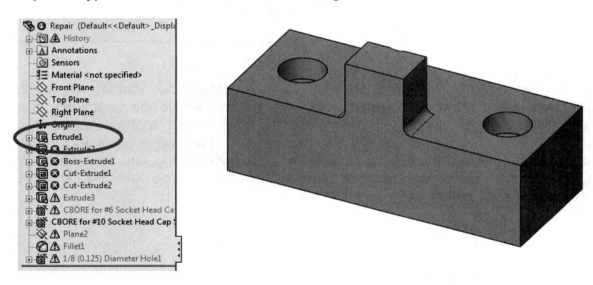

70. – Select the "Extrude2" feature, right-mouse-click and select "**What's Wrong?**" to view the error as before, *OR* rest the mouse pointer on top of it and wait to see the error in the pop-up bubble. Essentially we have the same problem as the first feature. Edit the *"Extrude2"* sketch to fix it.

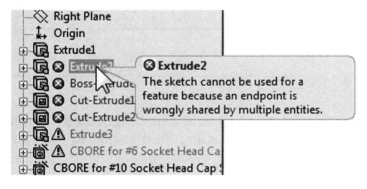

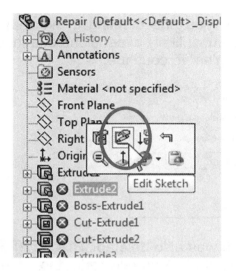

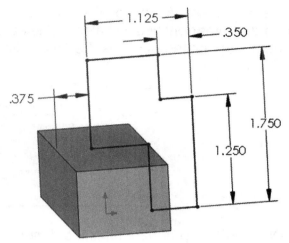

71. – Once we are editing the sketch, select the menu "**Tools, Sketch Tools, Check Sketch for Feature**" to find out what the problem is. Dismiss the error message to continue.

In this case, looking at all three problems listed in the "Repair Sketch" dialog (one overlapping entity and two with more than two entities at an endpoint), we can tell that we have two identical lines overlapping. Close the "Repair Sketch" dialog and delete one of the lines.

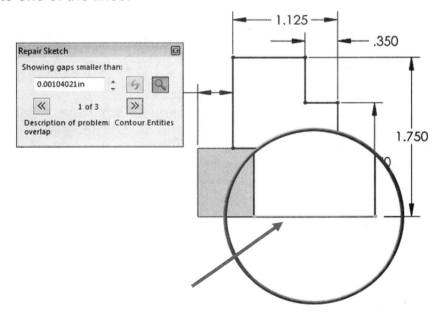

If we use window-selection we'll delete both lines; if we just click to select the line, only one line is selected.

72. – After selecting one of the overlapping lines and deleting it, we may be warned that other entities will also be deleted, most likely dimensions attached to the line being deleted. If this is the case, click "Yes" to continue.

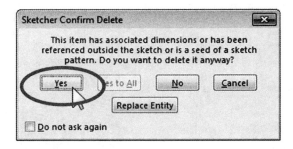

73. –Notice the top horizontal line's dimension was also deleted. This is because it was referencing the line we deleted. Manually add the missing dimension again and exit the sketch. Select "Continue (Ignore Error)" after we exit the sketch and dismiss the "**What's Wrong?**" message.

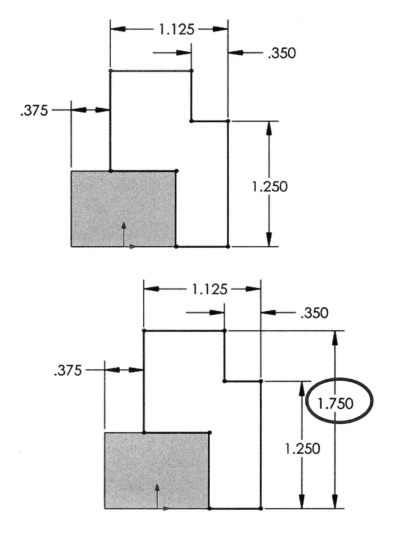

74. – Now our model is starting to look better. We have only fixed two errors and we have cleared many errors, better illustrating the importance of understanding parent/child relations.

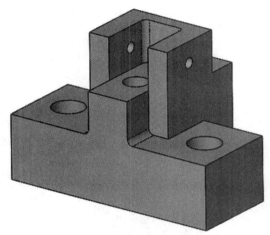

75. – After reviewing the next error with "**What's Wrong?**" we see that it's the same error as the two previous features. Editing the sketch we can see an extra diagonal line. We can either delete it, or convert it to **Construction Geometry**. The second option is usually safer, as we could lose dimensions and/or relations as in the previous step if we delete it, and if needed, the change be reverted. Select the diagonal line and convert it to construction geometry from the pop-up toolbar.

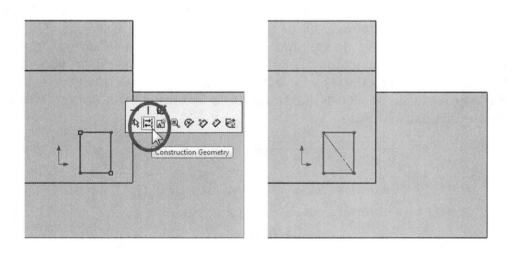

 Optionally we can change the line (or any geometry element) to construction geometry by turning on the "For construction" checkbox in the Property Manager after selecting it.

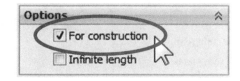

76. – Exit the sketch to rebuild the part dismissing the error dialog to continue. Since the newly created feature does not seem to be part of the original design, the logical step would be to delete it. Select the feature with the right-mouse-button, and after selecting "**Delete**" we get a "Confirm Delete" dialog:

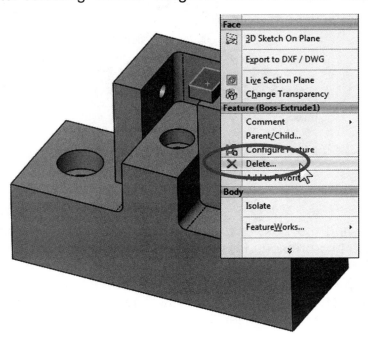

77. – In the confirmation we can see that if we delete this feature we would also delete its dependent features, all of which need to be in our part. The reason those features would also be deleted is because they are "children" features of *"Boss-Extrude1."* Activating the "Delete child features" checkbox shows all the features that would be affected. Select "Cancel." Since we don't want to delete any of these features, we'll edit those features to remove the dependencies *and then* delete this extra feature.

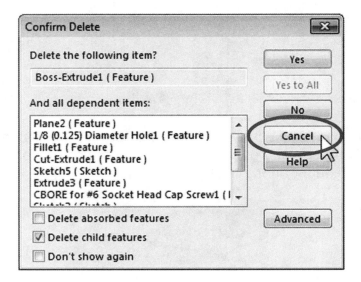

78. – Click in the *"Boss-Extrude1"* with the right-mouse-button and select the **"Parent/Child"** command from the pop-up menu.

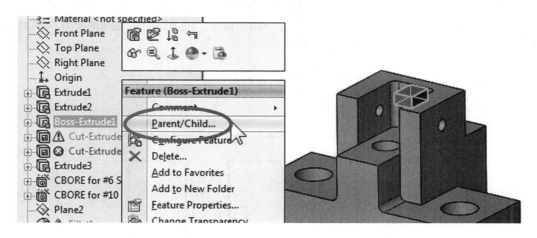

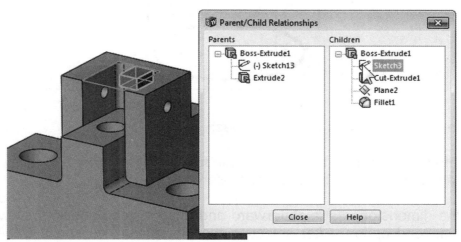

79. – We can see that the first dependent feature in the Children list is *"Sketch3"* (from the *"Cut-Extrude1"* feature). Close the **"Parent/Child Relationships"** window, select *"Sketch3"* in the FeatureManager, and edit it. At the same time, the error in this sketch says that a sketch element(s) or dimension(s) are referencing geometry that no longer is there, and so we need to fix it.

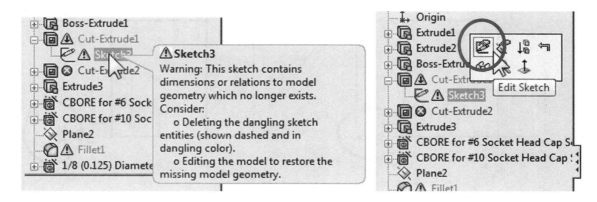

80. – Change to a Top view and "Hidden Lines Removed" mode for clarity. We can see a dimension colored in brown (0.603). The brown colored dimension (default color settings) tells us it is *'dangling,'* which means that it is referencing geometry that no longer exists. What we need to do is re-define what the dimension is referencing to fix this error.

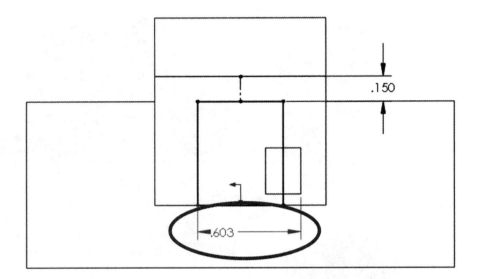

We can do this using one of two techniques:

    a) Delete the dimension and add it again to a valid reference, or
    b) Re-attach the dimension to a valid reference.

Deleting the dimension is straightforward and often a good solution, so we'll talk about the second option. After selecting the dimension, we see a witness line ending with a red dot; this is the witness line missing the reference. It *may* look as if it was referencing the existing edge, but in fact it's referencing geometry that is no longer in the model. To re-attach it, **drag the red dot** onto a valid reference; it can be any edge or vertex. For this dimension we'll use the right side edge. After re-attaching the dimension note that the geometry does not change, the value of the dimension is what updates.

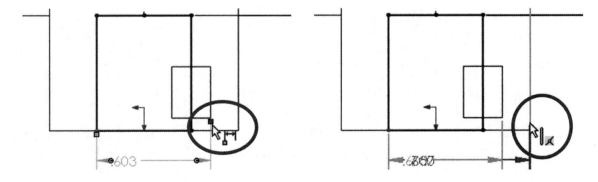

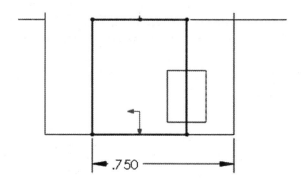

81. – After we exit the sketch dismiss the "**What's Wrong**" dialog. Notice that the error in "*Sketch3*" is fixed, and after reviewing the "*Boss-Extrude1*" "**Parent/Child**" relationships we can see, "*Sketch3*" and "*Cut-Extrude1*" are no longer listed as children features of "*Boss-Extrude1*" since we changed the dimension to reference a different edge, breaking the relation and fixing the error at the same time.

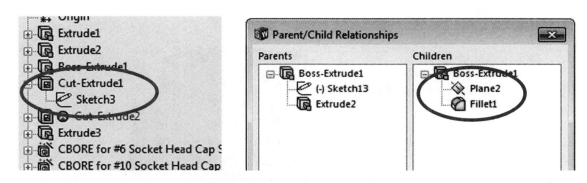

82. – From the relationships dialog we can see that the only children left are "*Plane2*" and "*Fillet1*." Close the dialog and review the "**Parent/Child**" relations of the "*Plane2*" feature. Here we can see the only child feature is the "*1/8 (0.125) Diameter Hole1*." A Hole Wizard feature has two sketches automatically made. The first sketch locates the hole's position, and is created in the plane/face where the hole is made; the second sketch is a profile for a revolved cut to create the hole.

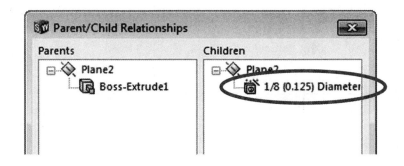

The most likely reason why the *"1/8 (0.125) Diameter Hole1"* feature is a child to *"Plane2"* is that it was most likely made in *"Plane2."* We can find out in which plane or face a sketch is made by editing its sketch plane. To do this expand the Hole Wizard feature, select the first sketch (*"Sketch12"*) which is the hole's location sketch, and from the pop-up menu select "**Edit Sketch Plane**." Since *"Plane2"* is listed, it confirms the sketch is located in it.

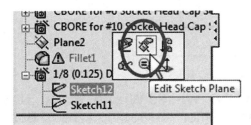 

83. – To change the sketch to a different plane (or face) and delete the relationship to *"Plane2"* (and by extension to *"Boss-Extrude1"*), select a new face to locate the sketch. Select the face indicated and click **OK** to finish. Close the "**What's Wrong?**" dialog to continue.

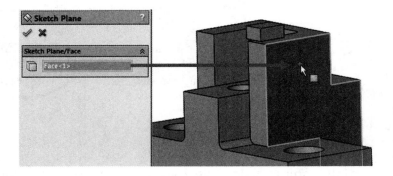

84. – Review the "**Parent/Child**" relationships again for *"Boss-Extrude1"* and *"Plane2."* Now the only children of *"Boss-Extrude1"* are *"Plane2"* and *"Fillet1"*, and *"Plane2"* has no children. Close the dialog to continue.

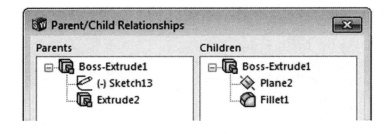

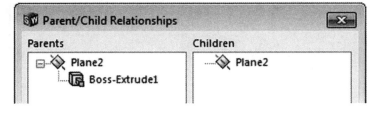

85. – Select *"Fillet1"* and check "**What's Wrong**?" Note that the icon is different; in this case the error is a warning. The fillet was built but it's missing one or more edges.

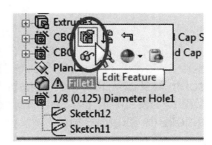

86. – Close the dialog and edit the *"Fillet1"* feature. When we edit the fillet we get a message letting us know that the fillet is missing three edges, this means that three edges that had been rounded, are no longer in the model, and the fillet command did not find them. Click OK to continue editing the fillet.

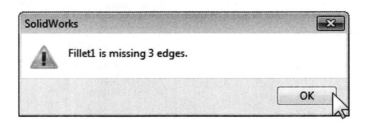

When we edit the fillet, the "Items to fillet" box is shown with a brown outline indicating that the fillet is missing one or more edges. In this case we can scroll down the list and delete the missing edges manually, or click OK. If we don't delete the missing edge references, we'll be asked if we want to remove the missing items and continue. Delete the edges from the list, *or* click OK to continue and select "Yes" to remove the missing references.

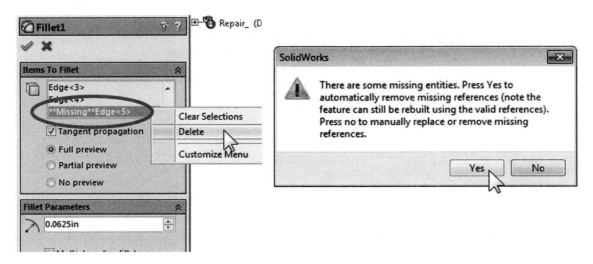

87. – After fixing *"Fillet1"* check the "**Parent/Child**" relations of *"Boss-Extrude1."* Since *"Fillet1"* is no longer a child feature of the *"Boss-Extrude1"* feature we can delete it.

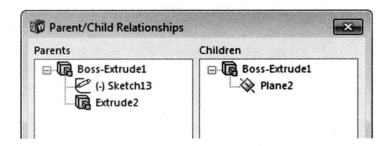

88. – Select *"Boss-Extrude1"* in the FeatureManager and delete it. This time the confirmation dialog only lists *"Plane2"*, which we don't need (or want now) and it's OK to delete it. Be sure to activate the option "Also delete absorbed features" to delete its sketch, too.

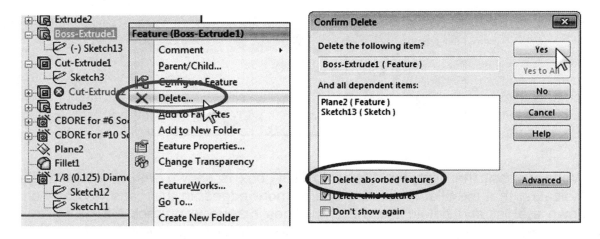

89. – The next error to fix is *"Cut-Extrude2."* The error we see using the "**What's Wrong?**" command means that the feature is not cutting the model.

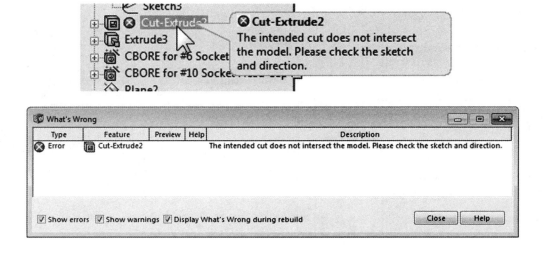

90. – Close the "**What's Wrong**" dialog and Edit the *"Cut-Extrude2"* feature. In the preview we can see that the cut is not going deep enough and we need it to make a 0.375" cut into the lower step.

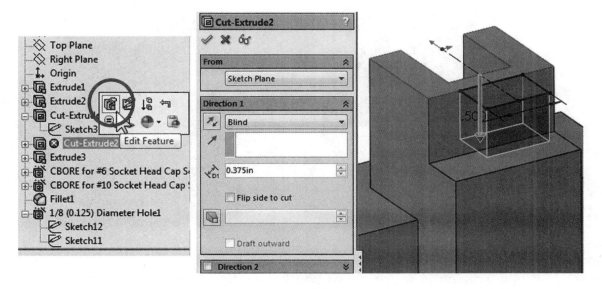

91. – Change the cut's end condition to "**Offset from Surface**" and select the lower step's face as indicated. This end condition will make the cut the offset distance going to either side of the selected face; if needed, turn on the "Reverse offset" checkbox to cut into the part as shown. Click OK to finish repairing the errors in this part.

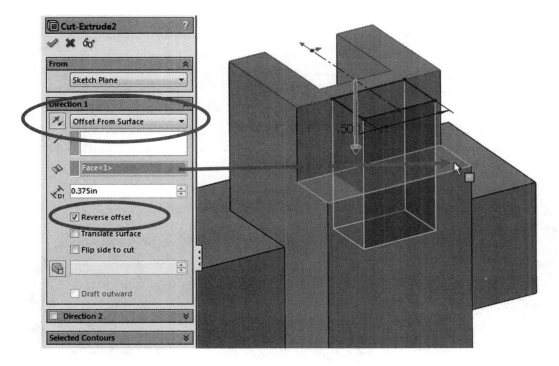

The finished part fully repaired looks like this. Save and close the part.

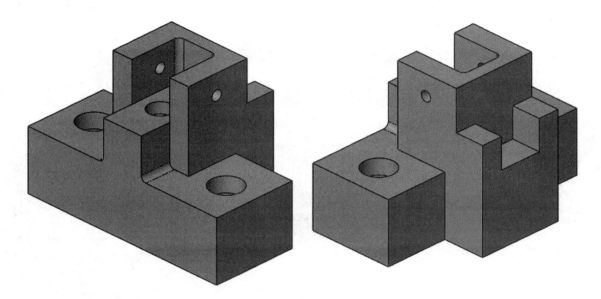

92. – When working in a sketch we can have different errors, so we decided to show them with a different part. Open the part *'Sketch Relations.sldprt'* from the included files. Like the part before, we have a number of errors in this part's sketch. After opening the file close the "**What's Wrong**" dialog and edit the *"Boss-Extrude1"* sketch. We'll show the use of diagnostic tools to help us correct these errors.

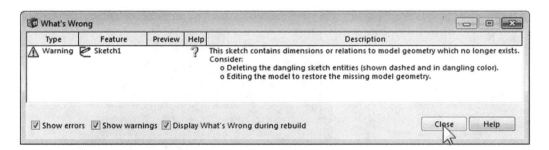

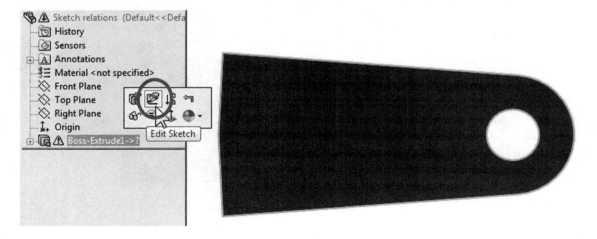

In a sketch we can have many types of geometric relations. What we are going to focus on is when relations are not solved correctly and generate warnings and errors.

An **Under Defined** or **Fully Defined** sketch can be used in a feature without a problem, the latter being the desired state. When we add conflicting relations that cannot be solved, a sketch's geometry can be in any of the following states:

State	Sketch Color	Description
Under defined	Blue	Not enough relations to fully define sketch
Fully Defined	Black	Enough information to fully define sketch
Over defined	Red	Conflicting relations cannot be satisfied.
Not Solved	Yellow	Relations cannot be solved; geometry cannot meet the required relations.
Dangling	Brown/Gold	Relations to geometry that no longer exists.
External	Can be under, fully or over defined, not solved or dangling.	Relations to geometry outside the sketch.
In Context *		Relations that reference other component's geometry added in an assembly.
Locked *	Can be any color	*'Frozen' In context* relations.
Broken *		*In context* relations that have been broken.

* Will be covered more in depth in the **Top Down Design** section later.

A sketch can become **Over Defined**, **Not Solved** or **Dangling** when geometry referenced by the sketch is deleted or modified, or when the user adds conflicting relations and/or dimensions to the sketch elements.

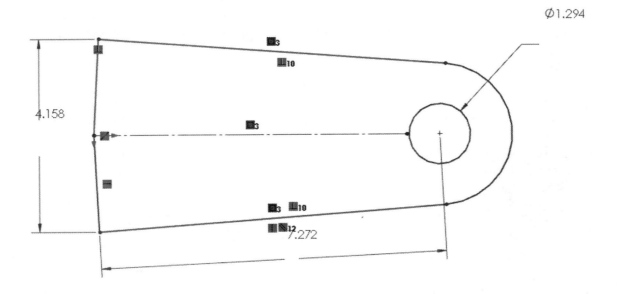

After we are done fixing the sketch it will look like the following image. In this case, while it is easier to erase the sketch and redo it, we'll take the other approach since it's a simple enough example to illustrate how to identify and make changes in the sketch.

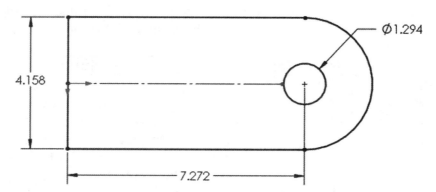

93. – After editing the sketch we see red and brown entities. There are two ways we can approach this: we can manually sort through the geometric relations and delete the conflicting ones, or we can use the **SketchXpert** tool to fix them. We'll show the manual process first, as this is a good starting point to understand what the **SketchXpert** does automatically. Select the "**Display/Delete Relations**" icon from the Sketch toolbar, or the menu "**Tools, Relations, Display/Delete**."

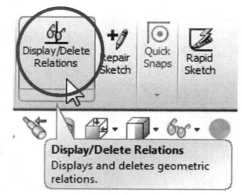

94. – In the PropertyManager we can see the list of existing relations in the sketch. We can see the relations highlighted with different colors, and the selected dimension's state under the list. At the top of the list we can filter the relations displayed by state, or if we select an entity, its relations will be listed.

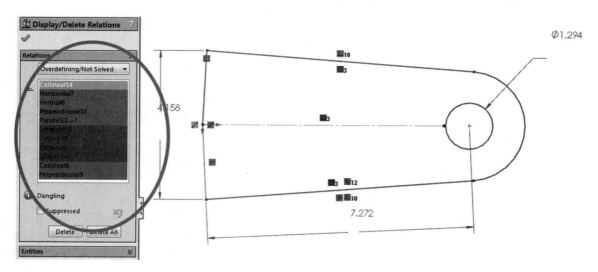

As the user can see, having a large number of conflicting relations can be difficult to sort manually. This is a good diagnostic option when we have problems with a few relations. On the other hand, this is also a good tool to identify and/or delete relations, especially "External", "In context", "Locked" or "Broken", if needed.

95. – Using this tool we'll remove the "**In Context**" relations, as we don't want our sketch to have relations to other part's geometry. An "**In Context**" relation is added when we modify a sketch inside of an assembly and add references to other components. By deleting this relation, our part will no longer be referencing geometry outside the part (more on this in the Top Down Design section).

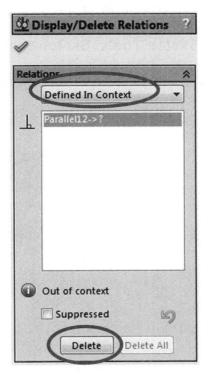

In the drop down list select "**Defined In Context**." In our sketch we only have one Parallel relation and its status is "Out of Context*." Click the "Delete" button at the bottom or select the relation and press "Delete" on the keyboard. Click OK to close the "**Display/Delete Relations**" dialog.

*More information will be provided later about relations in the context of an assembly in the "**Top Down Assembly**" lesson.

96. – After deleting this relation, our sketch is still in an undesired state, we have both unsolvable and over defined geometry, as we can see in the status bar and by the red and brown geometry.

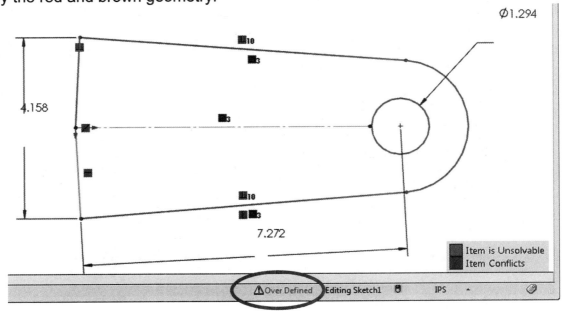

97. – The conflicting and over defined relations can be fixed using this tool, but it would take a long time; this tool is a good option when we have only a few relations that need to be fixed, or to find the relation that is causing problmes. In this case, the sketch has so many errors that we'll fix them using **SketchXpert**, which is a great tool to quickly review multiple possible solutions. To activate it click in the "**Over Defined**" message in the status bar, or use the menu "**Tools, Sketch Tools, SketchXpert**."

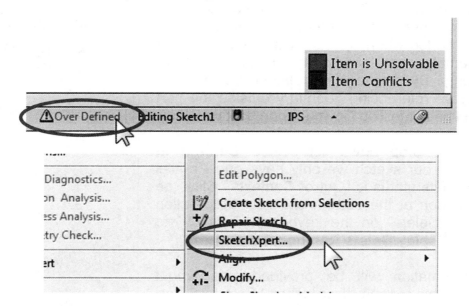

98. – In the **SketchXpert** dialog select "Diagnose" to automatically analyze the sketch and evaluate possible solutions.

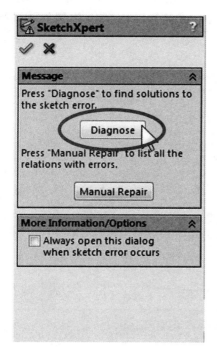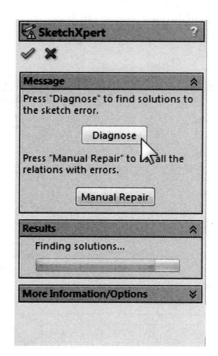

99. – SketchXpert quickly diagnoses the sketch and offers possible valid solutions. We can view each option by advancing in the "Results" box. In the "More Information/Options" box we can see the relations and/or dimensions that would be deleted if we accept the solution displayed. Scroll until you see the following solution and click "Accept" when done. This is not the optimum result, but it's the closest one that doesn't have an error.

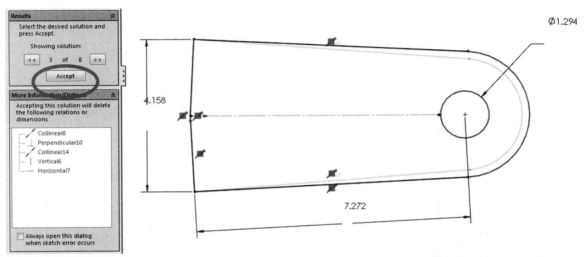

After accepting the solution we see the message "The sketch can now find a valid solution" with a green background to let us know that our sketch is error free. This does not mean that the sketch is what we want; it only says that there are no errors in it. Click OK to finish and continue.

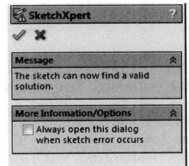

100. – We want to make the lines on the left to be collinear and horizontal (Remember the short red arrow at the origin indicates the horizontal direction in the sketch; here it is shown sideways to save space). What we need to do is to find what relations are keeping the lines fully defined, delete or edit them, and then make the lines horizontal. Turn on the display of sketch relations using the Hide/View drop-down icon or the menu "**View, Sketch Relations**" if not already activated; this way we can see all geometric relations in every geometric element.

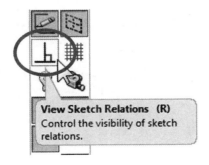

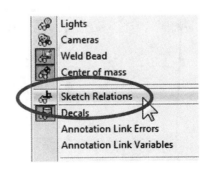

101. – We can see that there are only three relations in the two lines that we are interested in: a "Perpendicular," a "Coincident" and a "**Fixed**" relation. The "Fixed" relation is the equivalent of arbitrarily constraining an element in space. Think of it as putting a nail in a geometric element and hammering it in. Read: brute force. Select the "Fixed" geometric relation in the screen and delete it.

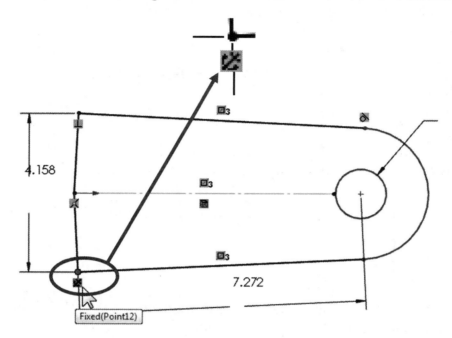

102. – If we pay attention, we'll see that we have a relation colored brown/gold in the centerline. This relation is "**On Edge**", this is the type of relation created when we use the "**Convert Entities**" command in a sketch. The brown/gold color means it is "dangling"; in other words, it lost the reference it was converted from, and we have to delete it. Delete the dangling relation and make both lines on the left "**Horizontal**". (Remember the sketch is rotated 90° in the picture.)

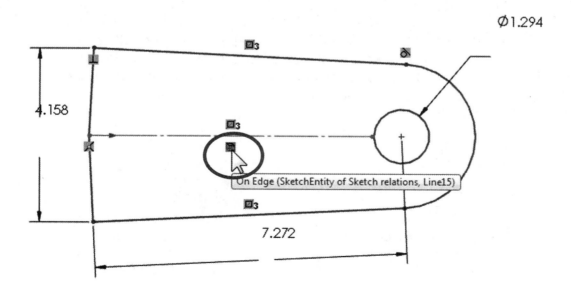

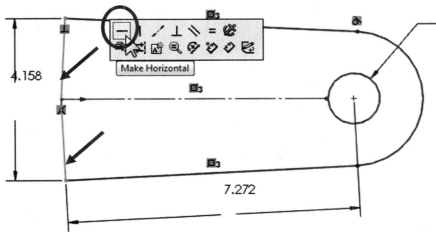

103. – Make one of the long lines "**Vertical**;" the other two lines will also become vertical as they have a "**Symmetric**" relation between them about the centerline.

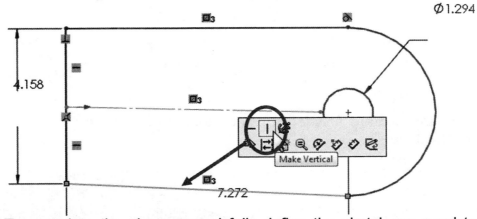

104. – To complete the changes and fully define the sketch we need to add a "**Tangent**" relation between the lower long line and the arc Press and hold the "Ctrl" key and select both entities; after releasing the "Ctrl" key select "Tangent from the pop-up toolbar. Note that the other side of the arc is already tangent to the top long line. To see the effect of having an under defined relation, click and drag the blue endpoint to see the effect of not having the tangent relation.

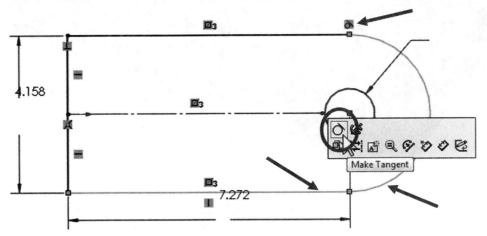

After adding these relations we can see the status bar indicates that our sketch is "Fully Defined."

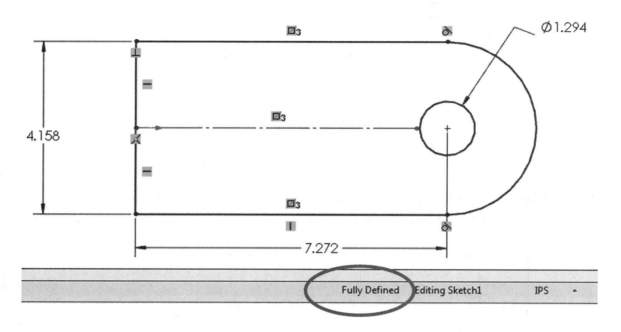

105. – Now that we are finished repairing the sketch, exit the sketch to finish. Save and close the file.

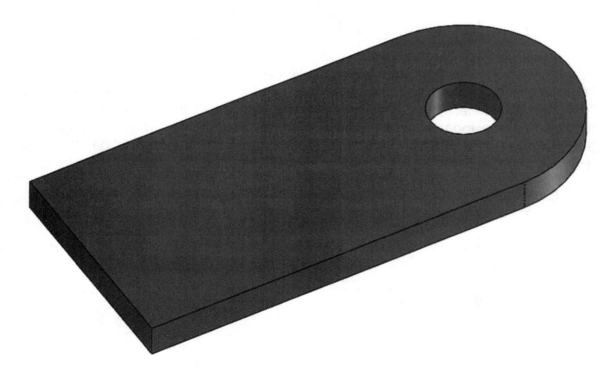

Exercise: Open the file *'Model Repair Exercise.sldprt'*. After opening it and rebuilding it, a list of errors will be displayed; however, the part has so many problems that no geometry will be available. Correct the errors in features and sketches. Use the following image of the finished part for reference. A high resolution image of this exercise is included for your convenience.

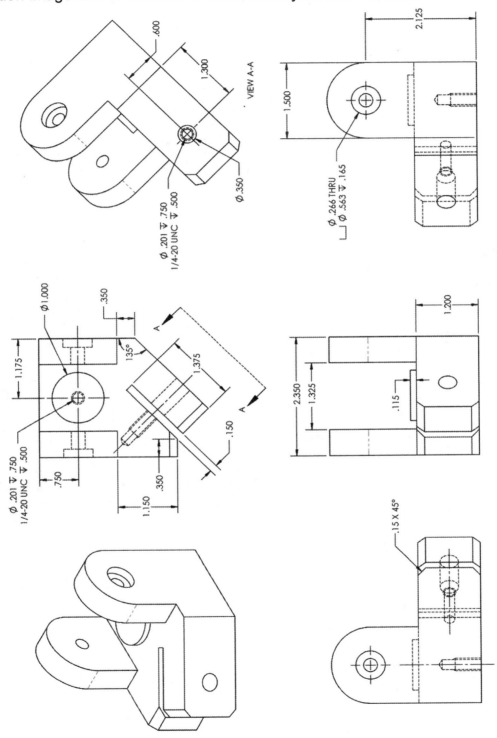

Notes:

Equations

One way to help us maintain design intent in our models is by using equations. Equations are commonly used to evenly space features, add or remove instances to a pattern's count, change dimensions when other features are modified, etc. All equations in SolidWorks have the following format:

> Variable = Expression
> Where:

Variable is the dimension/value to change (dependent value.)
Expression is the algebraic combination of other dimensions and values that will define the value of **variable**.

For example, if we have a part of length "L" where we want to evenly space a pattern of "N" number of holes spaced a dimension "S," the **variable** dimension will be "S," because that's the value we want to change when the number "N" and/or length "L" values change. Our equation would look like:

> S = L / N

A good practice when working with equations in SolidWorks is to rename dimensions and features, so instead of having an equation that reads:

> "D1@LinearPattern1" = "D3@Sketch1" / "D2@LinearPattern1"

It would look like this:

> "Spacing@Holes" = "Length@Base" / "Number@Holes"

This is more descriptive and easier to understand. If we have one, maybe two, equations in a part it may not be a problem to keep track of them, but with more equations it becomes more difficult to manage and modify them if needed.

106. – To show how equations work, we'll make a simple part, rename its features and dimensions, and finally add an equation. Draw the following sketch (any plane will do) and extrude it 0.5". Dimensions are in inches.

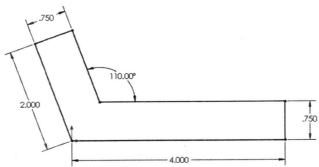

107. – Add a through hole in the corner...

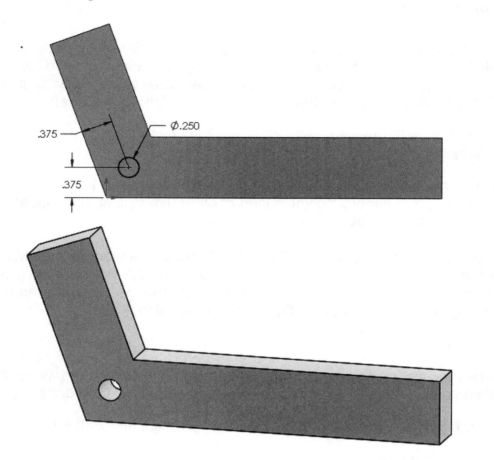

108. – Add a linear pattern. In this case the pattern will be made in two directions. Select a diagonal edge for Direction 1, three copies spaced 0.5″, and a horizontal edge for Direction 2, six copies spaced 0.5″ (don't finish it yet).

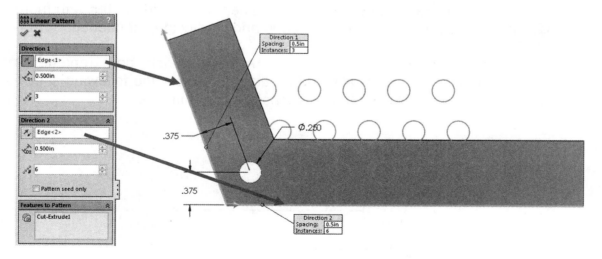

109. – Since we need a single row of copies in each direction, check the option "**Pattern seed only**" under the "Direction 2" options box. This option allows us to only copy the original hole, not the copies of the copy. Click OK to finish.

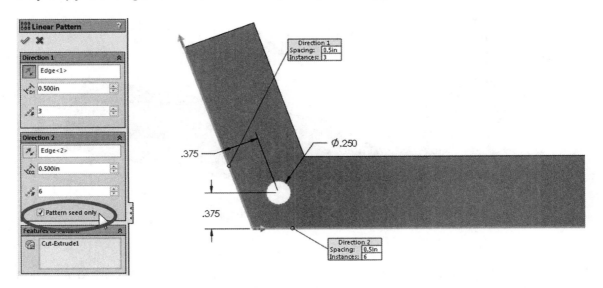

110. – Now we need to rename the dimensions. To show feature dimensions, right-mouse-click in the "**Annotations**" folder and select "**Show feature dimensions**"; to view dimension names, activate the menu "**View, Dimension Names**."

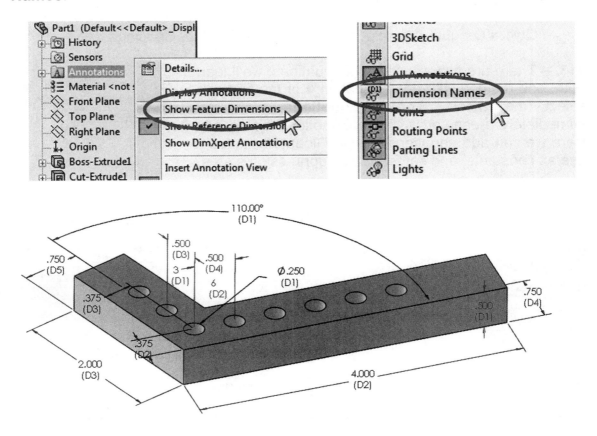

111. – Rename the dimensions for the side's length and linear pattern as shown; this way when we add the equations, it will be easier to identify the dimensions. Select each dimension and type the name in its PropertyManager.

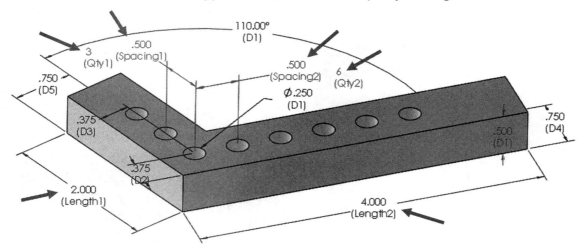

112. – Our design intent is to have a pattern of holes equally spaced to fill each side and update accordingly if the length or spacing changes. To accomplish this, we'll need two equations to calculate the quantity of holes (one for each direction). Our "dependent" dimensions will be *"Qty1"* and *"Qty2,"* and the driving dimensions will be *"Length1," "Length2," "Spacing1"* and *"Spacing2."* Based on this, our equation's general format has to be:

Qty1 = Lenght1 / Spacing1   Qty2 = Length2 / Spacing2

113. – There are two ways to add equations; we'll learn both, one method with each equation. For the first one, select the menu **"Tools, Equations."** We are immediately presented with the Equations dialog box. Here we can add equations, define Global Variables to use as constants, and set feature's suppression states.

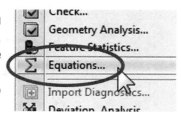

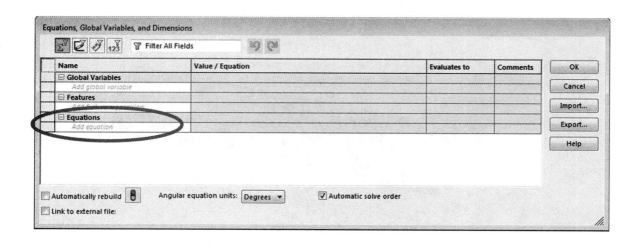

114. – To create the first equation click in the "Add equation" field in the Equations list; this is where the dependent variable will be listed. We can either type a dimension's name, or select it in the screen. Select the dimension *"Qty1"* in the screen to add its full name *"Qty1@LPattern1"* in the "Add Equation" field.

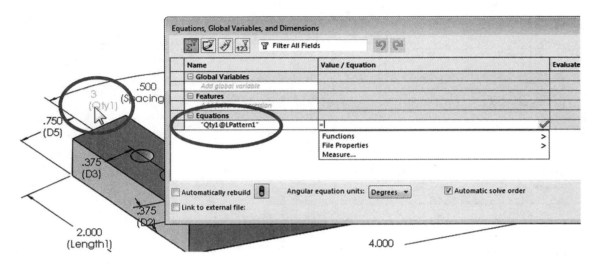

115. – After selecting the *"Qty1"* dimension, the equal sign is automatically shown in the Value/Equation field, waiting for us to type the rest of the equation to evaluate *"Qty1."* In order for our equation to calculate the correct number or holes, we have to subtract a small distance from the length in order to have a minimum space at the end. Otherwise we may get an extra hole at the end. After the equal sign, open a parenthesis, select the *"Length1"* dimension (its full name will be copied), type '- **0.375**', close the parenthesis, type '/' to divide and finally select the *"Spacing1"* dimension. The green checkmark next to the equation indicates that it is a valid expression. Our first completed equation looks like:

"Qty1@LPattern1" = ("Length1@Sketch1" - 0.375) / "Spacing1@LPattern1"

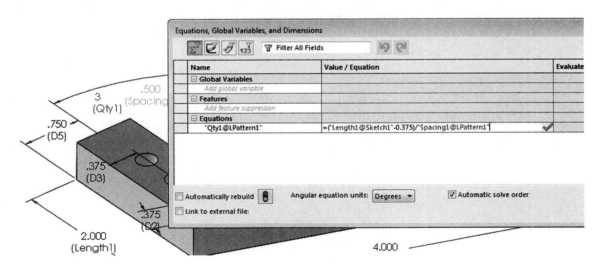

116. – In the "**Equations**" dialog we see the first equation is added, and after selecting the "Evaluates to" column the value is calculated. The value for the *"Qty1"* dimension evaluates to 3. Click OK to finish and continue.

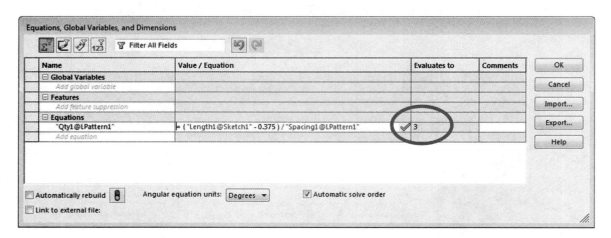

117. – After adding the equation, we see a red ∑ symbol next to the dimension, this indicates that its value is driven by an equation, and cannot be changed directly as other dimensions. It can only be changed when the dimensions driving it (*Length1* and *Spacing1*) change.

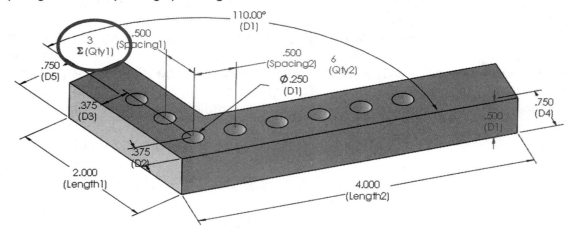

 After adding an equation to a model, a folder named *"Equations"* is automatically added to the FeatureManager.

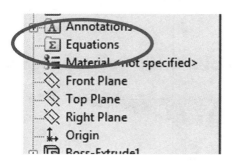

118. – To add the second equation we'll use a different approach. Double click in the *"Qty2"* dimension as if to change its value. In the second line, where we enter the dimension's value, type '**=**' (equal sign); now we are ready to enter the equation just as we did previously, but using the Length2 and Spacing2 dimensions. Remember we can select a dimension to copy its name into the equation.

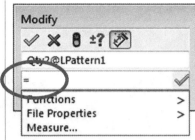

119. – Just as we did with the first equation, we'll open a parenthesis immediately after the equal sign, click in the *"Length2"* dimension, enter '**- 0.375**', close the parenthesis, add '*/*' and click in the *"Spacing2"* dimension. The completed second equation entered in the Modify dialog box looks like:

= ("Length2@Sketch1" - 0.375) / "Spacing2@LPattern1"

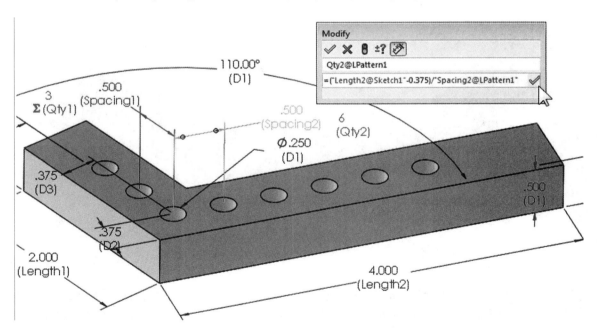

After clicking the green checkmark at the right side, the equation is created and the ∑ symbol is added. Click OK to finish adding the equation.

Now *"Qty2"* evaluates to 7. Both *"Qty1"* and *"Qty2"* have the red ∑ symbol added letting us know that these dimensions are equation driven.

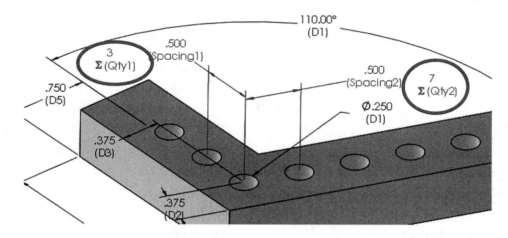

120. – To test the equations, change *"Length1"* to 2.75″ and *"Length2"* to 4.5″. Rebuild the model. Now we have 5 holes in the left side and 8 holes in the right side.

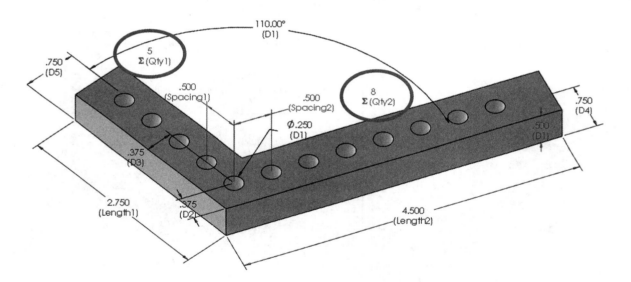

121. – Another way to maintain design intent is by using **Global Variables**. A Global Variable is essentially a constant with a name; after we create a variable it can be used in equations. Double click in one of the 0.75″ dimensions and enter an '=' (equal sign) followed by the variable name *"Width."* After entering the name, a Global Variable icon is displayed. Click in the icon to create a new Global Variable and at the same time make the dimension's value equal to it. Click OK to continue.

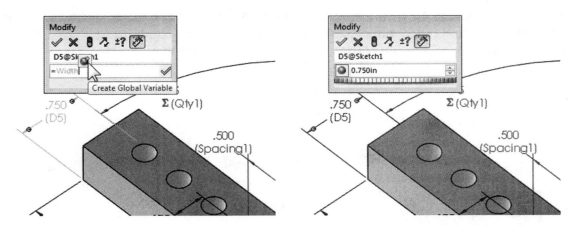

 We can switch from viewing the dimension's value or the Global Variable name using the toggle button on the left side of the Modify dialog box.

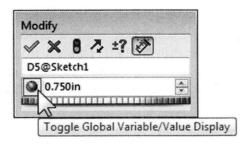

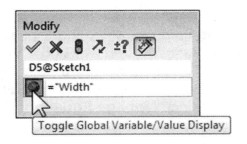

122. – After creating the Global Variable, the value is displayed with the red Σ symbol letting us know that the value is now driven by an equation. What happened was that the Global Variable was created, and at the same time an equation was added making this dimension equal to the *"Width"* variable.

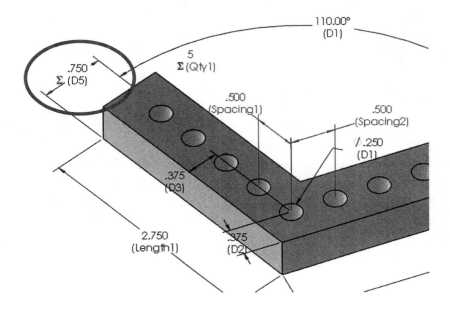

123. – To illustrate how to make a dimension equal to a Global Variable, double click in the other 0.75″ dimension, enter an '=' (equal sign), and from the drop down menu select *"Global Variables, Width"* and click OK to finish.

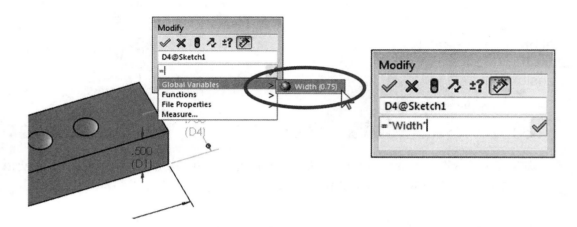

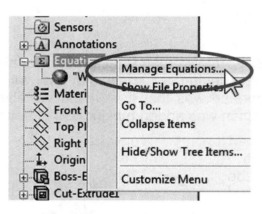

What we just did was to create a new equation making this dimension equal to the Global Variable *"Width."* Global Variables are automatically listed under the **"Equations"** folder in the FeatureManager and in the **"Equations"** dialog box. To change a Global Variable or an equation, right mouse click in the **"Equations"** folder and select **"Manage Equations."**

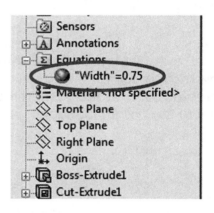

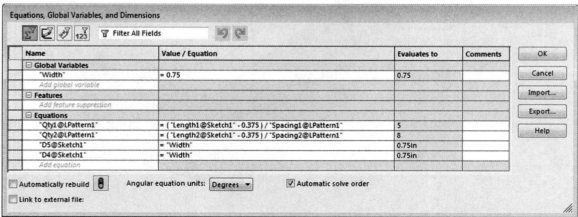

In equations we can't change the dependent dimensions directly (*"Qty1"* and *"Qty2"*), only the referenced dimensions (*"Length1," "Length2," "Spacing1"* and *"Spacing2"*). Global Variables allow us to update multiple dimensions at the same time, including when used in equations.

124. – In the **"Equations"** dialog we can Add, Edit, or Delete. To delete an equation, right-mouse-click in the equation, and select the option from the menu. Delete the last two equations in the list to learn a different way to make multiple dimensions equal to the same value.

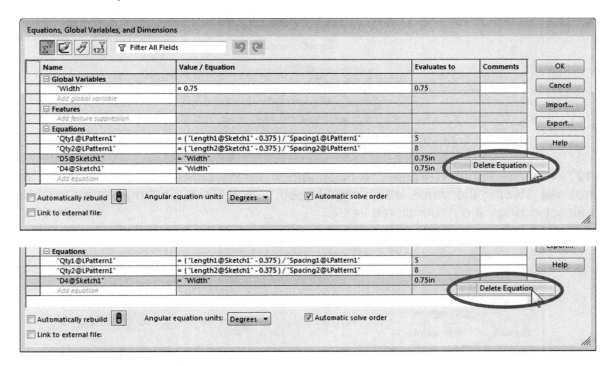

125. – Now our list only shows two equations. Click OK to close the equations dialog. After deleting the two previous equations, the dimensions are no longer preceded by the red Σ.

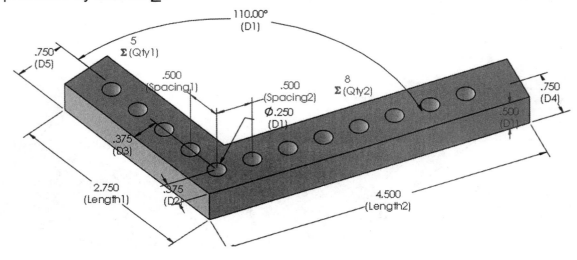

126. – A different way to make multiple dimensions equal to each other is by using the "**Link Values**" command. It is similar to making a dimension equal to a global variable as we did before; the difference is that we don't have to create a variable, and unlike using an equation where we need to change the Global Variable in the equations editor to modify its value, by linking values we can change it directly in any linked dimension. Right-mouse-click in one of the width dimensions, and select "**Link Values**."

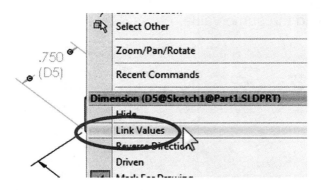

127. – In the "Shared Values" window enter *'Width_Link'* in the name field. (Do not use *'Width'*; the name was already used.) Click OK to continue. Note that we cannot change the dimension's value.

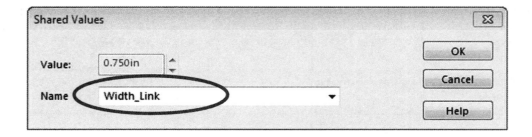

128. – After linking the dimension, its name is changed to *"Width_Link"* and a chain-link icon is added letting us know it is linked to a named value.

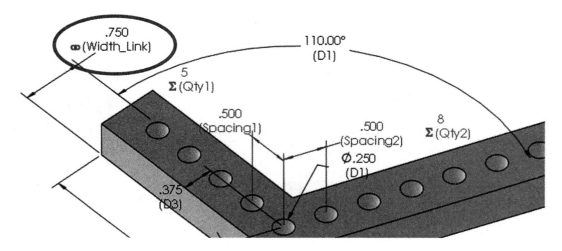

129. – To link the second dimension, right-mouse-click in it and select "**Link Values**" as we did for the first dimension, but instead of entering a name, select *"Width_Link"* from the drop-down list and click OK. Notice the previously made *"Width"* global variable is also listed as an option formatted as *"$VAR: var_name"*

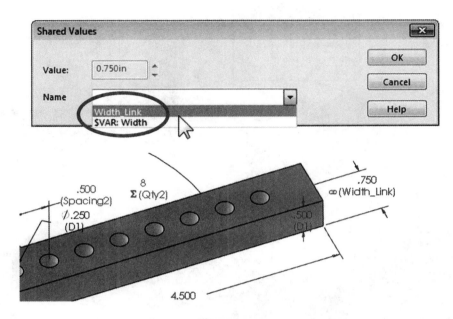

130. – Change the value of either linked dimension to 1″ and rebuild the part to see the other dimension update at the same time. Note the chain-link icon preceding the value alerting us that the dimension is linked to other dimensions.

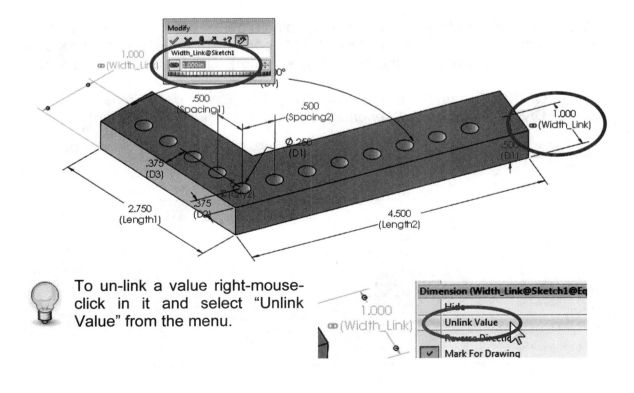

To un-link a value right-mouse-click in it and select "Unlink Value" from the menu.

131. – Add two new equations to center the holes about the *"Width_Link"* dimension. Optionally rename the dimensions. Select the dimension either in the screen or from the drop-down list. After rebuilding the part, the holes will be centered.

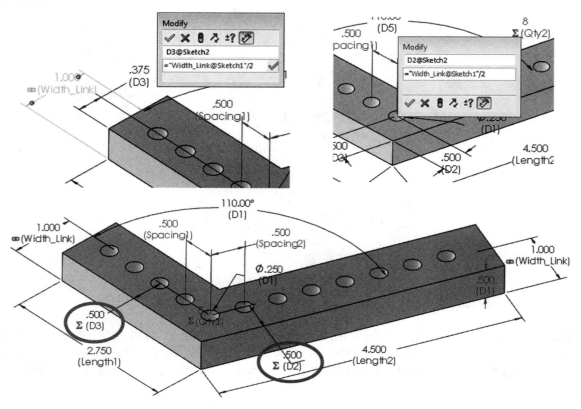

132. – To illustrate the effect of not subtracting 0.375″ from the length and show how to edit equations, remove the '**- 0.375**' portion from the two equations (it doesn't matter if we leave the parenthesis or not). Our equations should now read:

"Qty1@LPattern1" = ("Length1@Sketch1") / "Spacing1@LPattern1"
"Qty2@LPattern1" = ("Length2@Sketch1") / "Spacing2@LPattern1"

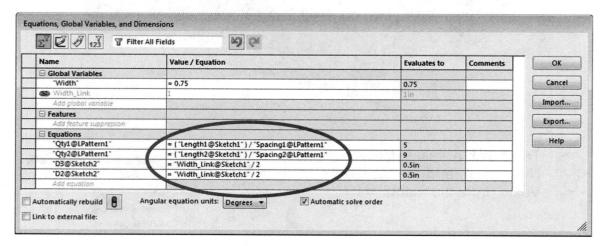

133. – Now the equations evaluate to 5 and 9 respectively. We can see the effects of the changes by rebuilding the model at the bottom of the dialog, or having them automatically rebuild as we make changes to them. Click OK to finish and rebuild the model to see the effect of the change.

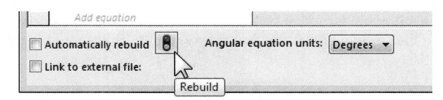

What happened with this change is that a sixth instance is added that breaks the edge of the part in the short side and gets very close to the edge in the long side; the equation resolves to 5.5 and is rounded up to 6. That is the reason why we subtracted a distance of 0.375″ from the length in the equations.

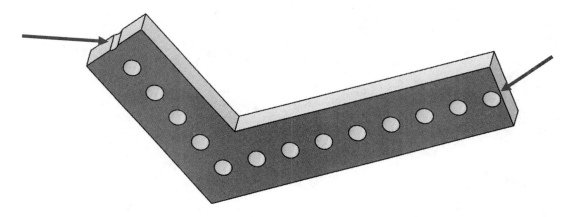

134. – Edit both equations again to subtract the 0.375″ distance from the length, returning the equations to their original value. Save and close the part.

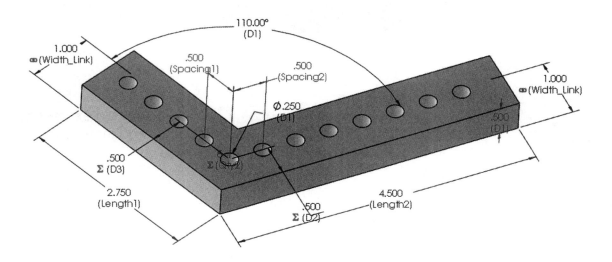

Exercise: Open the file *'Equations Exercise.sldprt'* from the included files and give all dimensions used in an equation a meaningful name. Add equations and use either Global Variables or Link Values to:

a) Make the bottom thickness equal to the wall thickness (10mm dimension).
b) Make the number of "Copies" equal to the outside diameter divided by 8.
c) Make the "*ShaftCut*" feature diameter ¼ of the "Body" inside diameter.

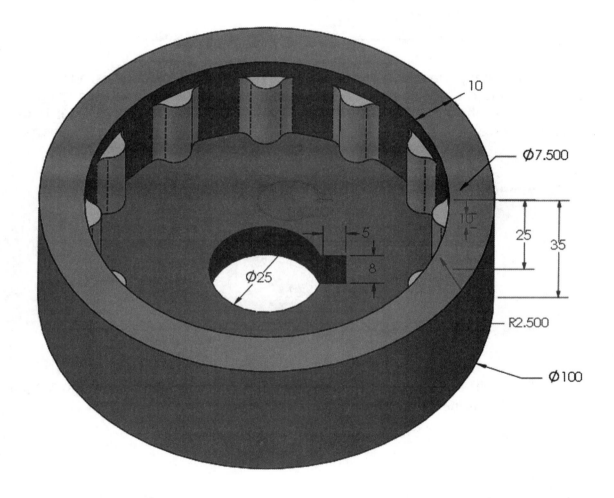

 In the SolidWorks equations, we can use almost any algebraic expression to evaluate values. As a general guide, you can write equations with the same algebraic format as you would in Excel formulas.

Top Down Design and Sheet Metal

Notes:

Understanding Top Down Design

In designing the power supply, we'll use different sheet metal and advanced assembly techniques including Top Down Design, and the first thing we need to learn is the process of designing components in the context of an assembly.

Top Down Design is generally referred to the process where we create and/or add features referencing other components in the assembly; in our example, we'll design a sheet metal enclosure for the power supply *around* the actual electronics of the power supply. In other words, we'll make an enclosure where the internal components fit, instead of building an enclosure and *then* trying to fit the components inside. That's the main difference.

One of the biggest advantages of Top Down Design is that when the referenced component (driving) is modified, the components designed around it (driven), also change. Looking at this example, the *"BASE"* component is the DRIVING component, the circular hole's edge is the driving feature, and the *"DRIVEN"* part will update when the *"BASE"* part updates. Here the hole's diameter is changed from 1″ diameter to 1.625″ and the pin's size changes accordingly.

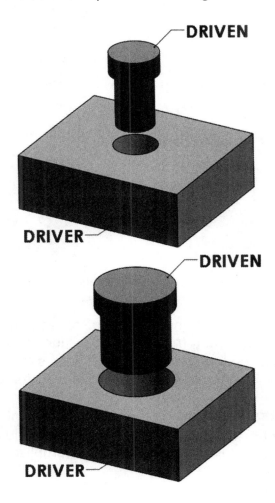

93

As a primer to **Top Down Design** before we start working with **sheet metal** parts, we'll make the components illustrated on the previous page as a simple example of how to work in the context of an assembly, or Top Down Design. In a **Bottom Up Design** approach (as we've done until now), we would make both parts and then assemble them together; the problem is that if a part is changed, the other has to be manually modified after.

In a **Top Down Design** approach, only one change is needed. For this example we'll make the *'Base'* part, add it to a new assembly, create the *'Driven'* part while in the assembly and finally modify the hole's dimension to propagate the changes to the pin (*'Driven'* part.) The steps we'll follow for the Top Down Design example are the following:

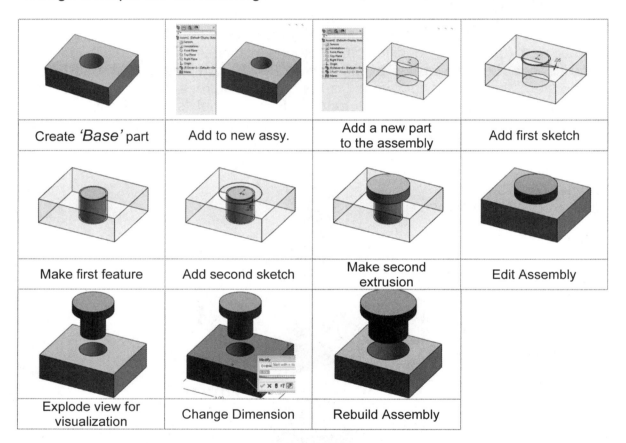

Create *'Base'* part	Add to new assy.	Add a new part to the assembly	Add first sketch
Make first feature	Add second sketch	Make second extrusion	Edit Assembly
Explode view for visualization	Change Dimension	Rebuild Assembly	

When designing components in the context of an assembly, you can reference other parts or assemblies (External Geometry), including geometry, sketches, planes, axes, etc., for sketch geometry, relations, sketch planes, feature end conditions, etc. The relations created this way are **External Relations** and are indicated in the Feature Manager with "**->**" both in the part's name and in each of the features that references external geometry. External relations are created every time dimensions, geometric relations, feature end conditions or sketch planes reference other components (parts or assemblies) while editing them in an assembly.

135. – Make the following part using the dimensions given and save it when done. Name this part *'Top Down Base.'* The hole is centered in the part.

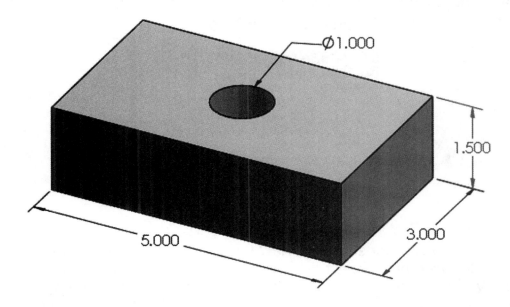

136. – Add the *'Top Down Base'* part to a new assembly. The location of this part in the assembly is not important for this example. Save the assembly with the name *'Top Down Assembly'*; saving the assembly is a requirement to work in the context of the assembly.

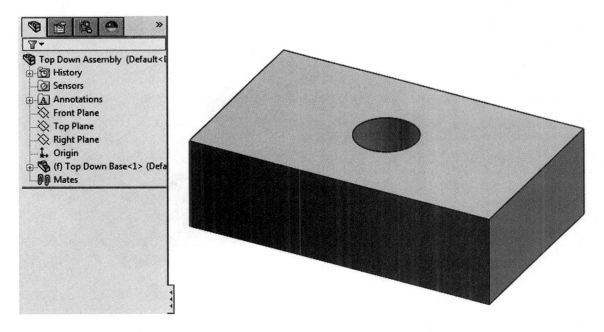

 It is important to save the assembly before external references can be added. The reason for this is because SolidWorks needs to know in which assembly the external reference was created.

137. – The next step is to add a new component. The new component will be created *inside the assembly* or *in context*. Click on the drop-down arrow from the "**Insert Components**" icon in the Assembly toolbar and select "**New Part**" or from the menu "**Insert, Component, New Part**."

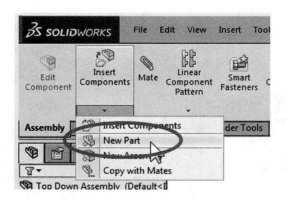

138. – Immediately after selecting "**New Part**", SolidWorks will ask us for a plane to locate the new part. Select the top face of the *'Top Down Base'* part in the assembly. By selecting the top face, we are creating the first external reference in the new part; the *"Front Plane"* of this new part will be mated coincident with the face selected using an "**InPlace**" relation that will fix the new part to the selected face. In other words, the new part will be immovable with all six degrees of freedom fixed.

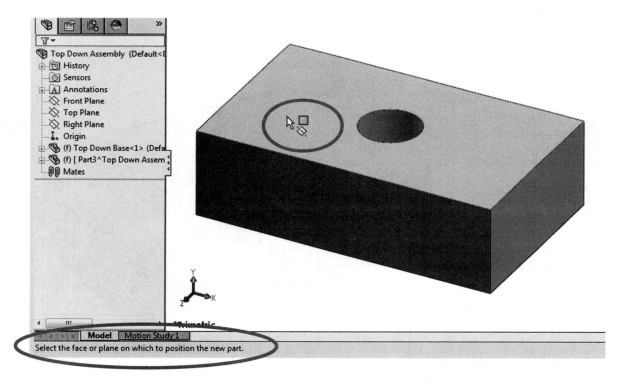

139. – After selecting the face in the assembly several things will happen:

A new part will be added to the FeatureManager tree and its name will be shown (by default) in blue. The name in blue indicates that we are *editing* this component in the assembly.

A new Sketch will be created in the *"Front Plane"* of the new part as indicated by the usual editing sketch indicators, plus an additional one: an "**Edit Component**" icon will be added to the CommandManager (or assembly toolbar) and will be activated. This indicates a component (part or subassembly) is being edited *in* the assembly. This icon will be visible in every CommandManager toolbar.

The most obvious indicator to know that we are editing a new part in the assembly is that the existing component(s) will become (by default) transparent.

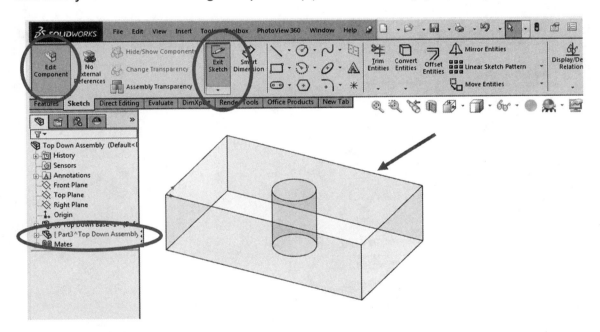

 If, after selecting the face to insert a new part, you are asked to name the new part, the System Option "**Assemblies, Save new components to external files**" is set. The default setting is *off*.

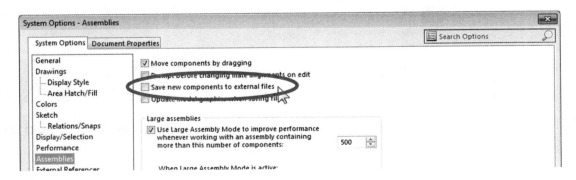

Making the other parts and/or sub-assemblies transparent is a System option in the "**Display/Selection**" section, and it has three different options:

- **Opaque Assembly** will force every component in the assembly to be opaque regardless of the component's transparency setting in the assembly.

- **Maintain Assembly Transparency** will leave components as they were; no changes to transparency will be made. Using this setting it will be more difficult to see that we are editing components in the assembly.

- **Force Assembly Transparency** (default setting) will make every component transparent except the one being edited. The level of transparency can be set with the slider.

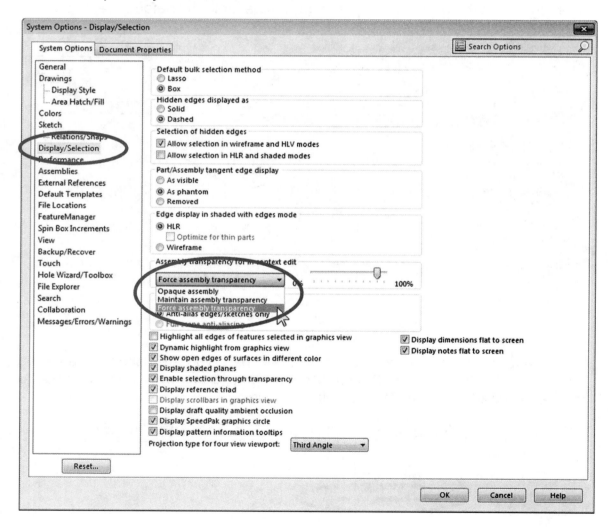

For this example we'll use the "Force assembly transparency" setting to help us better visualize if we are editing the assembly or a part.

140. – The next step is to design the driven part. What we want to do is to make the pin 0.05″ smaller than the hole. Since we are editing the first sketch in the new part, we'll use the **Offset Entities** command to offset the hole's edge. The only new thing here is that when we make the offset we are selecting the edge of a *different* component. Select "Reverse" if needed to offset the edge to the inside of the hole; click OK to finish.

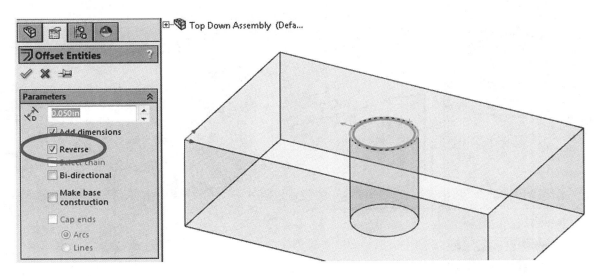

Your sketch will now look like this:

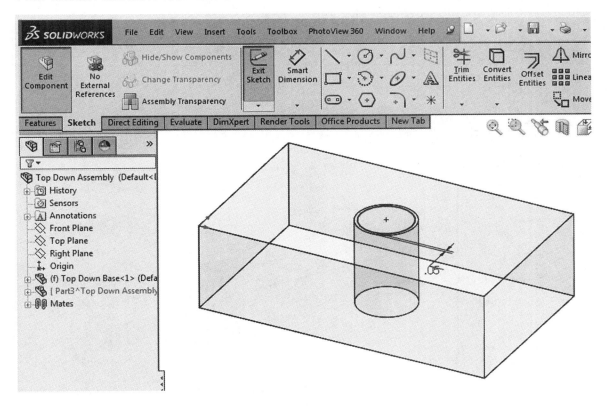

141. – Go to the Features tab, extrude the sketch using the "Up to Surface" end condition and select the bottom face of the *'Top Down Base'* part to make the new part's first extrusion the same thickness. Click OK to continue.

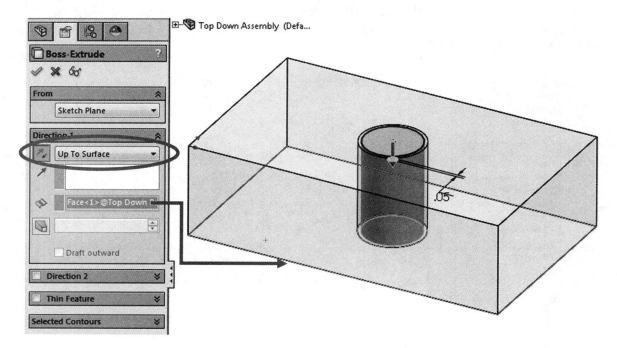

Our assembly should now look like this:

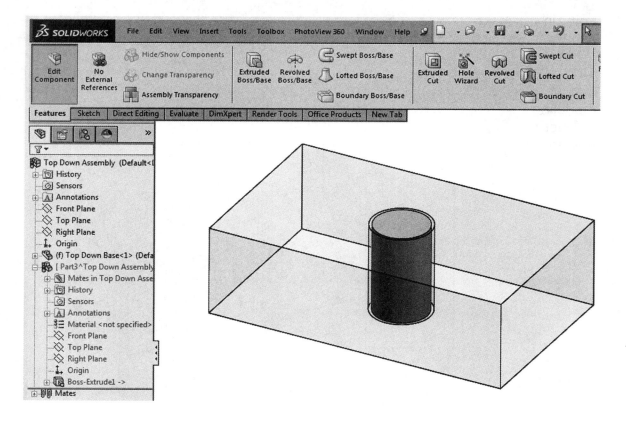

Let's take a look at the FeatureManager for a moment; there are a few things we have to pay attention to:

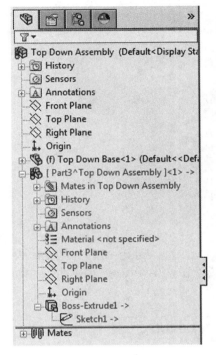

The part we added was automatically named, in this case **[Part3^Top Down Assembly]**. Notice the name is in rectangular brackets [ ]; this means that the part is *internal* to the assembly, meaning that while the part is a separate component, there is no external file for it (yet). In this case we only have 2 actual files on the hard drive: the assembly (*'Top Down Assembly.sldasm'*) and the Base part (*'Top Down Base.sldprt'*).

The name of the new part and all of its features are shown in blue; this means that the part (or subassembly if that was the case) is being edited.

At the end of the new part's name, we can see *"Boss-Extrude1"* and *"Sketch1"* names are followed by **->**; this means that the feature has *at least* one external reference. If you remember, *"Sketch1"* was made by offsetting the hole's edge and *"Boss-Extrude1"* was extruded up to the bottom of the *'Top Down Base'* part.

Later in the lesson we'll go more in depth about external references, and internal and external components.

142. – The second feature of the new part will be the head of the pin. Remember that we are editing the new part in the context of the assembly, and EVERY part modeling command is available to us. Add a new sketch on top of the extrusion we just made. Using the **Offset** command, make a sketch 0.5″ bigger than the hole in the *'Top Down Base'* part, just as we did for the first sketch.

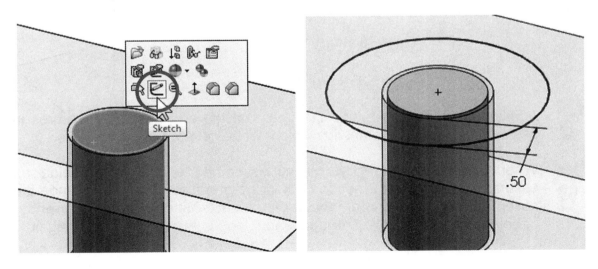

Extrude the sketch 0.5″ up to finish the part.

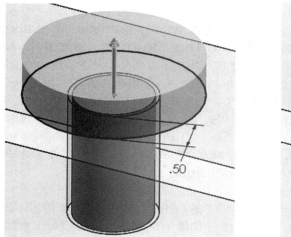
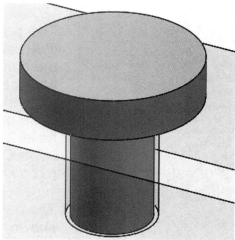

143. – We have finished the new part in the context of the assembly. What we need to do now is return to editing the assembly. As a reference, the different editing levels within SolidWorks are:

Icon	Editing Level
	Part Sketch
	Part Feature
	Part
(Yellow & Blue)	Part in Assembly
(Yellow & Blue)	Subassembly
* As many Subassembly levels as needed.	
(Yellow & Blue)	Subassembly
(Yellow & Green)	Main Assembly

 Additionally, SolidWorks has assembly sketch and assembly features in assemblies and subassemblies which will be covered later.

 When we are at the part (or subassembly) editing level inside an assembly, we can right mouse click in the graphics area, and change the assembly transparency from the pop-up menu "**Assembly Transparency**" or select the "**Assembly Transparency**" icon in the CommandManager.

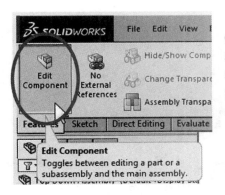

Currently we are editing the part *inside* the assembly. To return to editing the main assembly, select the "**Edit Component**" icon in the CommandManager (or Assembly toolbar if visible) or,

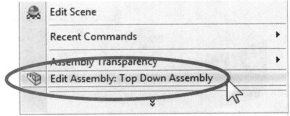

...right mouse click in the graphics area and select "**Edit Assembly:** *name*" or,

...select the **Edit Assembly** confirmation corner (just like when we close a Sketch).

Now that we are editing the assembly, everything should look as we were used to, and the status bar at the bottom will read "**Editing Assembly**." Notice that when we are editing a part in the assembly, the assembly tools are disabled.

Editing part   Editing Assembly

144. – Now that we are back in the assembly editing level, add an exploded view and pull the pin out for visibility.

145. – Change the hole's diameter from 1″ to 1.625″ and rebuild the assembly to see the changes propagate to the driven component.

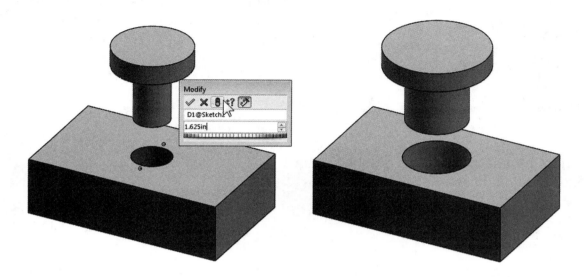

146. – Save the assembly and when asked select "**Save All**," and "**Save internally**" for now.

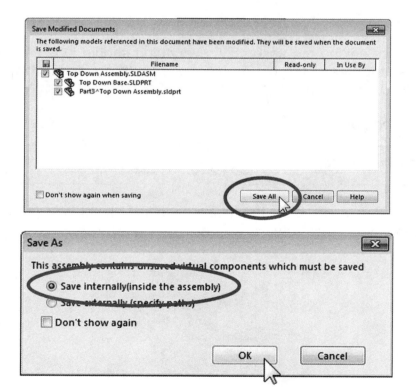

 It may seem a little confusing at first, but it's very simple; we are either editing the part or the assembly. The important thing is to *be aware* if we are editing a part or an assembly.

147. – Saving the part "internally" means that there is not an actual file that we can reference, and therefore we cannot make a drawing of it while it remains an internal part. To *externalize* the part and make a file for it, we have to (optionally) rename it like a regular feature (with a slow double click or in its properties) and then save it. Rename the internal part as *'Top Down Driven.'* Right-mouse-click on it and select "**Save Part (in External File)**." Select the path to save the file and click OK.

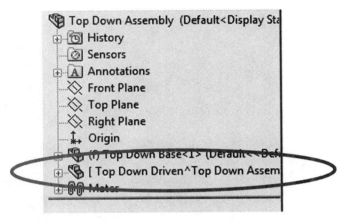

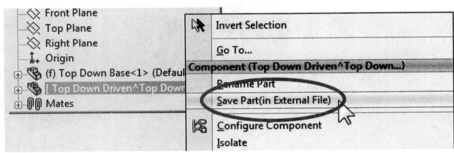

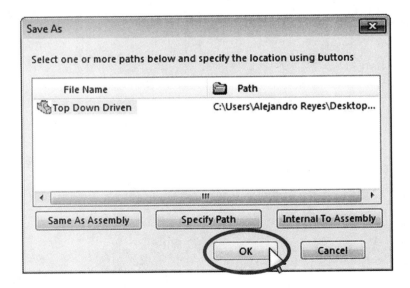

Now that the part has been "externalized" we have a file for this part and the FeatureManager shows the part's name as any other assembly component, except that it shows that it also has external references (->). Save the assembly to continue.

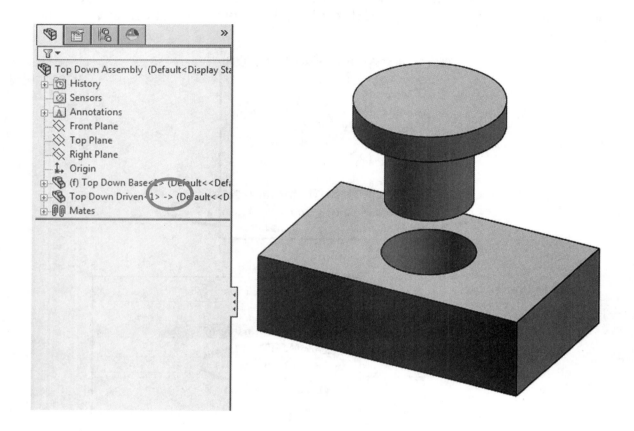

More about external references

External references can have four different states: In Context, Out of Context, Locked and Broken. Here is a description and example of each.

In Context:
When a part with external references and the assembly where those references were created <u>are open in SolidWorks</u> (loaded in memory), if a referenced part (driving) is modified, the driven part updates automatically, as in the previous example. **In context** external references show "**->**" in the FeatureManager.

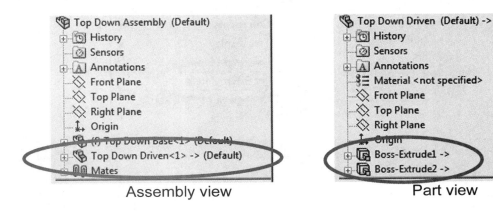

Assembly view   Part view

Out of Context:
Out of Context is when a part with external references is open, but the assembly in which those references were created is not (not loaded in memory.) If a part referenced (driving) is modified, the driven part **WILL NOT** update until the assembly where the external references were made is opened.

Open the *'Top Down Driven'* and *'Top Down Base'* parts but **not** the assembly. Change the *'Top Down Base'* part's hole size and then go back to the *'Top Down Driven'* part; notice the changes do not propagate to the driven part. Finally, open the assembly to see the changes propagate. **Out of Context** references are displayed with "**->?**" in the FeatureManager.

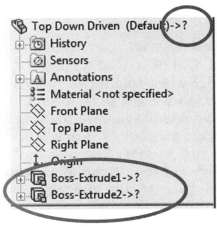

Locked:

When external references are locked changes will not propagate, even if the assembly is open (loaded in memory or In context.) To lock a part's external references, open the *'Top Down Driven'* part, right mouse click at the top of the part's FeatureManager and select "**List External Refs...**"

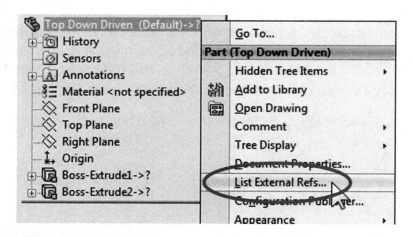

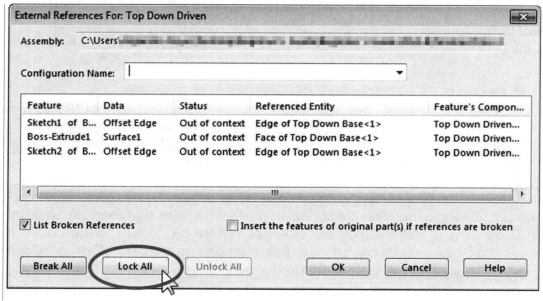

In the "**External References**" list we can see the features with external references, status (In context, Out of context, Locked or Broken) and which part and entity is being referenced. To lock references means that we can temporarily *freeze* them. This is useful when we are working in an assembly and we don't want to propagate changes immediately. After opening the *'Top Down Driven'* file, list the external references and select "**Lock All**." When locking external references, ALL references are locked. There is no way to lock individual references selectively; it's all-or-nothing. When locking references, we'll see the following message; click OK to continue and close the external references window.

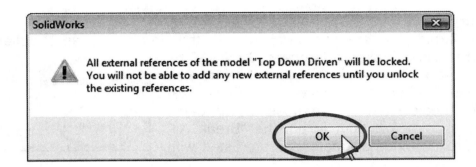

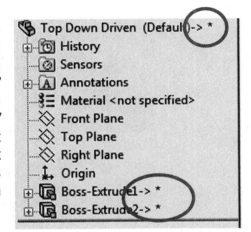

When the external references are locked, we cannot add more external references until they are unlocked. Open both parts and the assembly, with the *'Top Down Driven'* external references locked. Make a change to the hole's diameter in the *'Top Down Base'* part. Return to the *'Top Down Driven'* part and confirm that changes have not propagated, even when the assembly is open. Locked external references are listed in the Feature Manager followed by "**->***."

To unlock external references, list the part's external references again, and this time select "**Unlock All.**" After unlocking, changes will propagate following the rules when parts are "**In Context**" or "**Out of Context.**"

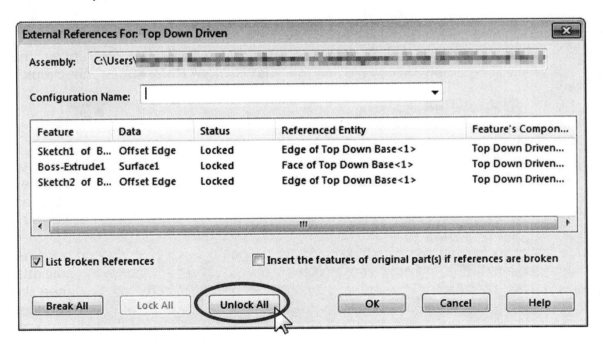

Broken:

Broken external references work exactly the same way as "**Locked**"; the difference is that Broken references **CANNOT** be re-established. Once a part's external references have been Broken, there is no way to recover them. To break external references list the part's references as before, but select the "Break All" option. Broken external references are listed with "**-> x**" in the Feature Manager.

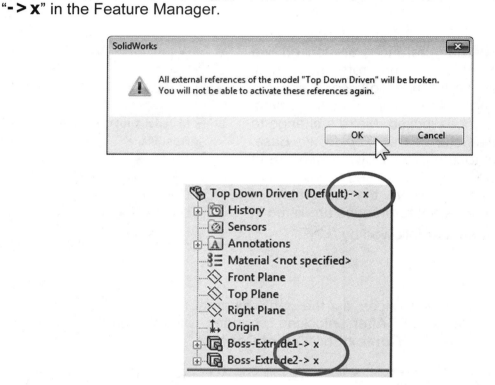

 When listing External References for a part with Broken references, the list will be empty unless we use the "**List Broken References**" checkbox.

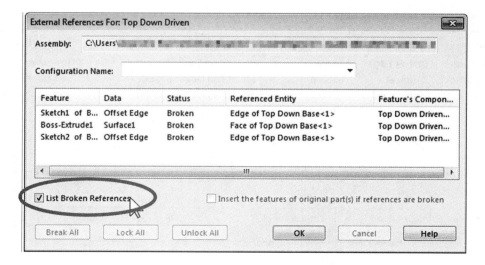

Sheet Metal and Top Down Design

When we talk about a sheet metal process, we are talking about trimming and bending sheets of metal beyond their yield strength causing a permanent deformation. Sheet metal components are made from steel, aluminum, bronze or pretty much any malleable metal, and can produce components with medium to high tolerances; this manufacturing process is well suited for high production volumes where the components manufactured need to/can have flexibility in an assembly process, and/or the final product's appearance does not negatively affect the cosmetic requirements of the final product, like the power supply or locker we'll be making in this lesson.

Due to the inherent nature of the process of cutting and bending metal, this manufacturing technique is used mostly for internal components that are "out of sight" or don't have a high aesthetical requirement, since the resulting components are not always as pretty or good-looking as a plastic injection molded part, and have a more "industrial" look. A very good example is a computer's case. If you look at the back of your PC's case, you'll see that most likely it is made from sheet metal, and the front (more visible parts) are likely made of plastic to give it a more attractive look as a consumer product.

Sheet metal tends to be a relatively cheap manufacturing process, and lends itself very well for mass produced components using multiple stations or progressive dies, where a long sheet metal strip is fed in one end, and is progressively trimmed, cut, punched and bent until a finished part is completed at the other end. Other sheet metal manufacturing techniques include cutting the sheet metal with computer controlled torch, plasma, laser or water jet machines and bending the part afterwards using presses and bending tools; this process is often used with larger components.

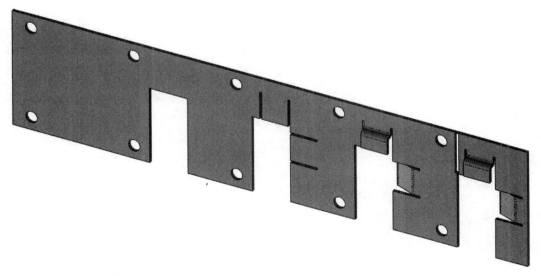

Example of a progressive die sheet metal part

Notes:

148. – Now that we have a basic understanding of what **Top Down Design** is and how it works, we'll design a computer's power supply enclosure using a Top Down design technique, while at the same time we learn how to make sheet metal parts. For our example we'll start with the assumption that the electronic components are already designed and the enclosure will be designed *around* the power supply. From the included files open the *'Power Supply.sldprt'* part file and open it.

*If you have SolidWorks Professional or Premium you *may* be asked to run a feature recognition, if so, click "No" to continue. Feature recognition attempts to create parametric features from an imported feature-less solid body using the SolidWorks' add-in FeatureWorks.

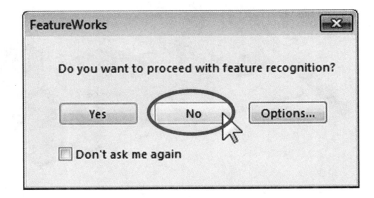

Our electronics board looks like this:

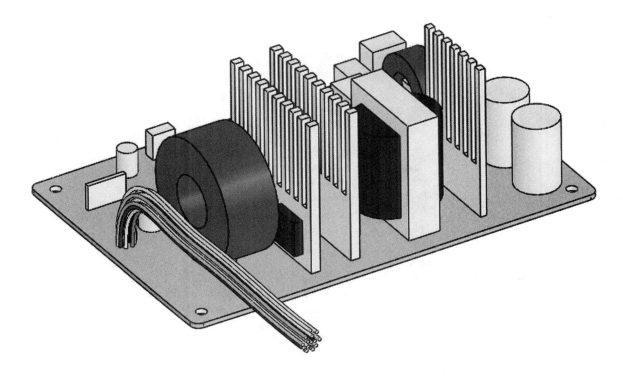

149. – The first thing we need to do is to make a new assembly, add the electronics board at the assembly's origin, and save the assembly as *'Power Supply Assy.'* Remember that the assembly must be saved to be able to edit a part in the context of the assembly.

150. – The first component we are going to design in the context of the assembly is the bottom half of the enclosure. Select "**New Part**" from the drop down list in the "**Insert Components**" icon in the Assembly toolbar, or from the menu "**Insert, Component, New Part**."

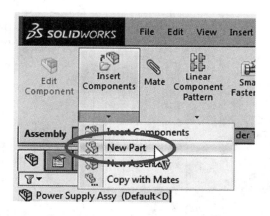

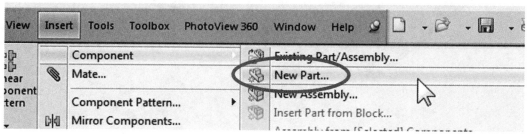

151. – Select the assembly's *"Right Plane"* in the FeatureManager to locate the new part's *"Front Plane"*; remember that when we locate the plane, at the same time we'll be adding an "InPlace" mate.

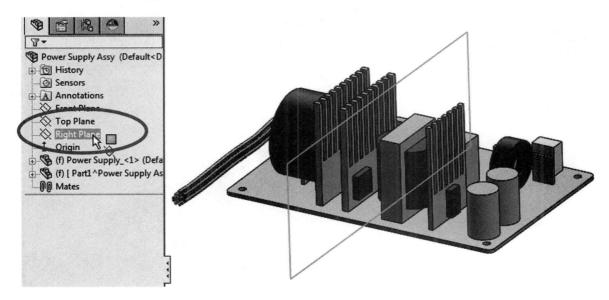

Immediately after selecting the assembly's *"Right Plane"* in the FeatureManager, this is what will happen:

- The electronics board will become transparent (if the option is set in "**Tools, Options, Display/Selection, Force assembly Transparency**").

- The "**Edit Component**" icon in the CommandManager will be enabled.

- A name will be assigned to the new part in the FeatureManager tree if the option "**Tools, System Options: Assemblies, Save New Components to external files**" is NOT checked; if set, you will be asked for a new part name.

- The new part's FeatureManager will be blue (default setting).

- A new sketch will be created in the new part's *"Front Plane."*

- The status bar will read *"Editing Sketch."*

- The title bar will read:

    Sketch1 of Part#^Power Supply Assy -in- Power Supply Assy.SLDASM

    Which means:

    Sketch1: We are editing the new part's first sketch (Sketch1)

    of Part#^Power Supply Assy: of **internal** *Part#* in the *'Power Supply Assy.'*

    -in- Power Supply Assy.SLDASM: in the *'Power Supply Assy.sldasm'* file.

- An "**InPlace**" mate will be added automatically, fixing the *"Front Plane"* of the new part to the selected plane or face, in our case, the assembly's *"Right Plane"* restricting all six degrees of freedom of the part.

As we can see, there are plenty of visual clues to alert us when we are editing a part in the context of an assembly.

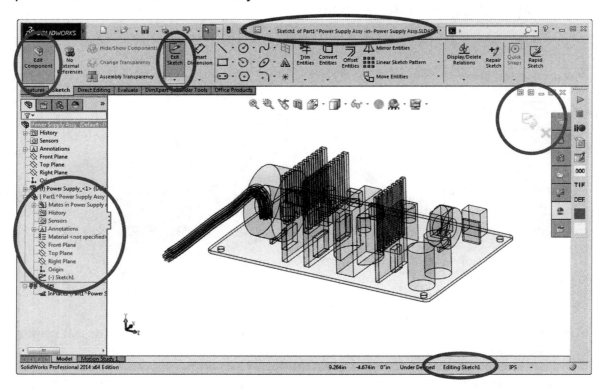

152. – Now we can start the sheet metal work. If not already visible, activate and select the Sheet Metal toolbar in the CommandManager and/or individual toolbar by either a right-mouse-click in the CommandManager's tabs and selecting "**Sheet Metal**" from the pop-up menu, or in the menu "**View, Toolbars, Sheet Metal**."

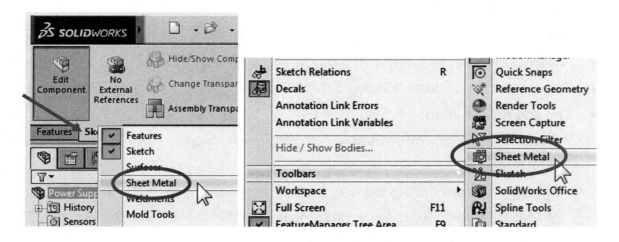

The Sheet Metal toolbar has most of the commands needed, and looks like this:

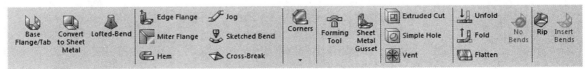

Command Manager

Floating Toolbar

153. – When working with sheet metal components we have multiple tools available to us, as well as different techniques to start a sheet metal part. We can:

- Make a bent part (using a Base Flange) and add more sheet metal features.

- Start with a flat pattern (unbent sheet metal) and add features and bends.

- Convert a solid model to sheet metal by adding bends and rips.

In this example we'll start making our sheet metal part using a "**Base Flange**," and will cover multiple different new features and commands while completing the first component. Before we start the first part, let's take a look at some of the most common sheet metal features and the terminology used.

Icon	Feature Description	Example
Sheet-Metal	Definitions of default bend parameters including sheet metal thickness, bend radius, relief type and related options.	N/A
Base Flange/Tab	Base Flange/Tab will be the first sheet metal feature (or additional tabs) in the part. If it's the first feature, it will create the required Sheet Metal features automatically (Sheet-Metal, Base-Flange and Flat-Pattern).	

Base Flange/Tab	Tab: Tabs are used to add more material to an existing sheet metal part using a sketch. Tabs can be bent later.	
Flat-Pattern:	Represents the sheet metal part in its unbent state. If this feature is suppressed, the part is bent; if it's unsuppressed, the part is unbent (flat).	
Convert to Sheet Metal	Used to convert a solid, surface or imported model to Sheet Metal by adding rips and bends to allow the part to be flattened.	
Edge Flange	Used to add flanges to model edges. Flange shapes can be modified and do not need to be the full length of the edge. Circular edge flanges are allowed.	
Miter Flange	Used to automatically add flanges to a series of adjacent edges, including non-planar flanges.	
Hem	Used to add a hem to a linear edge. Mostly used to remove sharp edges and reinforce a sheet metal part.	

Jog	A jog will add two bends on a model using a single sketch line.	
Sketched Bend	A sketched bend will add a bend to a flat area defined using a single sketch line.	
Corners	Used to close corners after relief cuts are made.	
Forming Tool	Used to create indentation forms using forming tools like louvers, lances, flanges, etc.	

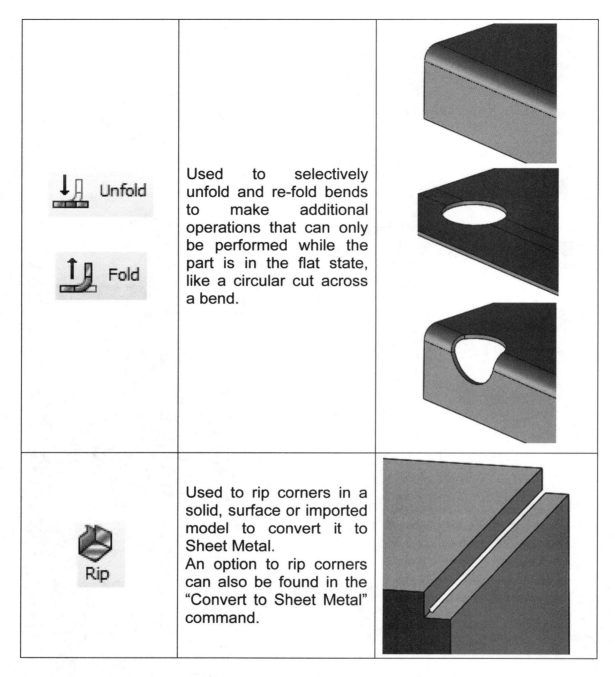

	Used to selectively unfold and re-fold bends to make additional operations that can only be performed while the part is in the flat state, like a circular cut across a bend.	
	Used to rip corners in a solid, surface or imported model to convert it to Sheet Metal. An option to rip corners can also be found in the "Convert to Sheet Metal" command.	

154. – Now let's design the lower part of the sheet metal enclosure. Select a Left view orientation (Assembly view orientation), and draw the following sketch. Remember that we are already working in a new sketch in the new part inside the assembly. Note that this is an open sketch and is dimensioned referencing the lower and outer edges of the electronic board. Both vertical lines are equal length.

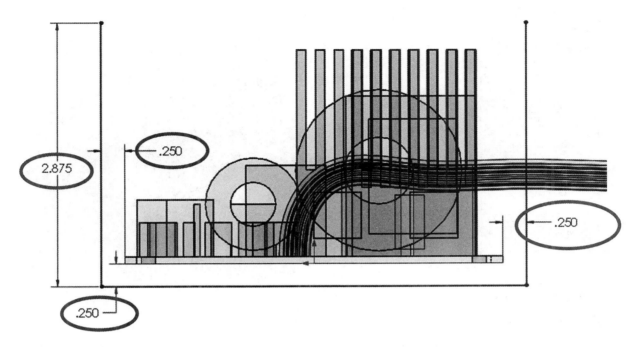

A sheet metal component is essentially a thin extrusion with a few more parameters that we'll explain in the next step. With our sketch complete, create the Base Flange feature selecting the "**Base-Flange/Tab**" icon from the Sheet Metal toolbar in the CommandManager. Remember, we are designing the enclosure <u>in the context of the assembly using a Top Down design approach</u>.

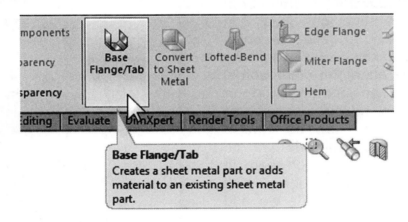

155. – Extrude the "**Base Flange**" 7″ deep using the "**Mid plane**" end condition. In addition to the end condition for the extrusion we see more options to control the sheet metal's thickness, bend radius, bend allowance and auto relief type. For our example, use a **thickness** of 0.030″ with a **bend radius** of 0.05″ and a **K-Factor** of 0.5 in "Bend Allowance." **Auto Relief** will be set to "Tear." In this case we want to have a 0.25" space between the part and the electronic board, click in the "Reverse Direction" checkbox to make sure the material is added outside, and not inside.

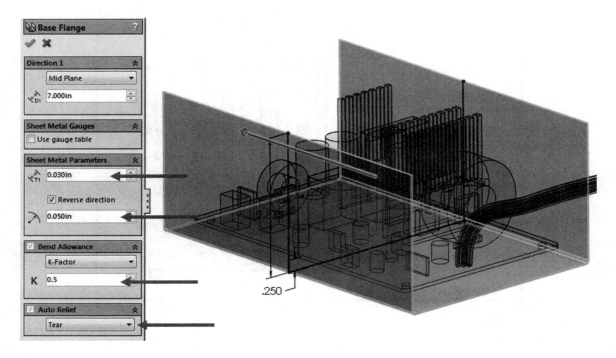

What the options in the "Sheet Metal Parameters" mean:

Thickness of the material: This is the uniform thickness of the material that will be used to make the part.

"Reverse direction" checkbox: Since a sheet metal part is a Thin Feature, we need to define on which side of the sketch we want the material to be added; in other words, if the given dimensions are *inside* or *outside* dimensions. In our example we want them to be *inside* dimensions and **the material will be added *outside*.** It may not seem to be important at first, but if we don't pay attention to which side we add the material in our designs, the resulting part will be one material-thickness bigger or smaller, and that *may* ruin your design. Remember to pay attention to this very important detail when designing sheet metal components and not ignore it.

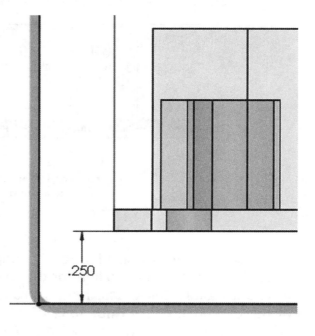

Default bend radius: The sheet metal bend radius measured on the inside of the bend.

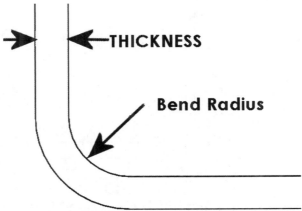

The "**Bend Allowance**" section refers to the way SolidWorks calculates how much material will be consumed by a bend. When we bend a sheet metal component, the material on the inside face of the bend is compressed and the material outside is stretched. For example, if we take a 4″ long strip of metal, and bend it 90 degrees in the middle, the length of the flat portions plus the bent region will not necessarily add to 4″. SolidWorks has three methods to calculate the flat pattern length: **Bend Allowance**, **Bend Deduction** and **K-Factor**.

- **Bend Allowance:** When we use the Bend Allowance method we specify a length of metal that will become the bent region and is calculated by <u>adding</u> a specified length to the flat regions.

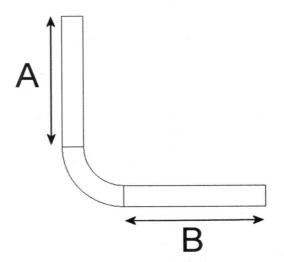

Flat Pattern Length = A + B + Bend Allowance (BA)

- **Bend Deduction:** If we use a Bend Deduction, we <u>subtract</u> a specified length to obtain the flat pattern length.

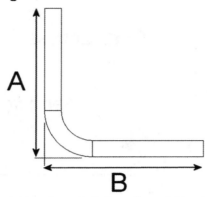

Flat Pattern Length = A + B – Bend Deduction (BD)

- **K-Factor:** As we explained before, the material inside the bend is compressed and the material outside is stretched. The line between these regions is known as the **neutral axis**, which is the line that does not change length. The ratio between the distance of the neutral axis to the inside face and the sheet metal's thickness is known as the K-Factor.

K-Factor = t / T

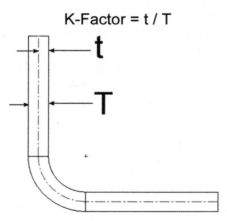

The bend allowance (BA) is then calculated by:

$$BA = \pi (R + KT) A/180$$

Where:

R = inside radius (Bend Radius)
A = Bend angle in Degrees
T = Material Thickness
K = K-Factor (t/T)

The method to calculate a sheet metal's bend allowance varies by preference, experience, availability of data, material, etc. There are numerous tables, books, references and guides to help us calculate it, and since these parameters vary based on material, thickness, bend angle, material grain direction, process and even temperature, we will not go into those details as they are well beyond the scope of this book; to simplify our examples we'll direct the user to use a generic K-Factor value for our examples.

The "**Auto Relief**" section refers to the relief cuts generated by SolidWorks for sheet metal fabrication processes. These cuts are added as needed for bending purposes. During the bending process, it's common to make small cuts (reliefs) at the end of bends to allow the metal to bend properly and not deform the material unexpectedly.

The types of Auto Relief available are:

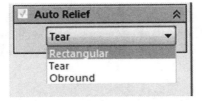

- **Rectangular:** In a rectangular relief, a cut is made using a ratio or by specifying dimensions. The ratio is the cut's size (Dim) calculated based on the sheet metal's thickness as follows:

$$Dim = (Material\ Thickness) \times (Relief\ Ratio)$$

- **Obround:** Is just like the Rectangular relief, but the relief's cut is rounded.

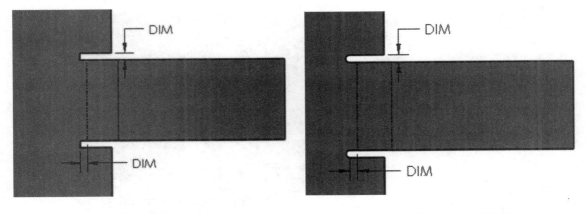

Rectangular Relief   Obround Relief

- **Tear:** In a tear the smallest possible relief cut is made, just enough to allow the metal to bend as intended. This is similar to ripping the material at the time of bending during fabrication.

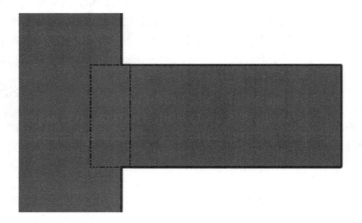

156. – After extruding the *"Base-Flange1"* with the defined parameters, our part looks like this. Keep in mind that we are still editing it in the context of the assembly, and that's the reason why the electronics board is still transparent.

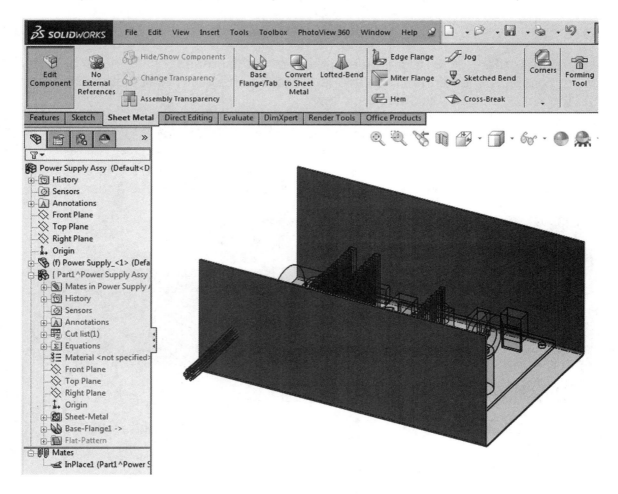

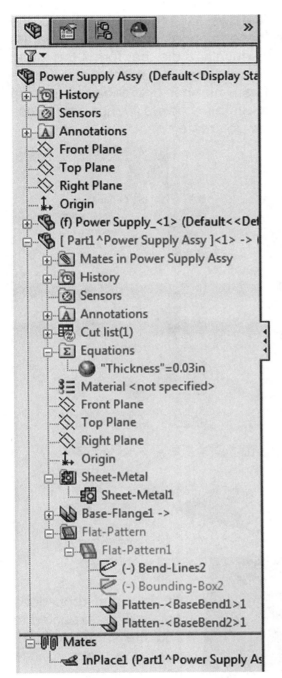

157. – The FeatureManager now shows the new part and the Sheet Metal features.

In *"Part1^Power Supply Assy"* (in blue because we are editing it in context of the assembly), we see the new sheet metal features added by the "Base-Flange" feature.

"Cut list": Shows the number of bodies in the part. If multiple sheet metal bodies are added to a part, they will be listed under this folder.

"Equations": Lists the equations and Global Variables in the part. For sheet metal components, a Global Variable is added automatically called *"Thickness"* which is used to make dimensions equal to the material's thickness.

Inside the *"Sheet Metal"* folder we see **"Sheet-Metal1,"** which is the feature that defines the default sheet metal parameters of the part including material thickness, default bend radius, bend allowance and relief type.

"Base-Flange1" is the first feature of the sheet metal part. Just like other sketch-based features, it includes the sketch used to make it and the bends created when the feature was made. Notice the "In Context" icon (**->**) next to the feature, letting us know that the part has external relations in the context of the assembly, in our case the "InPlace" mate and the dimensions to the board.

The *"Flat-Pattern"* folder contains the *"Flat-Pattern1"* feature, and is suppressed when the part is bent (default). To show the part in its flat (unbent) state we have to unsuppress it. The *"Flat-Pattern(1)"* feature includes a sketch (*"Bend-Lines2"*) automatically generated with the part's bend lines and a list of all the bends in the part, as well as a *"Bounding-Box2"* sketch, which is a sketch with the outline of the smallest square in which the unbent part can fit, useful to calculate the minimum material size required for fabrication.

127

158. – The next step is to add mounting features to screw the board to the enclosure. The way we are going to create these supports is by using a feature called "**Forming Tool**." A Forming Tool is essentially a die that forms the sheet metal by bending, cutting and stretching as if it was pressed into the part, hence the *"Forming"* part of the name. SolidWorks has a number of pre-loaded forming tools to make flanges, louvers, lances ribs, etc. In this step we'll learn how to use an existing forming tool and later we'll learn how to create one.

Rotate the model to approximately this orientation, expand the Design Library, from the "**forming tools**" folder select the *"lances"* subfolder and select the *"bridge lance"* forming tool. To apply it to our model, drag it onto the bottom face of the model, and before dropping it make sure the tool is going 'into' the part (pressing the metal from the outside in), as in the following image. To reverse the direction of the forming tool press the "Tab" key before releasing the mouse button.

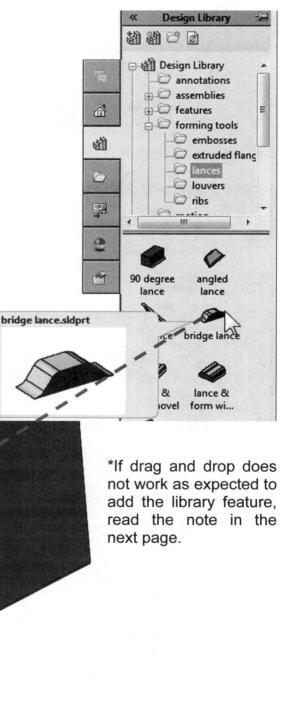

*If drag and drop does not work as expected to add the library feature, read the note in the next page.

If after dropping the forming tool into the sheet metal part SolidWorks asks *"Are you trying to make a derived part?"* click "**No**." What happens is that the *"forming tools"* folder is not defined as such. To fix it, right-mouse-click in the "**forming tools**" folder in the Design Library and turn on the "**Forming Tools Folder**" option. Now you can drag-and-drop the *"bridge lance"* form tool into the sheet metal part.

159. – After dropping the forming tool in the part, we use the "**Form Tool Feature**" PropertyManager to define the face where we want to locate it, if we want to rotate it, flip the form tool to go in or out, etc.

160. – If needed, we can rotate the form tool and change the direction the lance is going. For the "Rotation Angle" make sure we have "90deg" for the alignment we need. If the lance is not going in the correct direction, press "Flip Tool" to change it and match the next image. Here we can also change the face to add the form tool in the "Placement Face" selection box.

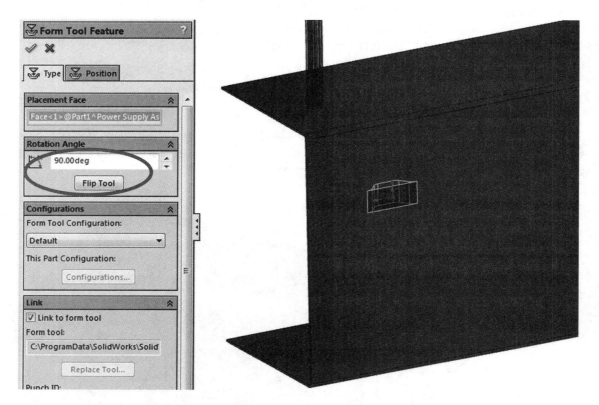

161. – After making sure the form tool has the correct orientation and direction we have to locate it accurately. Select the "Position" tab at the top of the properties and change to a bottom view with "Hidden Lines Visible" mode for easy visualization.

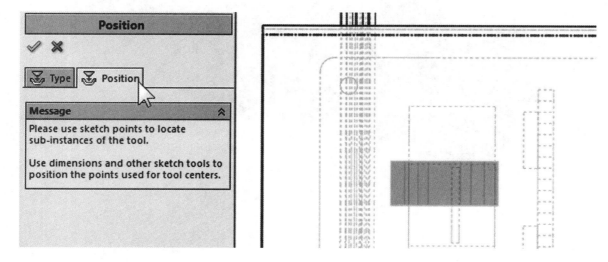

162. – The process to locate the form tool feature is similar to the hole wizard. Notice we have the sketch point tool active, and we can add all four locations needed at the same time. Since we want to locate the form tool features concentric with the mounting holes in the board, touch the edge of each of the other three holes in the electronics board to reveal their centers, and add a point concentric to each one.

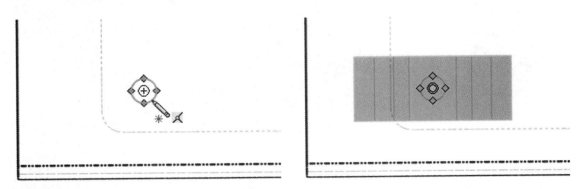

163. – Remember we still have one point close to one corner where we dropped the form tool at the beginning. To locate the last form tool, turn off the "Sketch Point" tool and add a concentric relation between the point and the edge of the last mounting hole.

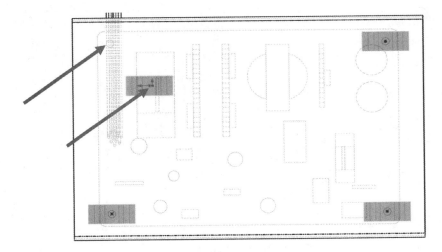

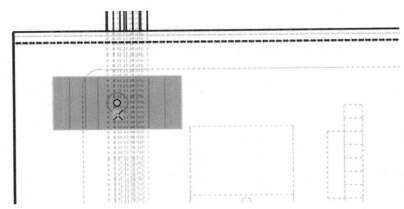

164. – After locating the last form tool, click OK to finish. The finished part now looks like this.

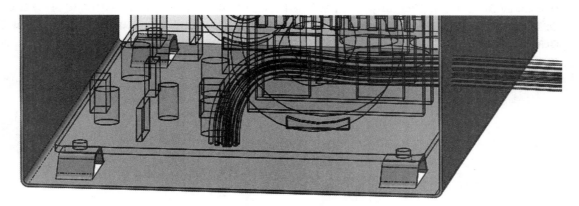

165. – After adding the lances we need to make holes for the screws that will hold the board to the lances. Switch to a top view, select the top of any of the lance form features just added, and add a new sketch. Draw four circles concentric to each of the electronics board mounting holes, make them all equal, and dimension one of them 0.1″ in diameter.

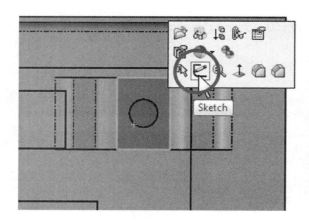

166. – After drawing the four equal circles, select the "**Extruded Cut**" icon to make a cut using all circles at the same time. Notice the extra options in the "Direction 1" box. Use the "Link to Thickness" option to make the cut exactly as deep as the sheet metal thickness, essentially making the depth equal to the "Thickness" global variable created when we added the first sheet metal feature.

The second new option is "**Normal Cut**." A normal cut means the cut will be made perpendicular to the sheet metal, even if the cut is being made at an angle. The reason for this is because in a sheet

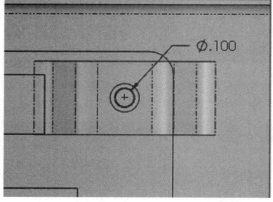

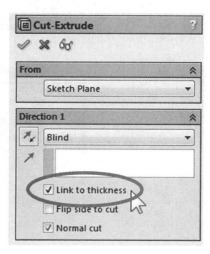

metal manufacturing process is very difficult to make cuts at an angle in thin metal, it's not practical, and very unlikely needed. Therefore, this option exists by the very nature of the manufacturing process itself. In our example it makes no difference because the cut's direction is perpendicular to the part. Click OK to add the holes. If the cut had been made oblique (at an angle) to the surface, this is what happens:

Oblique cut in sheet metal (section view)

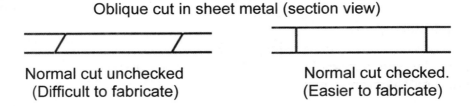

Normal cut unchecked   Normal cut checked.
(Difficult to fabricate)   (Easier to fabricate)

167. – After adding the mounting holes we need to add a fan to help us cool the power supply. In many cases we'll need to include purchased components in our designs, like electrical components, hardware, etc. Many suppliers provide 3D models of their products, saving designers the trouble of modeling purchased parts, reducing the risk of design errors and at the same time improving the chances of having their products selected for our designs, making it a win-win situation. There are multiple sources of 3D models, from manufacturer's web sites to portals that consolidate multiple suppliers and user submitted models, like SolidWorks' 3D Content Central. 3D Content Central is a free portal that is also integrated in SolidWorks, and can be accessed directly from within SolidWorks or on the web at www.3dcontentcentral.com.

Remember that at this point we are editing the sheet metal part *in* the assembly, and now we need to add a new component to the assembly. To do this we need to stop editing the part, and return to editing the assembly. Select the "**Edit Component**" toggle button in the toolbar or CommandManager, *or* the confirmation corner *or* right mouse click and select "**Edit Assembly: xxxx**" to stop editing the sheet metal part and return to the assembly level.

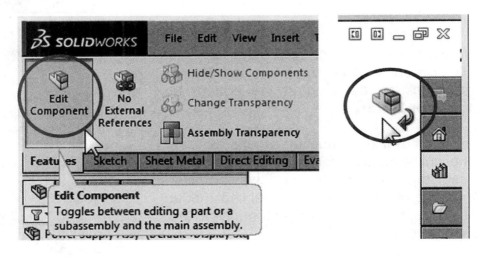

133

When we stop editing the part and return to editing the assembly:

- The sheet metal part's FeatureManager is no longer blue.
- The Power Supply electronics will become opaque again (following the transparency rules set in the options.)
- The status bar will read "Editing Assembly."
- The "Edit Component" button will be disabled (since no component is being edited.)
- The confirmation corner will be off.
- The title bar will read *'Power Supply Assy.sldasm.'*

Basically, the assembly looks like any other assembly and now we are ready to add more components to it.

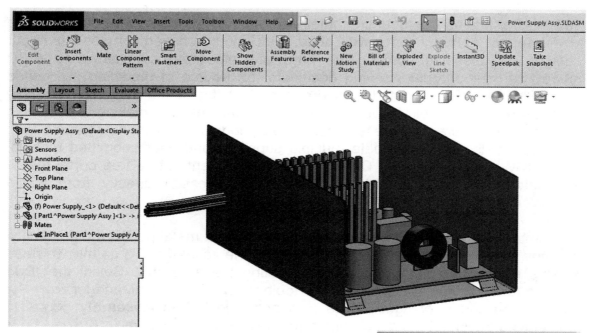

168. – In SolidWorks, select the Design Library, go to "**3D ContentCentral**, **User Library, Home Page**" and click on the icon at the bottom to display the **User Library**.

A web browser will be launched where we can search the online catalog for components. Users need to register (free) in order to download files. In our example we'll use a user supplied model instead of a supplier certified component for the power supply we need.

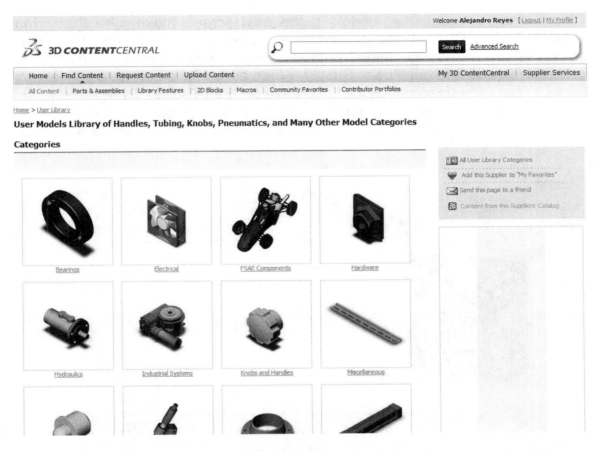

In the search box, type "*PABST 512F.*" This is the exact model we'll use; however, users would typically search by model or part number, or make a general search for "Fan", and select a suitable model from the available options. The search now shows the part we are interested in.

169. – After selecting the model, in the next screen make sure the format selected is SolidWorks 2014 (or earlier) and download the 3D model. After we are presented with the download window, we can drag the model's link into the assembly to add it to our Power Supply. When asked for a part name, select the folder to save the downloaded part and save as **"Fan-PABST 512F."** The model will be automatically added to your model. Save it and return to the assembly. Optionally save the part to your PC and add it to the assembly as we have done before.

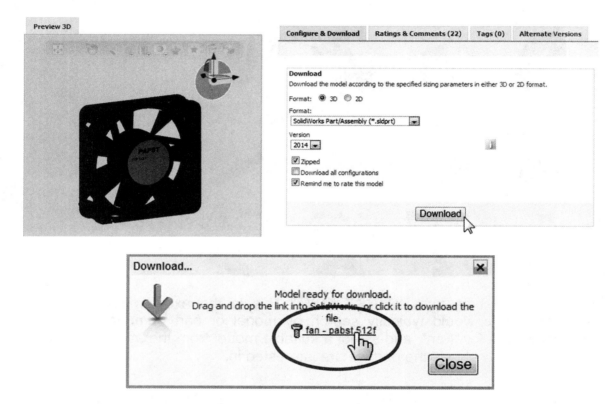

IMPORTANT: Drag and drop *may* only work when using Internet Explorer, for other browsers click in the link to save the file to disk and manually add to the assembly.

Select the folder to save the model...

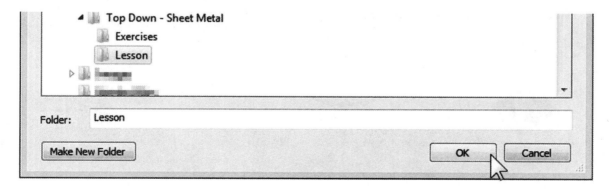

… and add the fan to the assembly, locating it approximately as shown.

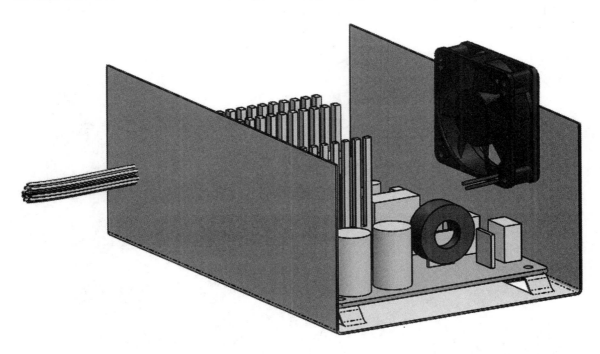

170. – Add a Coincident mate between the Fan and the inside wall of the enclosure.

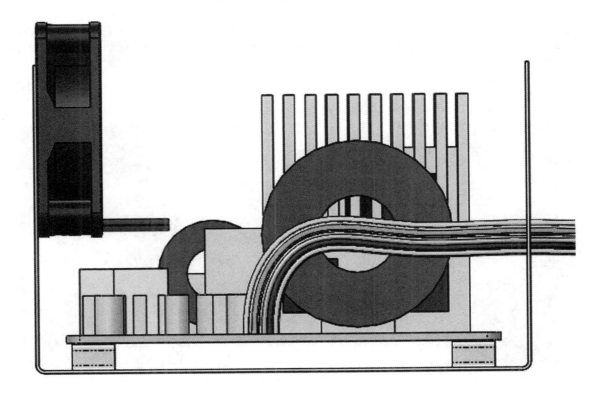

171. – Then add a second Coincident mate between the *"Right Plane"* of the Fan and the face indicated in the electronics board, which places the fan about half way between the heat sinks.

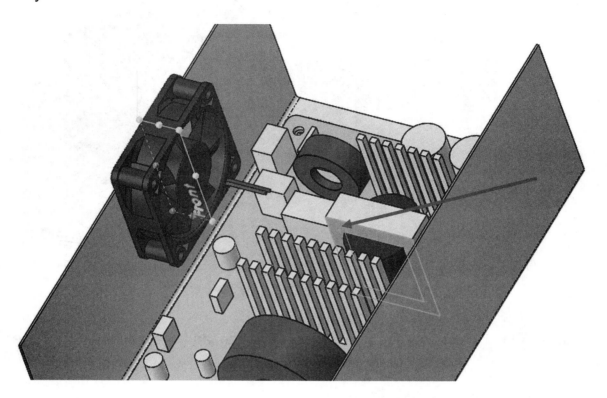

172. – To finish locating the Fan, add a 0.2″ Distance mate from the top face of the Fan to the top of the sheet metal part. Our Fan is now fully defined in the assembly.

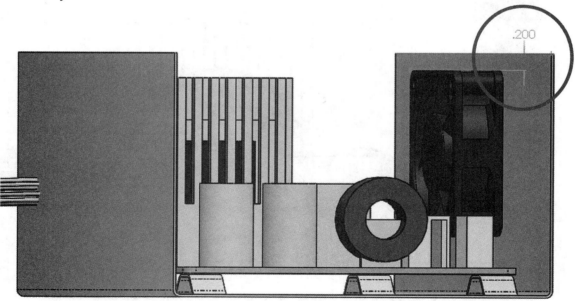

173. – Now with the fan fully defined in place, we need to edit the sheet metal part to make a ventilation cut. Select the sheet metal part in the FeatureManager or in the graphics area and select "**Edit Part**" from the pop-up menu or the "**Edit Component**" command in the CommandManager.

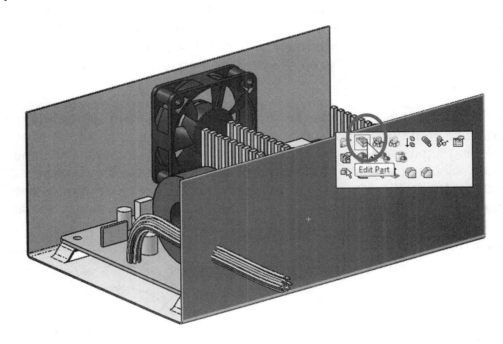

174. – We are now editing the sheet metal part in the assembly; the electronics board and the fan are transparent and the box is opaque. Switch to a Back view, change the display mode to "Hidden Lines Visible," and hide the electronics board component to improve visibility.

 While editing a component in the assembly, we can hide, show and change the transparency of other components in the assembly at will.

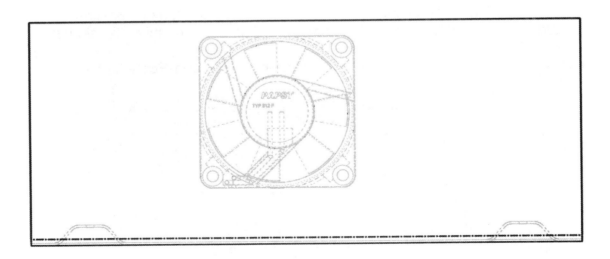

175. – To add ventilation to our power supply, we'll make a cut using a feature called "**Vent**" using the fan's location for reference. Select the face in the back of the sheet metal part and make the following sketch.

By matching the vent's location to the Fan, if the Fan's position needs to be changed the vent location will change as well; that is the idea behind designing in the context of an assembly. Notice the sketch used for the vent has intersecting lines, and the only reference to the Fan is a Concentric relation to the Fan's center.

First make the first circle concentric to the Fan, then either change to a "Hidden Lines Removed" view mode and/or hide the Fan to finish the sketch and avoid adding unintended geometric relations to the Fan. We only need one concentric relation to the Fan for the entire sketch. To make things easy, add one circle concentric to the fan, change to Hidden Lines Removed mode, and finish the sketch, this way is easier and we won't capture unwanted external relations.

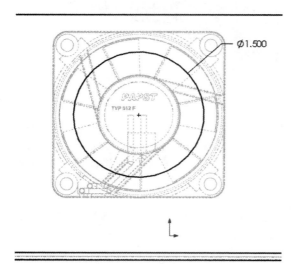

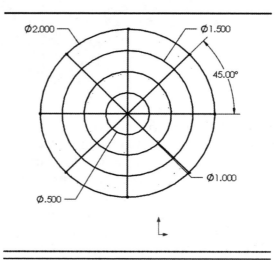

Hidden Lines Visible
(to add concentric relation to first circle)

Hidden Lines Removed
(for clarity to finish the sketch)

176. – To add the vent select the menu "**Insert, Fastening Feature, Vent**."

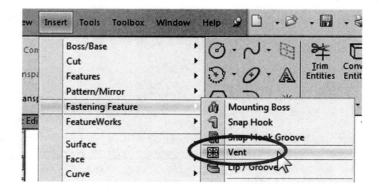

140

In this step the fan was hidden for clarity. Select the outer (2″) circle in the "Boundary" and set the "Radius for the fillets" value equal to 0.05″ (this will be the fillet radius for all inside edges). In the "Ribs" selection box select all the straight lines and make their width 0.1″. For the "Spars" selection box select the 1″ and 1.5″ diameter circles and make them 0.1″ wide. Finally, select the inner circle in the "Fill-In Boundary." The Fill-In Boundary is an area that will be covered. Note the "Flow Area" section lists the total area and the open area percentage for reference. Click OK to complete the Vent feature.

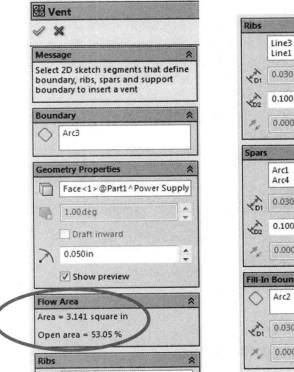
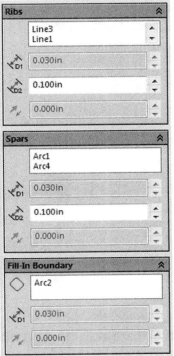

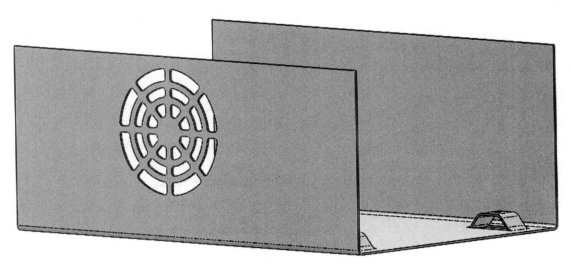

Fan and electronics board hidden for visibility

177. – So far the sheet metal component has been an *'internal'* component in the assembly (unless, if we were asked for a name when the component was added). If you remember, an internal part means that it only exists inside the assembly; there is no file on the hard disk that we can make a reference to. Since we need to have a *part file* in order to be able to make a drawing of it, we need to *externalize* the part. (It's not possible to make a drawing of a part internal to an assembly.) Since we never gave the part a (file) name, it was automatically assigned by SolidWorks. In this step we'll make the part external. To externalize the part, select it in the FeatureManager with a right mouse click, then select **"Rename Part"** from the pop-up menu and change the name to *"Box Bottom."*

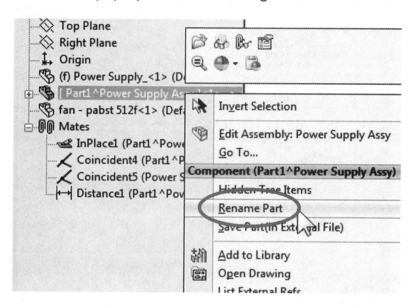

178. – After renaming the part, select it again with a right mouse click and select **"Save Part (in External File)"** from the pop-up menu.

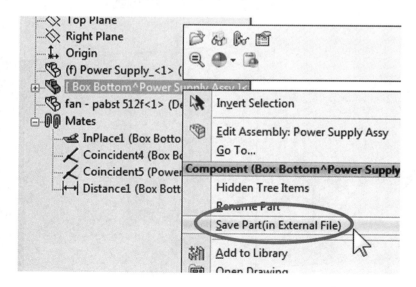

Make sure the path is correct and click OK. The *'Box Bottom'* has been saved as a part file and now looks the same as any other part in the assembly's FeatureManager. Note that we are still editing it inside the assembly.

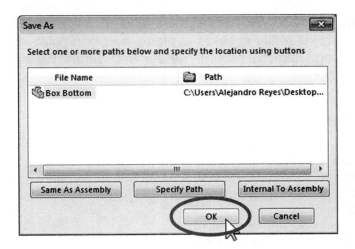
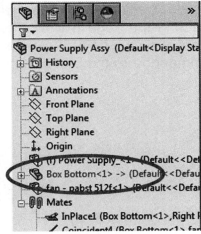

179. – At this point we can stop editing the part in the assembly and open it in its own window, or continue working in the context of the assembly. We'll edit the part in its own window as much as possible to make it easier to visualize the process, and revert to editing in the assembly as needed to take into consideration other components. At this point we can optionally stop editing the part in the assembly or not before opening the part in its own window. For our example turn off the **"Edit Component"** icon…

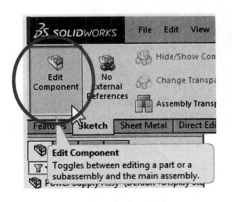

or

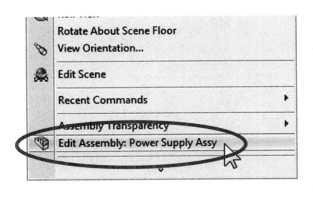

… and then open the *'Box Bottom'* part in its own window.

180. – The next step is to reinforce the sides to make our sheet metal part stronger by adding a miter flange. To do this we need to make a sketch perpendicular to an edge. One way to do this is to select the thin face at the end of the part and insert a sketch in it (which is difficult unless we zoom in really close), or select the **Miter Flange** icon and click near the end of the edge. By selecting the edge SolidWorks automatically creates a plane perpendicular to it where we can draw the miter flange sketch profile. Select the "**Miter Flange**" command from the Sheet Metal tab,

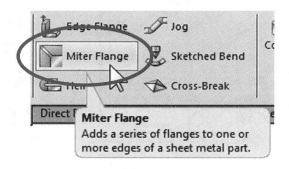

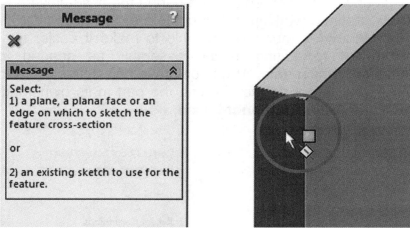

Zoomed in VERY close to select face (harder)

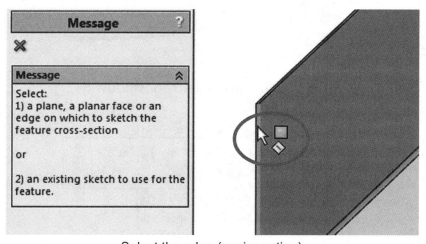

Select the edge (easier option)

181. – Select an edge (a plane and a sketch will be created automatically) and draw a single line perpendicular to the edge as shown; dimension it 0.25″ long. This will be the profile of the miter flange. Notice the new plane created at the end of the edge where our sketch is made.

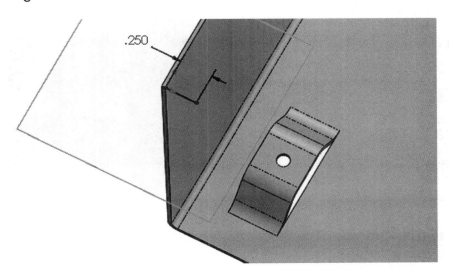

182. – Click on the "**Miter Flange**" command icon (or Exit Sketch) to continue. Notice the miter flange preview in the first edge. To continue the flange along the bottom and other side, either select the "**Propagate**" icon to automatically select these edges, or manually select them in the screen. Leave the "**Use default radius**" checkbox selected to make the flange's bends the default radius as defined when the sheet metal parameters were defined. "**Gap Distance**" is the gap at the corners of the miter. Leave this value at 0.01." The "**Trim side bends**" option removes material when a bend touches an existing bend (in this case there is no difference in the result.) In "Flange Position" select the "Material inside" option for our part. See the differences for flange position in the next page. Click OK to complete the miter flange.

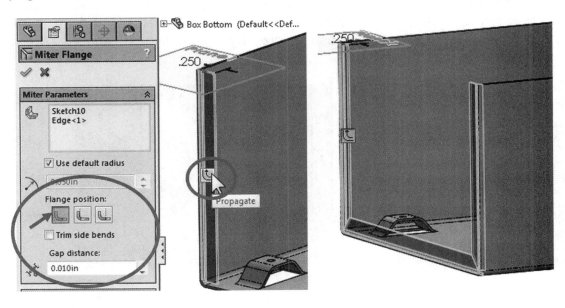

The different "**Flange Position**" options mean (Looking at the part from above):

- **Material Inside**: The Flange added will not extend beyond the existing edge of the part (indicated by the arrow), maintaining the original part dimensions.

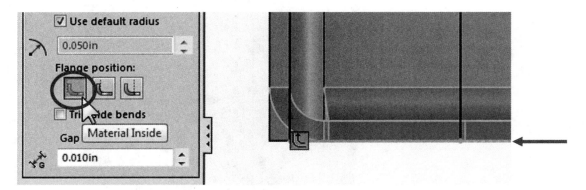

- **Material Outside**: The flange added will be one material thickness outside the existing edge (indicated by the arrow).

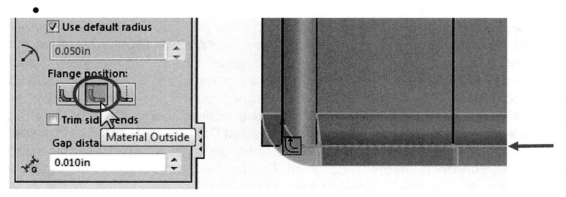

- **Bend Outside**: The flange will leave the existing edge in its original position (indicated by the arrow) and will add the material and the bend outside. The part will increase in size by one thickness and one bend radius.

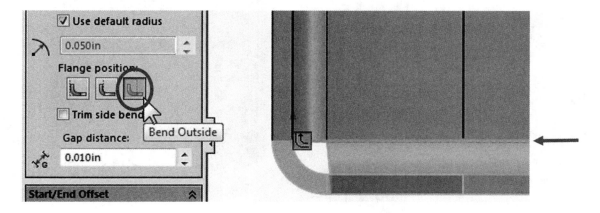

183. – Add a second "**Miter Flange**" to the other side using the same parameters as the first one. If we select the edge, a new plane will be added.

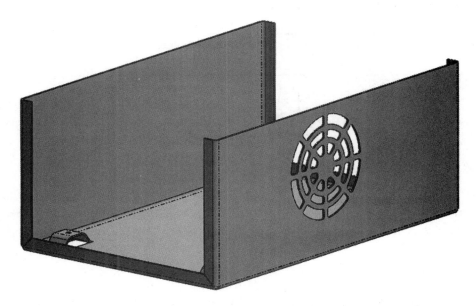

184. – Now we need to add a cut on one side to allow the power supply wires to exit the enclosure. Go back to the *'Power Supply Assy'* and make the electronics board visible. Select the *'Box Bottom'* part either in the FeatureManager or the assembly and click in "**Edit Part**" to

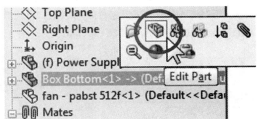

continue, this way we can add features in the context of the assembly and reference other components. Hide the fan if wanted or needed.

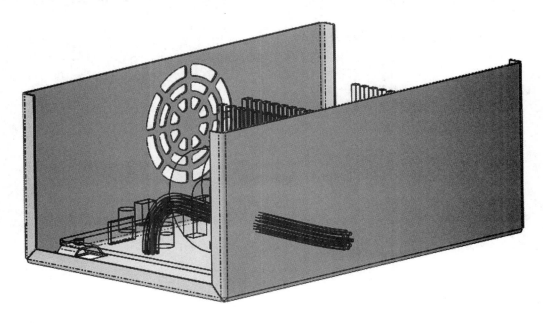

185. – To make the cut, first we need to unfold the miter edge closer to the wires. We'll use the "**Unfold**" command from the Sheet Metal tab, or from the menu "**Insert, Sheet Metal, Unfold**."

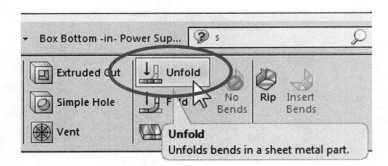

 The cut can be made without unfolding the flange, but we chose to do it this way to show the reader a different approach and cover more functionality.

In the "Fixed face" selection box select the model's face that will remain fixed; in other words, when we unfold the miter flange, the flange will move out instead of the other way around. In the "Bends to unfold" select the bend region of the flange. Click OK when done.

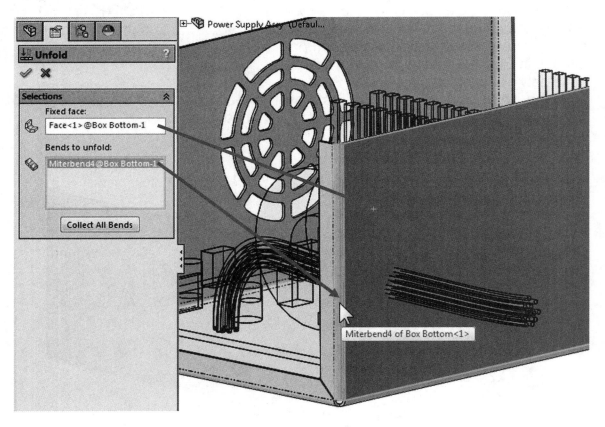

After selecting OK the miter flange unfolds.

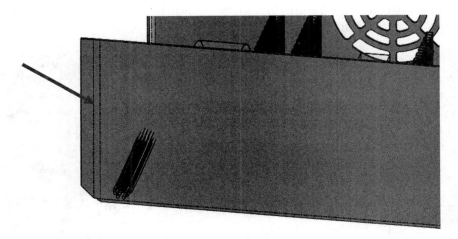

186. – While unfolded we can add the cut for the wires and then re-fold the flange. Switch to a Front view and make the next sketch around the wires. Make the cut using the "**Link to thickness**" option. Remember we are editing the '*Box Bottom*' part in the assembly (Hidden Lines Removed for clarity.)

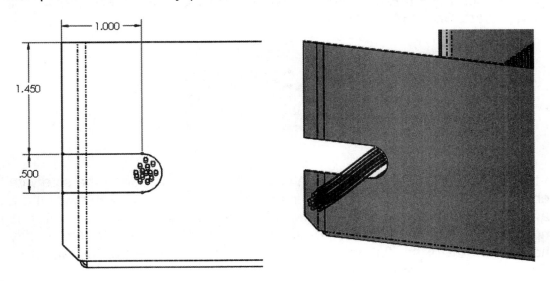

187. – To re-fold the flange after making the cut, select the "**Fold**" command from the Sheet Metal toolbar or from the menu "**Insert, Sheet Metal, Fold**."

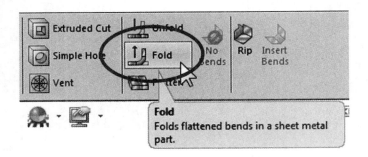

The "Fixed face" is automatically selected; all we have to do is select the bend region (SolidWorks only allows us to select bends). Clicking in the "Collect All Bends" button will add all unfolded bends.

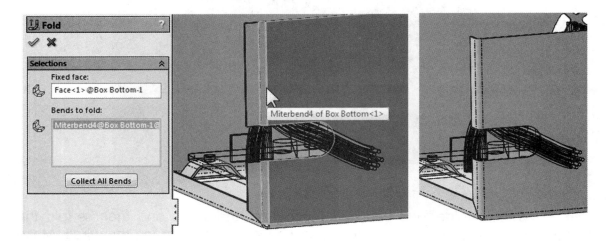

188. – The next step is to add flanges at the top of the box to screw the cover in place. Select the "**Edge Flange**" command from the Sheet Metal toolbar, or the menu "**Insert, Sheet Metal, Edge Flange**."

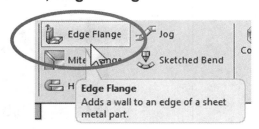

In the "Flange Parameters" selection box, select the top edge of the side with the vent. As soon as the mouse pointer is moved, you can see a preview of the edge flange. Click towards the center of the box to add the flange inside; the size is not important at this time. Now click on the "Edit Flange Profile" button to modify the shape and size of the flange. (If we want the flange to be the full length of the edge, there is no need to edit the profile.)

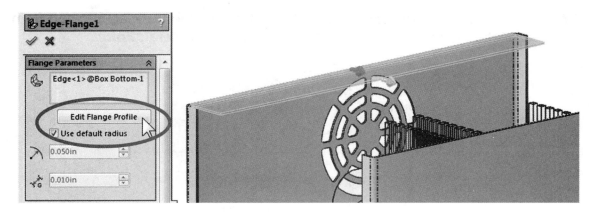

We can modify the sketch as much as we want, as long as we end with a single closed profile. Drag the lines on the sides towards the middle, delete the short line in front and replace it with a tangent arc. The lines can be dragged even if they appear fully defined. Add dimensions as needed to complete the sketch.

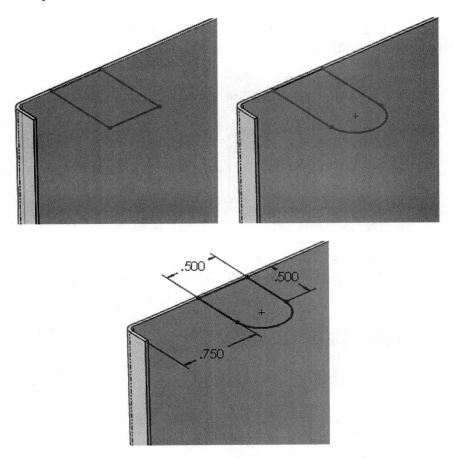

189. – Click the "Back" button to return to the "Edge Flange" command when the sketch is complete.

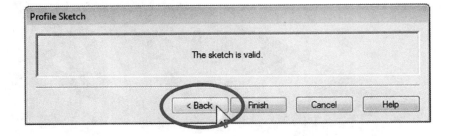

 If we select "Finish" the flange will be completed with the parameters entered before editing the sketch. This message will let us know if the sketch profile is valid or not.

In the "Flange Parameters" set the angle to 90 degrees. In the "Flange Position" selection box, select the "Material Inside" option, and turn on the "Custom Relief Type" checkbox. Select an Obround relief for the flange with a 0.5 ratio to override the default settings. Click OK to add the edge flange.

 For rectangular or obround relief types we can also define the relief by specifying the width and depth.

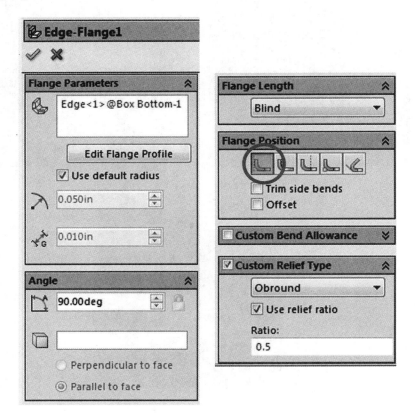

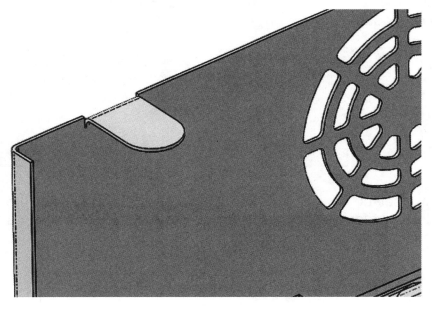

190. – Now we need to add a hole for a screw to the tab we just made. In this case we'll use one of the built in Form Tools to do it. Open the "Design Library" in the task pane, and scroll down to "forming tools, embosses." From the list of tools select "extruded hole." Drag-and-drop it in the top face of the tab we just made. Make sure it is going *into* the part, and not out. We can reverse the direction of the form tool by pressing the Tab key before releasing the mouse when we drop it or using the "Flip tool" button later.

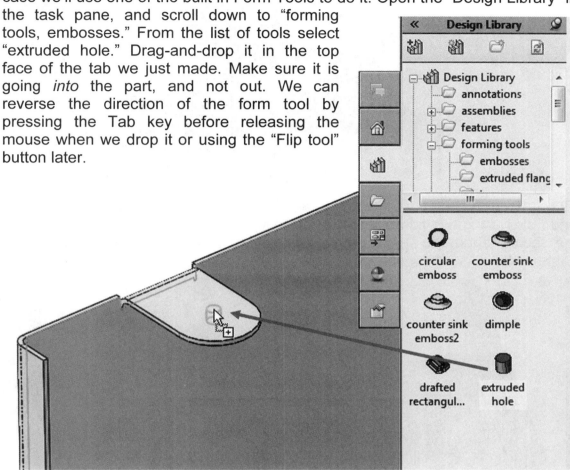

191. – In the "Position" tab of the form tool properties, make the locating point concentric with the edge of the round tab and click OK to finish.

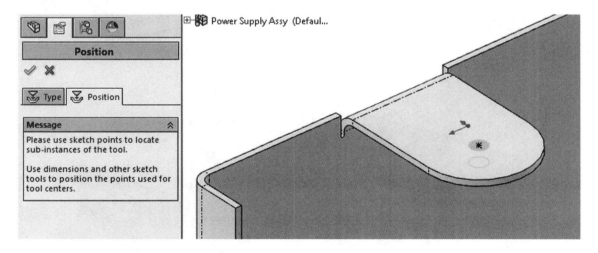

Our part should now look like this:

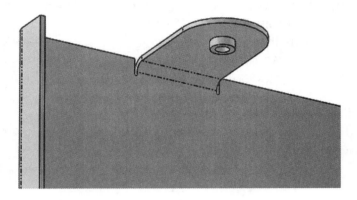

192. – The part requires four flanges (and holes). Add the other three flanges and holes using two mirror features: the first about the part's *"Front Plane"*, and the second about the *"Right Plane."* Remember that we are still working in the context of the assembly at this point.

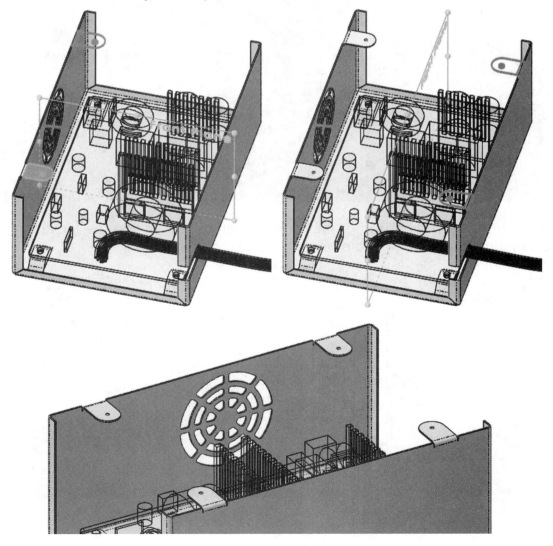

193. – Form tools can be used only in sheet metal parts, but a **library feature** can be used in any part. To make a library feature that we can reuse in our designs, first we need to make a new part (outside the assembly), add the feature(s) we want the library to have and save it as a Library Feature.

In this example we'll make a Library Feature of a cutout for a power supply connector, and add it as a feature to our sheet metal part next to the vent.

To build a library feature, make a new part (NOT in the assembly), add a new sketch (any plane) with a 3″ x 2″ square starting at one corner, and make an extrusion 0.125″ thick. After extruding it, add the following sketch in a face and make a cut using the "Up to next" end condition. The first feature's size and thickness are not important; it just has to be big enough to hold the feature we want to make a library with.

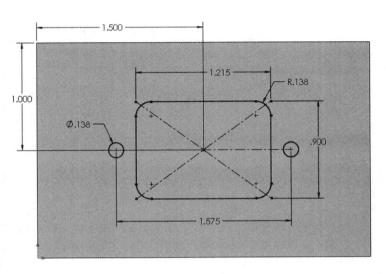

In this sketch we are only adding two dimensions to the edges of the first extrusion, essentially adding only two external references in this sketch. Note that the origin is in a corner and not in the center. If we had made the rectangle's center at the origin, and the second sketch also in the origin, adding the 1″ and 1.5″ dimensions would have over defined the sketch.

194. – Once the cut is made, select the *"Cut-Extrude1"* feature in the Feature-Manager, select the menu **"File, Save as…"** and from the "Save as type" drop down selection box select "Lib Feat Part (*.sldlfp)." Name it *'Power Connector Cutout.'* Notice that as soon as we select the library feature type, we are switched to the "Design Library" folder. Save the library in this folder.

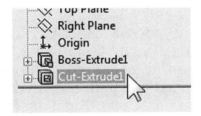

If the feature(s) to be saved as a library is not pre-selected in the FeatureManager, saving a library feature will not work.

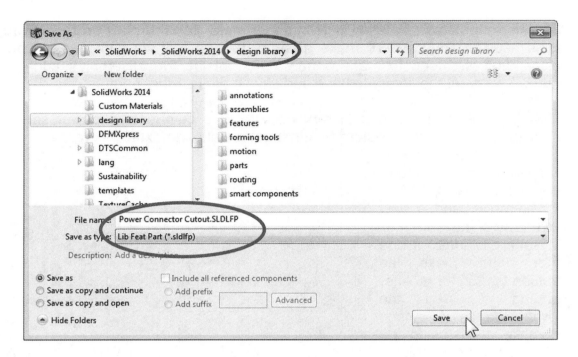

After saving the feature as a library, the FeatureManager changes to display it as such. A green letter "L" is superimposed in the feature(s) that are the library, and the icon at the top is changed to a library feature.

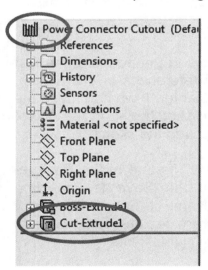
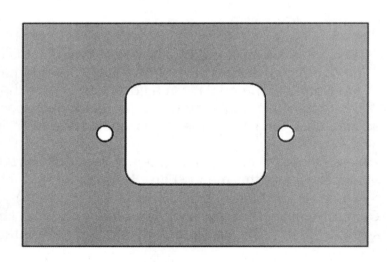

195. – After the library is saved, we can see two new folders added to the FeatureManager: *"Dimensions"* and *"References."* The *"Dimensions"* folder defines which dimensions can be changed by the user, and which will remain hidden when the library is added to a part, and which dimensions will be used to locate the library when added to a part. Right-mouse-click in the *"Annotations"* folder and turn on the "Show Feature Dimensions" option; also turn on the option to display dimension names in the menu "**View, Dimension Names**."

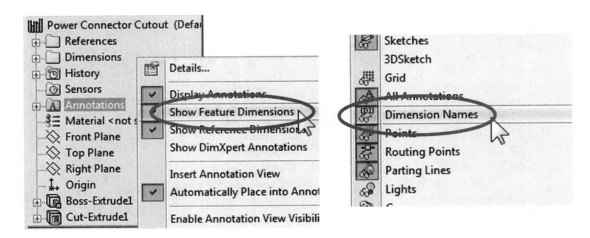

196. – Rename the two indicated dimensions as shown.

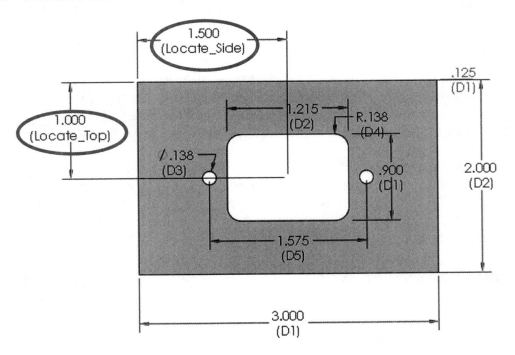

197. – After renaming the dimensions, expand the *"Dimensions"* folder, where we can see all of the part's dimensions listed. Drag-and-drop the *"Locate_Top"* and *"Locate_Side"* dimensions to the *"Locating Dimensions"* folder, and the rest to the *"Internal Dimensions"* folder.

Locating Dimensions are used to position the feature when we add the library to a part. **Internal Dimensions** cannot be changed by the user and the dimensions not added to either folder can be changed at the time of adding the library to a part. After adding the dimensions to the different folders, turn off the dimension display, save the changes and close the library file.

The library file we just made is now available to us under the main *"Design Library"* folder in the task pane (or the folder it was saved).

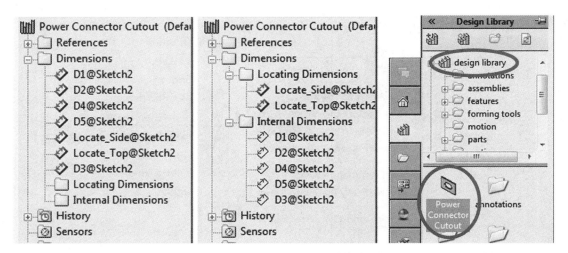

 Library features can be organized by dragging-and-dropping them to the folders in the Design Library; also, more folders can be added as needed.

 Library features can have multiple features; however, to make these features as easy to use as possible, be sure to add subsequent features referencing only the features included in the library, making the first feature a "parent" to all other features in the library. By doing this, the number of references needed when inserting the library is kept to a minimum.

198. – We are now ready to add the *'Power Connector Cutout'* library to the *'Box Bottom'* part, either in its own window or while editing it in the assembly (your choice). Open the task pane; select the *"Design Library"* folder, and drag-and-drop the *'Power Connector Cutout'* library onto the part. In this image we are editing the sheet metal part in the assembly.

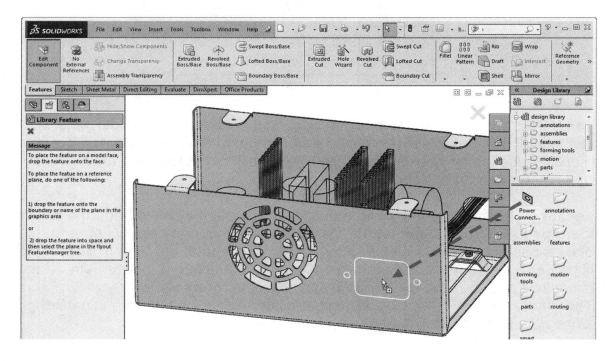

After dropping the library, a preview pop-up window will show us the edges that we need to select in the part for each of the references in the library to locate it. When an edge is selected, the "References" selection box will show a checkmark next to each reference listed. In this case the first reference needed is *"Locate_Side"* (yours may be *"Locate_Top"*); select the edge indicated to continue. The library's locating dimension will then be referenced to this edge.

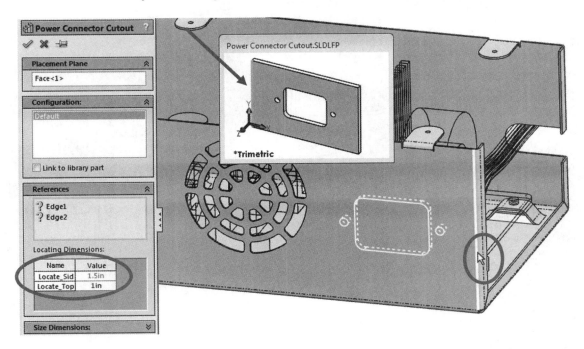

199. – After selecting the first reference we are asked for the second reference in the preview window. Select the top edge of our part as reference to continue.

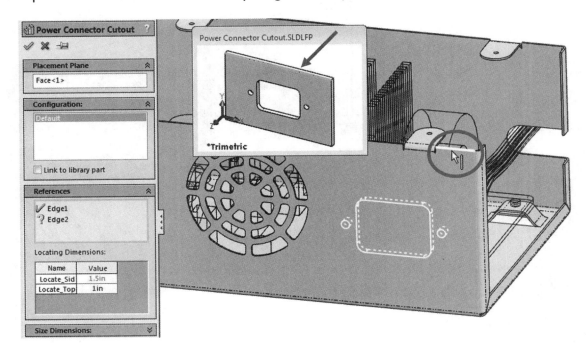

200. – After the required references are selected they are listed with a green checkmark, letting us know that we have selected valid references. Now we need to change their values. Click inside each dimension's value in the "Locating Dimensions" box, and change their values to "Locate_Side" = 1.25" and "Locate_Top" = 1.5". Click OK to finish adding the library.

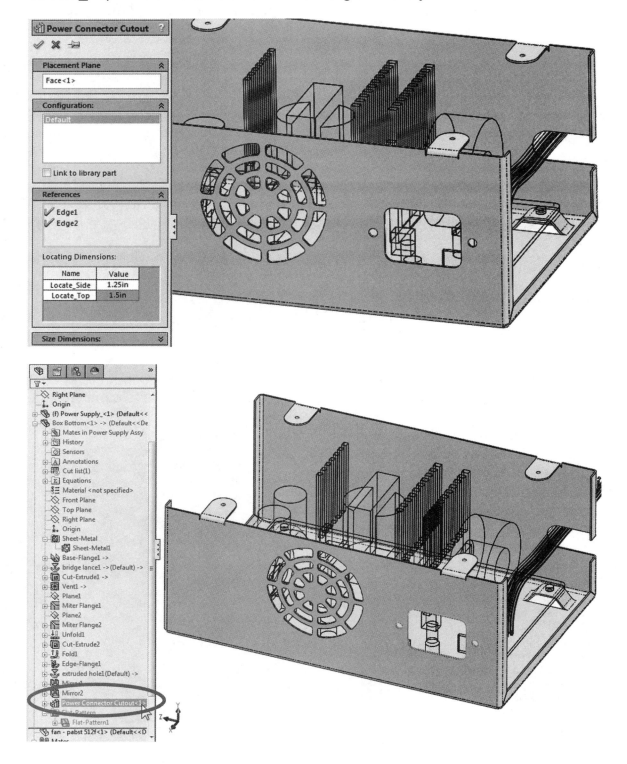

If a dimension is not designated as *"Internal"* or *"Locating"* in the library, its value can be changed in the "Size Dimensions" box, after selecting "Override dimension values". Also, if the library has no locating dimensions, we'll see an "Edit Sketch" button to locate the library instead of the "References" selection box.

Library features may have configurations, and every time it's used, a different configuration can be selected.

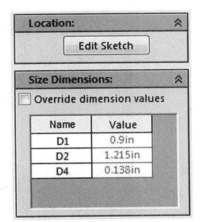

201. – Just as we can make libraries of individual features, we can also make libraries of parts and even assemblies. In the next step we'll create a library part and use it in our assembly. The part we want to make a library of is the Power Connector for which we just made a cut out. A library part can be any SolidWorks part, and can be used as such just by putting it in the *"Parts"* folder of the *"Design Library."* What makes library parts useful, more than keeping them handy, is that we can add **"Mate References"** to them, so that when we drag them into an assembly, the mate references will attempt to find a matching entity to mate with. Open the part *'Power Connector.sldprt'* from the included files for this example. Optionally, as a challenge, feel free to open the part's drawing and build it yourself.

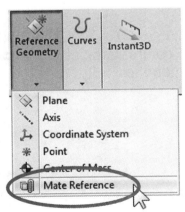

202. – Once the connector is open, go to the menu **"Insert, Reference Geometry, Mate Reference,"** or the drop-down menu in the **"Reference Geometry"** command in the Features tab of the Command-Manager.

We can add up to three mate references to a part; for our example we'll only use a single reference.

In the "Primary Reference Entity" select the circular edge in the left mounting hole of the connector; the "Default" mate alignment is the only option for circular edges, and leave Alignment to "Any." Optionally give a name to the mate. Click OK to add the mate reference.

161

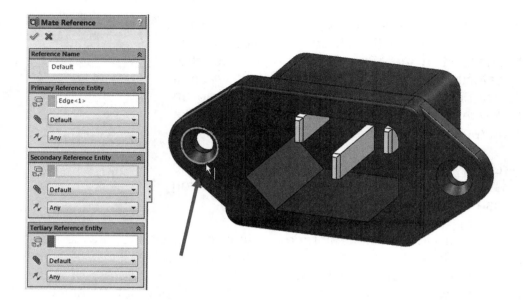

 Mate References can be added to any entity that can be mated, including faces, edges, vertices, axes, sketch entities, points, planes, etc.

Mate Alignment refers to how the mated faces will be oriented. For flat faces imagine an arrow (Vector) going away from each face. "**Aligned**" means both arrows are pointing in the same direction; "**Anti-Aligned**" means they are pointing at each other. There is no analogy for cylindrical surfaces, just flip them as necessary. A good practice is to position parts as close to the desired orientation before adding mates; SolidWorks will not rotate a part more than 180 degrees to add a mate.

203. –. After adding the mate references, a new folder is added to the FeatureManager listing the mate reference added. Save the *'Power Connector'* part, now it's ready to be added to the *"Design Library."*

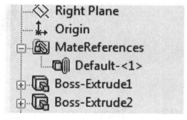

204. – To add the part to the library, right-mouse-click at the top of the FeatureManager in the part's name, and select "Add to Library" from the pop-up menu, *or ...*

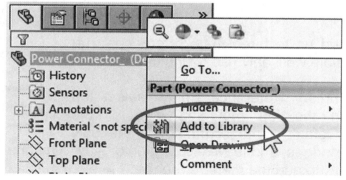

… expand the *"Design Library"* and drag-and-drop the part <u>from the top</u> of the FeatureManager to the task pane in the desired folder, in this example we'll select the "parts" folder.

 Before adding the part to the library using drag-and-drop, we need to keep the *"Design Library"* visible by clicking on the auto-show pin.

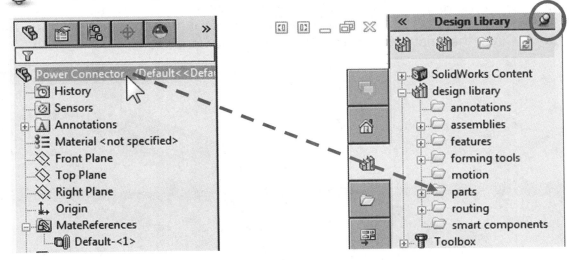

205. – Now we are asked to select the folder to add the library to, and optionally give it a different name. Click OK to add the *'Power Connector'* to the library and close it. If we don't close it we'll be asked if we want to use the open part when adding it to the assembly.

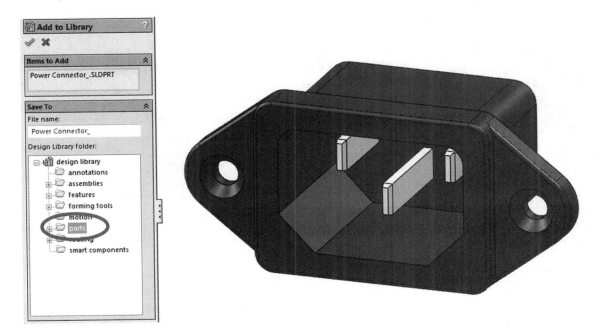

206. – Now, go back to our assembly. After making the library part, it's time to add it to our assembly. <u>Be sure we are editing the assembly</u> before adding the library part; if needed turn off the "**Edit Component**" command.

From the *"Design Library,"* drag-and-drop the *'Power Connector'* in the hole of the cutout. Notice, as we move it close to a hole or a flat face, the part automatically snaps, trying to find a match using the defined mate reference. When we drop the part in the hole, one Concentric and one Coincident mate are added. Close the "Insert Components" message to continue.

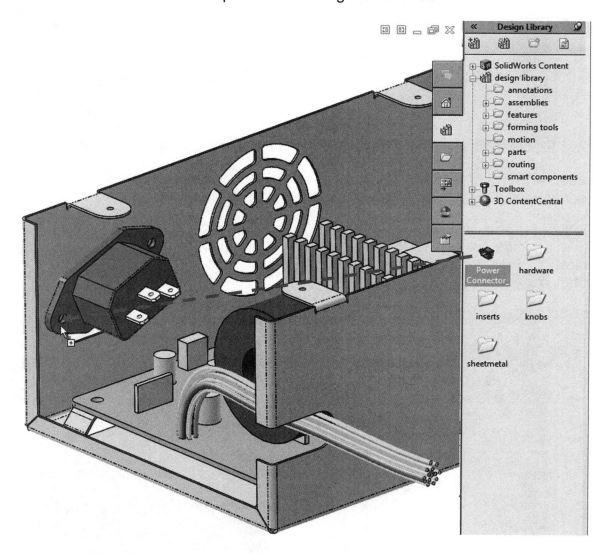

 When adding library parts with mate references, if needed, we can press the Tab key once to flip the alignment and reverse the direction of the part before dropping it in place.

207. – The only thing left for us to do is to add a Parallel mate to align the connector using the faces indicated.

208. – In order for the fan to properly cool the power supply, we need to add a series of slots to let air into the enclosure. Open the *'Box Bottom'* part in its own window, add the following sketch using the "**Slot**" tool and make a cut using the "Link to thickness" option.

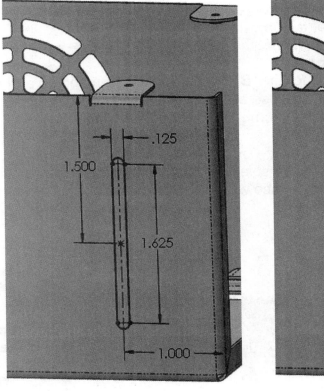
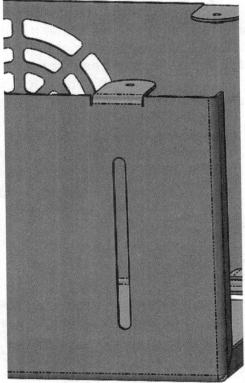

209. – Make a linear pattern of the slot with 15 copies spaced 0.25″.

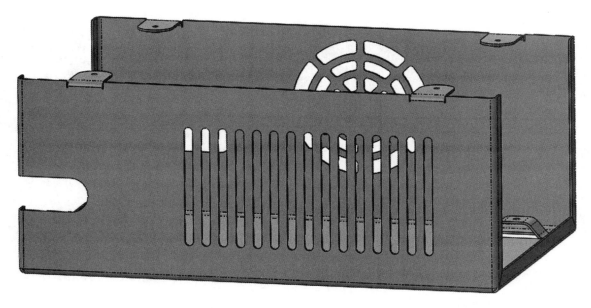

210. – The last operation to complete our first sheet metal part is to eliminate all the sharp edges. When fabricating sheet metal components, it's common to have sharp edges due to shearing and punching operations, and eliminating them will prevent accidents. The **"Break Corners"** command is in the Sheet Metal toolbar, under the drop-down **"Corners"** command, or the menu **"Insert, Sheet Metal, Break Corner."**

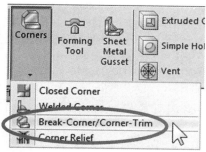

There are two options for the **"Break Corners"** command: to add a chamfer or a fillet. For this part we'll use a 0.050″ chamfer. Select all edges that could potentially have a sharp corner and press OK to finish the part.

 When using the **"Break Corner"** command a filter to select small edges is automatically enabled to make selection easier. Small edges can also be window-selected. Selecting a face will add all of its small edges, too.

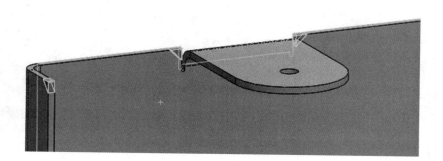

211. – Save the *'Box Bottom'* part. Our finished component now looks like this:

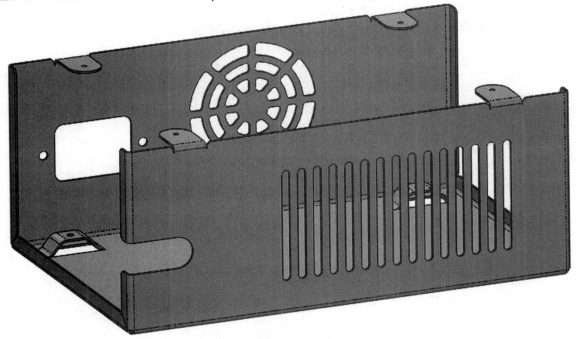

212. – Sheet metal components can be flattened at any time. The *"Flat-Pattern"* feature at the end of the FeatureManager is suppressed while the part is in the bent state, and when we select the **"Flatten"** command from the Sheet Metal toolbar, it is unsuppressed and the sheet metal part is unfolded.

A flat pattern is required by manufacturing to know what size the sheet metal needs to be cut and how to bend it. Using the correct bend allowance is extremely important, otherwise, when the calculated flat pattern is bent, the resulting part will not have the expected design dimensions. To see the flat pattern for the *'Box Bottom'* select the **"Flatten"** command.

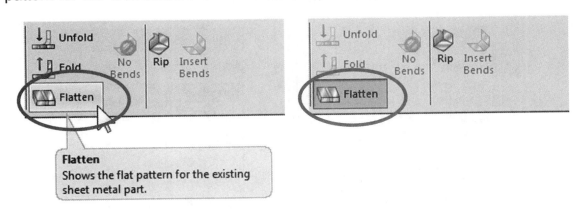

The *"Flat-Pattern"* folder and the *"Flat-Pattern(1)"* feature are now unsuppressed, and we can see the flattened sheet metal part and the *"Bounding-Box"* sketch that represent the minimum material size needed to make this part.

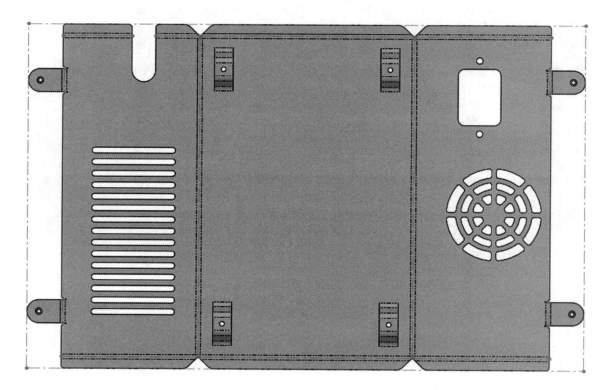

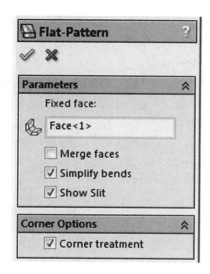

Remember to turn off the "Flatten" command before returning to the assembly. If we don't, the part will be shown flattened in the assembly and cause errors because other parts are mated to the sheet metal, and in-context features were created in the assembly.

213. – Editing the *"Flat-Pattern1"* feature will show the following options:

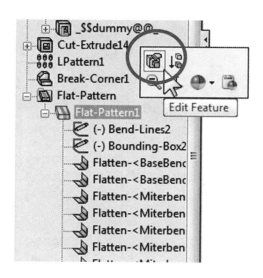

Flat Pattern Options Description

- **Merge faces**: When this option is checked, no bend regions are shown in the flat pattern, merging every flat face together. If left unchecked, all the bend regions will be displayed.

 Merge faces is useful when the flat pattern is exported to a computer numerical control (CNC) software, this will help in programming by avoiding multiple broken lines and faces in every edge.

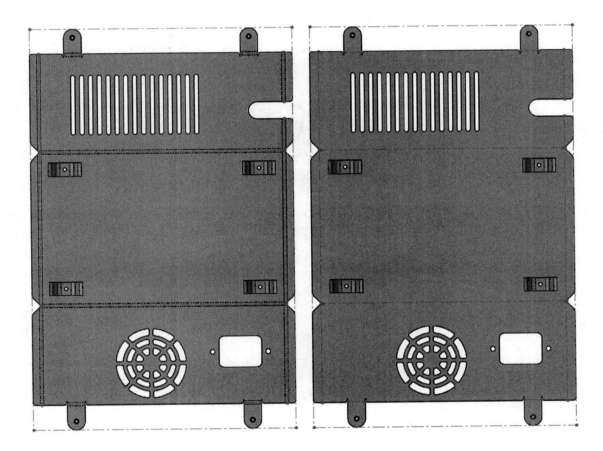

Merge faces unchecked   Merge faces checked

- **Simplify bends:** When flattening complex bends (like bends across a curved area), this option will make a straight edge in the bend region, making the manufacturing process easier.

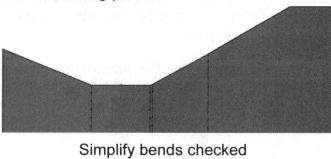

Simplify bends checked

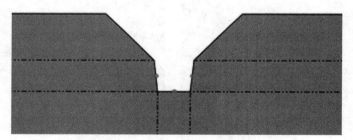

Simplify bends unchecked

- **Corner Treatment**: Will make a flat pattern easier for manufacturing to fabricate by eliminating complex cuts in bent corners.

No corner treatment

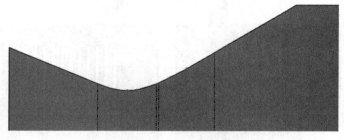

With corner treatment

214. – After finishing the *'Box Bottom'* part, the second component to be made is the power supply's cover. Just like the bottom part, we'll design the cover in the context of the assembly. Change to the assembly window, if not already there, and make sure we are editing the assembly and not a part.

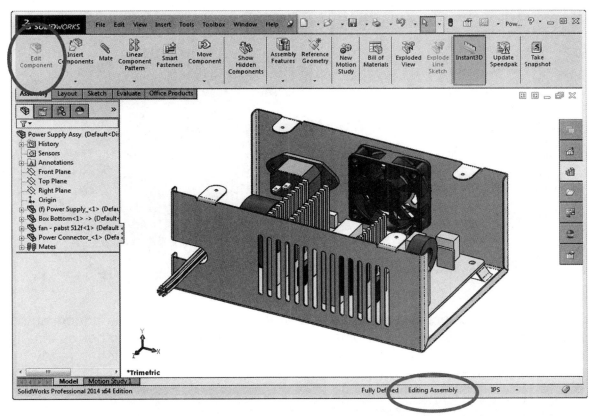

Add a new part to the assembly. From the "**Insert Components**" drop-down icon select "**New Part**," or from the menu "**Insert, Component, New Part**."

215. – In the FeatureManager, select the assembly's *"Front Plane"* to add the new part. Remember, the *"Front Plane"* of the new part will be aligned and constrained to the selected plane or face with an "**InPlace**" mate.

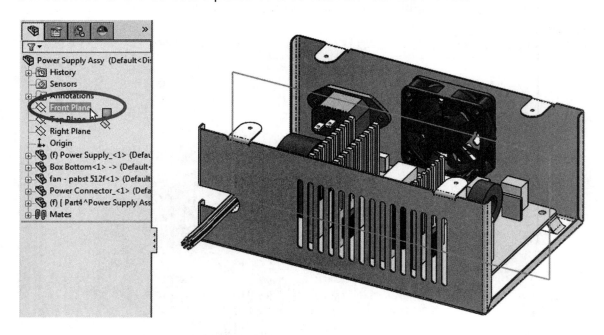

As soon as we select the assembly's *"Front Plane"*, a new part is added to the assembly and it's displayed in blue in the FeatureManager. The other assembly components become transparent (if the option is set), and a sketch is ready for us to start working in the new part's *"Front Plane."*

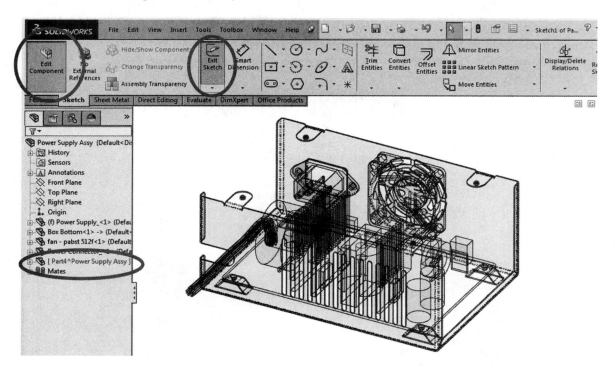

216. – Switch to a Front view, hide the electronics board, the fan and the connector; they will not be needed to design the cover and keeping them visible makes the screen unnecessarily busy. Draw the next sketch using three lines; add the necessary coincident relations to make the lines coincident to the top, bottom and side edges.

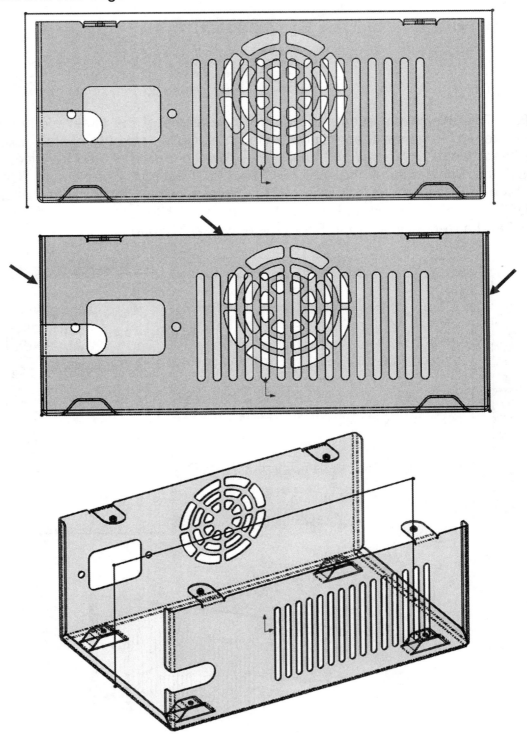

217. – After completing the sketch, select the "**Base-Flange**" icon from the Sheet Metal toolbar.

218. – This sketch represents the inside dimensions of the sheet metal cover; therefore, we'll change the settings to add material *outside* the sketch. Extrude the base flange in *"Direction 1"* using the "**Up To Surface**" end condition and select the front face of the *'Box Bottom'*; in *"Direction 2"* also use "**Up To Surface**" and select the rear face of *'Box Bottom.'* Make the thickness equal to 0.030″, the default Bend Radius = 0.015″, in "Bend Allowance" use a *"K-Factor"* of 0.5 and a Rectangular "Auto-Relief" relief with a 0.5 ratio.

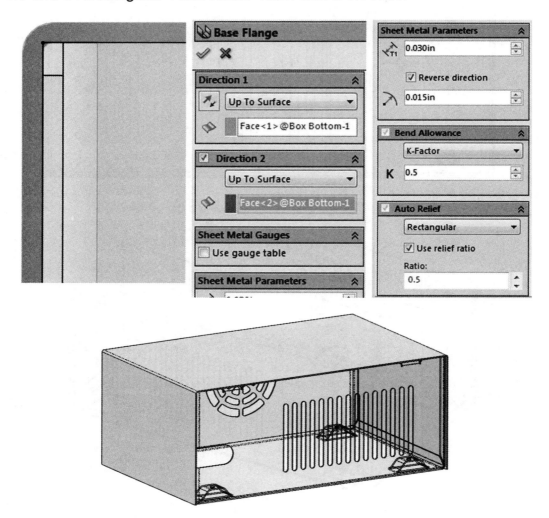

219. – The next step is to add jogs at the bottom of the cover; these jogs will help us assemble the cover to the bottom. To add a jog we need a sketch with a line; we can draw the sketch before or after selecting the "**Jog**" command.

The first jog will be done adding the sketch first, and then making the jog. Select the side of the cover, and draw a single line sketch. (Remember we are editing the cover in the assembly.)

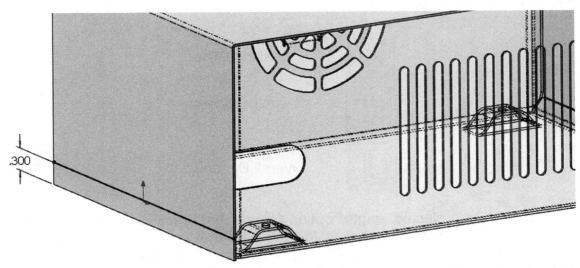

The sketch line does not need to cross the face that we intend to bend, but the jog will.

220. – Select the "**Jog**" command from the Sheet Metal toolbar or from the menu "**Insert, Sheet Metal, Jog.**"

In the "**Jog**" command, under the "Fixed Face" selection box, click anywhere in the sketch face above the line; this is the part of the face that will not move. Leave the option "Use default radius" checked to make the bends the same radius as the rest of the part.

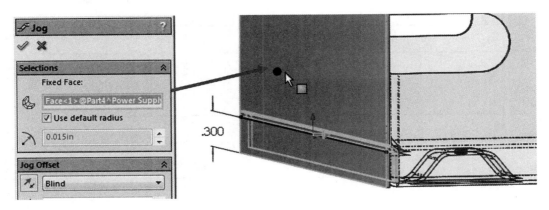

In the "Jog Offset" section we'll define the parameters for the jog bends. We can make the jog a fixed distance, up to a surface, vertex or offset. The purpose of the jog is to bend the cover inside the miter flange of the base part to assemble them together.

When using the "Blind" end condition, there are three different ways to define how the jog distance is measured, and the icons for each type illustrate the difference between them:

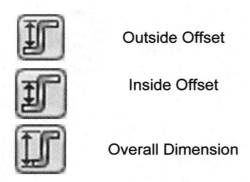

Outside Offset

Inside Offset

Overall Dimension

The "Fix projected length" option means that, if checked, material will be added to the flat pattern equaling the jog plus the jog bends, making the flat pattern bigger as seen below. If it's un-checked no material will be added, the flat pattern will remain the same, and the bends will be made with the existing material.

Fixed projected length option:

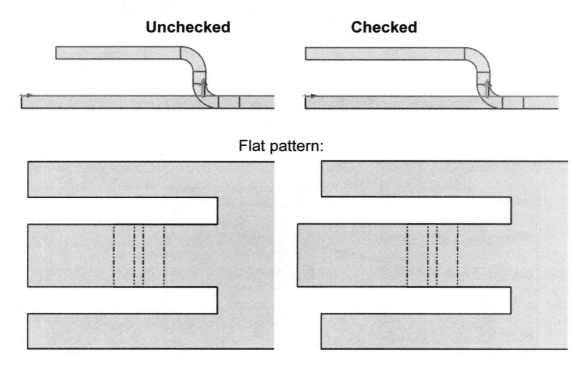

Unchecked   **Checked**

Flat pattern:

The "Jog Position" is the same definition as when making flanges. Change to a Front view and zoom in the jog area for visibility. Select the "Blind" end condition and make it 0.1″ going into the enclosure (use reverse direction if needed); use the "Overall Dimension" and "Material Inside" options, and make the "Jog Angle" 90 degrees. Notice the option "Fix projected length" is not checked; if checked, the jog would extend and go through the bottom part. Click OK to finish.

 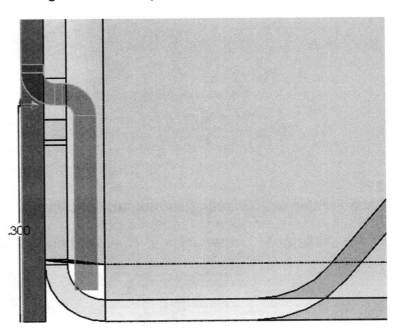

221. – For the other side, select the "**Jog**" command first. When asked to select a face on which to draw the sketch, select the cover's opposite side face, draw the same sketch and add the second jog with the same settings as the first one, also going inside.

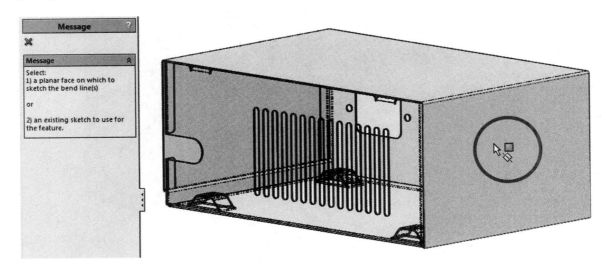

177

The second jog should now look like this when finished:

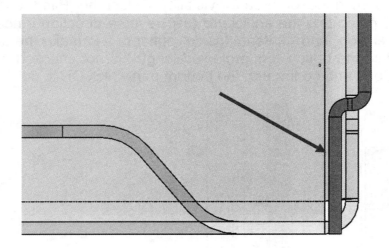

222. – With the jog we bent the entire length of the cover to fit behind the miter flange of the *'Box Bottom'* part, but now we have interferences at the corners. To see the overlapping areas better we need to make all components opaque. Click in the "**Assembly Transparency**" command and select the "Opaque" option.

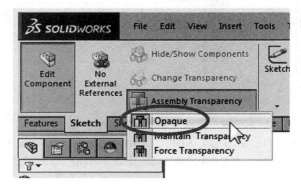

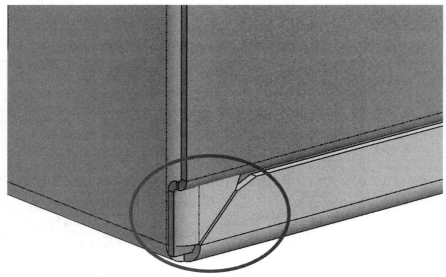

223. – To correct this interference we'll have to make a cut in the corners. Use the "**Assembly Transparency**" command and select "Force Transparency" to make the rest of the components transparent again. Add the following sketch on either side of the cover (left or right will work). Make the top of the rectangle coincident with the jog's bend line and the opposite corner coincident with the lower outside corner of the jog. To fully define the sketch,

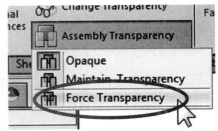

dimension the inside of the rectangle 0.050″ past the miter flange corner. (Note the miter corners were trimmed.)

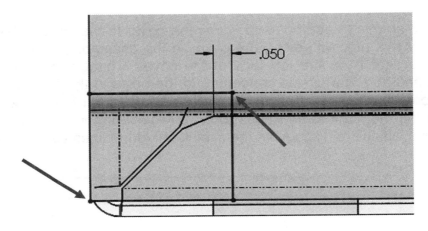

Mirror the sketch about the part's center and make a cut using the "Through All" end condition to cut all four corners at the same time (image shown using "Shaded" mode for clarity, no edges).

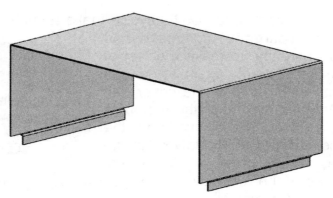

Cover part by itself to show the corner cuts

224. – The next step is to add holes to screw the cover to the *'Box Bottom'* part. In the assembly (we are still editing the cover in the assembly), switch to a top view and add a sketch on the top face of the cover. Change to "Hidden lines visible" mode and add four circles concentric to the holes in the tabs of the *'Box Bottom.'* Dimension one circle 0.150" diameter and make all four circles equal. Make a cut using the "Link to thickness" option. (The "Shaded" view was added for clarity, as it's difficult to distinguish the sketch in hidden lines visible mode.)

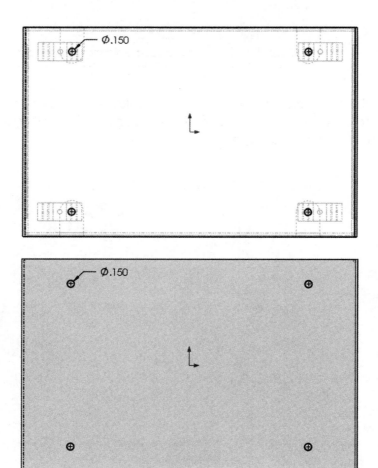

225. – Now add ventilation slots to the cover. Add a new sketch in one side of the cover and dimension it as shown. Just as with the corner cuts, make a cut using the "Through All" end condition to cut both sides of the cover.

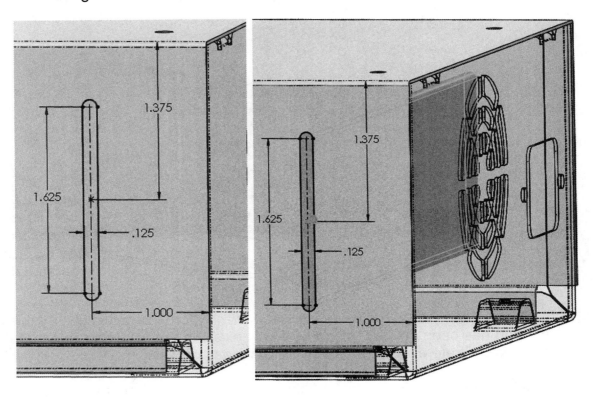

226. – Make a linear slot pattern with 12 instances spaced 0.250″.

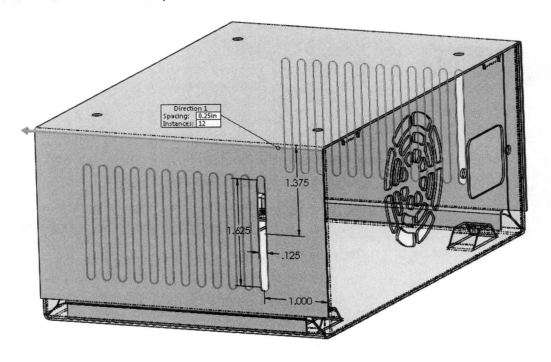

227. – As a final step, break all sharp corners using the "**Corners, Break-Corners**" command from the Sheet Metal toolbar. Add a 0.1″ chamfer break. Select the side and the jog faces to automatically select all corners. Click OK to continue.

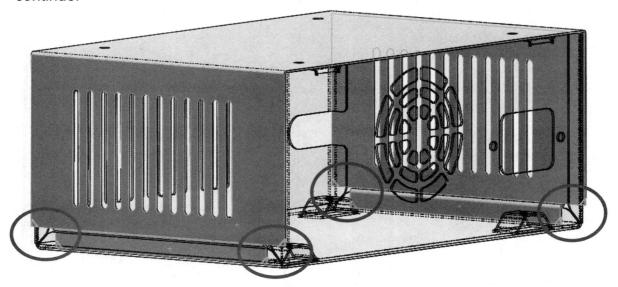

228. – Now that the part is finished we'll *externalize* it from the assembly. All this time the part has been internal, meaning it only exists in the assembly and there is no *'file'* we can reference for a drawing or other assemblies.

Stop editing the cover in context of the assembly and return to editing the assembly. Turn off the "**Edit Component**" button in the CommandManager, *or* right-mouse-click in the graphics area and select "**Edit assembly: Power Supply Assy**."

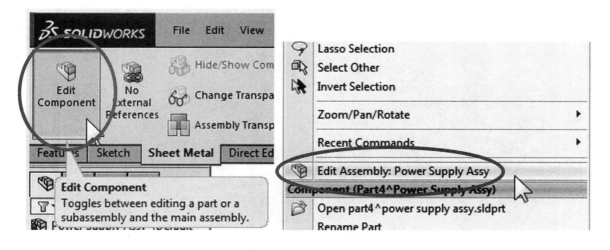

229. – Rename the part as *'Box Cover'* with a slow double-click, or from the right-mouse-click menu select "**Rename Part**." Then right-mouse-click in it again and select "**Save Part (in External File)**" from the pop-up menu. Make sure the path is correct and click **OK** to save the file.

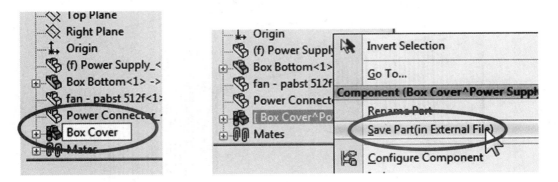

230. – Open the *'Box Cover'* part in its own window and view the flat pattern by selecting the "**Flatten**" command in the Sheet Metal toolbar. After reviewing the flat pattern, turn it off (re-fold the part) and return to the assembly.

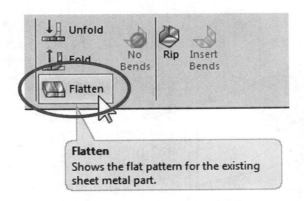

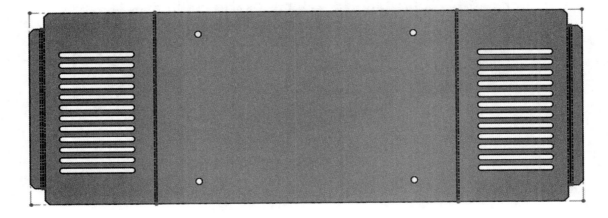

231. – Make all the components visible in the assembly, and if wanted, make an exploded view to finish. We'll work on the sheet metal drawings next.

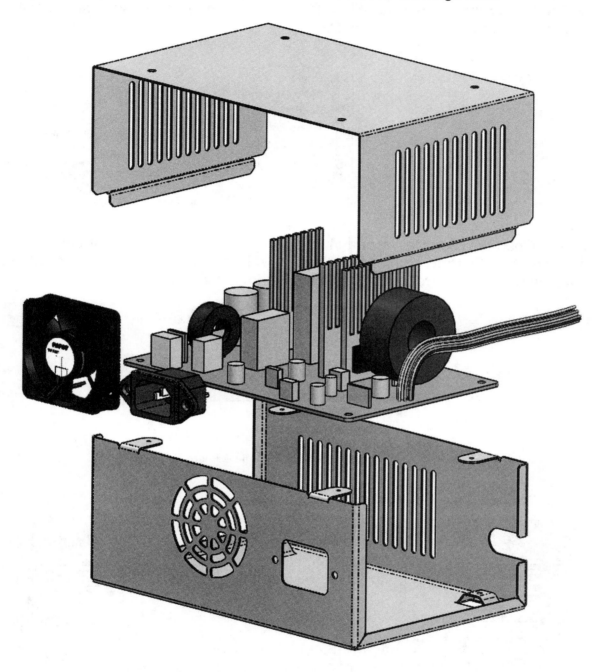

Sheet Metal Drawings

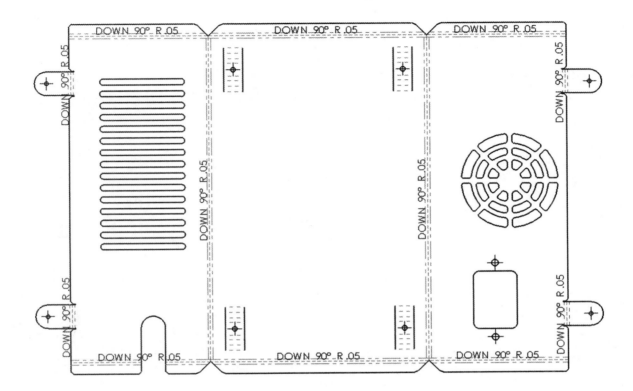

Notes:

232. – When we talk about a sheet metal part's drawing, there are a few differences compared to regular drawings. When we make a sheet metal part drawing, the first thing to notice is that we have a new option to add a **"Flat Pattern"** view besides the standard views we are used to. Open the *'Box Bottom'* part and make a new drawing. Drag the "Flat pattern" from the View Palette into the drawing.

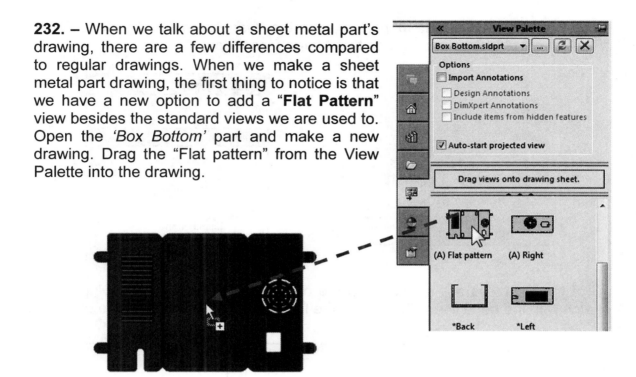

As soon as we drop the flat pattern in the drawing, the sketch with bend lines and bending notes is added automatically. These notes will help you bend the part correctly. Adjust the drawing view's scale and/or bending note's font as needed.

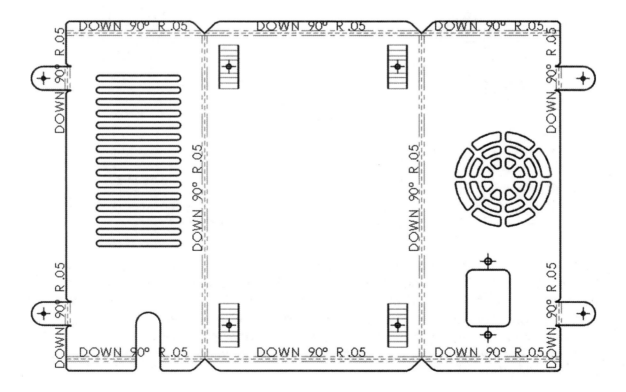

233. – Open the *'Box Bottom'* part, expand the *"Flat-Pattern"* folder and edit the *"Flat-Pattern1"* feature and turn on the "Merge faces" option.

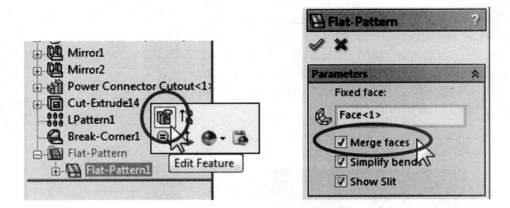

All of the flat faces are merged and the only thing visible in the drawing is the sketch with the bend lines. Go back to the flat pattern drawing to see the effect.

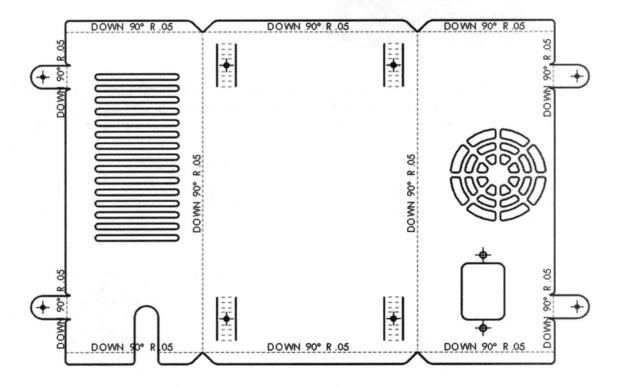

234. – It's common to export a 1:1 scale drawing in DWG or DXF format with the "Merge faces" option turned "On" for manufacturing (CNC) purposes, and you may or not want the bend lines and notes in this file, as a CNC program only uses the contours to trace the part. To turn off all notes and bend lines in the drawing, you can right mouse click in the *"Annotations"* folder of the drawing and turn off "Display Annotations," or hide the sketch with bend lines in the *"Flat-pattern1"* feature.

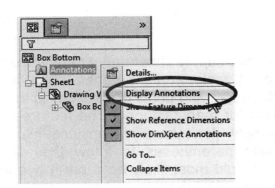 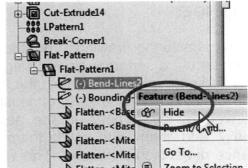

If needed, flat patterns usually have to be manually dimensioned.

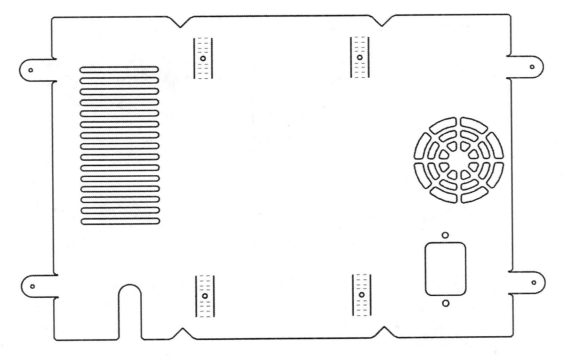

 To enable or disable the option to add sheet metal notes in the flat pattern when making a drawing, go to the menu "**Tools, Options, Document Properties, Sheet Metal**" and turn off "**Display sheet metal bend notes**."

 One more thing to know: the flat pattern is a part sub-configuration where the *"Flat-pattern1"* feature is unsuppressed showing the part unfolded. Preferably show the flat pattern using the "**Flatten**" button instead of manually changing configurations, since we can accidentally suppress or unsuppress features that need to be On or Off in a configuration.

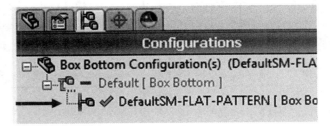

CHALLENGE:

Make manufacturing drawings from the power supply sheet metal parts including a flat pattern in a separate sheet using a 1:1 scale. In a sheet metal drawing we usually find general dimensions in the folded part and detail dimensions in the flat pattern drawing using ordinate dimensions for manufacturing. The features that are driven by other parts in the assembly don't have dimensions; we can manually add them and mark them with parentheses to know they are a reference dimension if needed. The vent would most likely be a pre-made tool; we just need to locate its center and possibly orientation. Add centerlines if needed to locate its dimension.

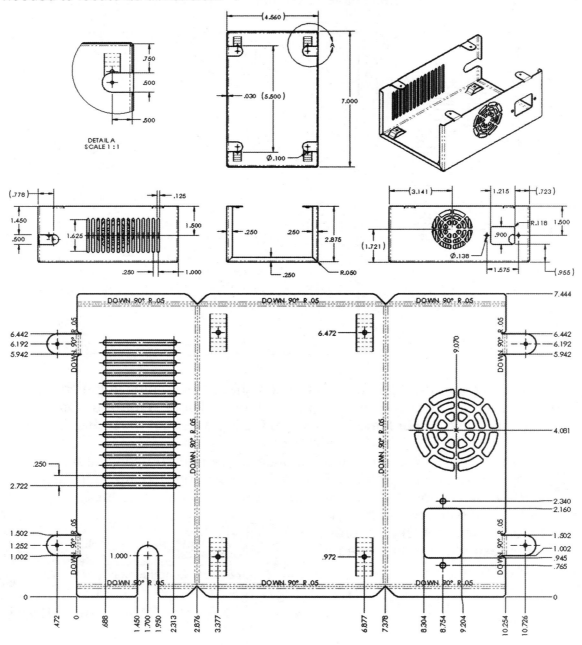

Sheet Metal Locker Top Down Design

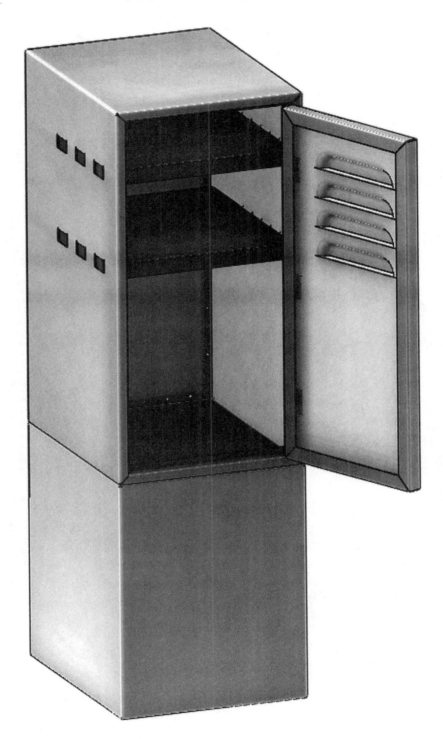

Notes:

235. – For our next sheet metal project we'll build a metal locker also using a top down design approach. Our locker will have a base to raise it from the floor. In this example, we'll learn how to convert solid parts into sheet metal. A solid or shelled part can be converted directly as shown here:

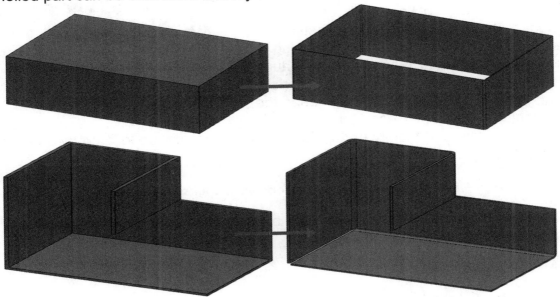

What happens when we convert a solid part into sheet metal is that the face selected as "fixed" will remain static, the faces connected to it by edges defined as "Bends" will be part of the sheet metal component and the rest of the faces will be deleted. If the part is made of surfaces or a shelled part before converting it to sheet metal, some of the faces not connected by "Bend" edges will also be removed. The size of the sheet metal locker we are going to design is smaller than a real one, in order to better see the part's thickness and features.

236. – The first part of the assembly will be the locker's base. Make a new part in inches, draw the following sketch in the *"Top Plane,"* and extrude it 8" up. Note the square is centered about the origin to maintain our design intent.

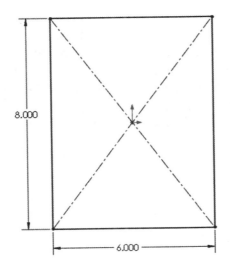

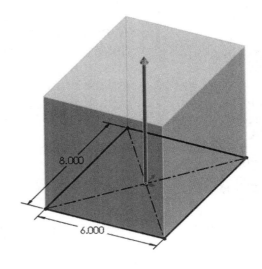

237. – From the Sheet Metal toolbar select "**Convert to Sheet Metal**."

238. – For our locker, we'll use an 18 gauge* steel sheet metal, but instead of entering the values directly, we are going to use a "Sheet Metal Gauge" table. Gauge tables are Excel files that list the material's Gauge number, its thickness, and the available bend radii for each thickness. We can add, delete and modify existing gauge tables, which are located by default in the folder:

C:\Program Files\SolidWorks Corp\SolidWorks\lang\english\Sheet Metal Gauge Tables

The default gauge table location can be changed in the SolidWorks options under "File Locations."

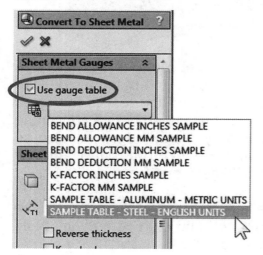

Turn on the "Use gauge table" option, and from the drop down list select "SAMPLE TABLE–STEEL-ENGLISH UNITS."

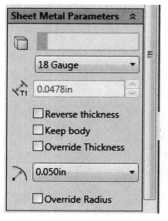

After loading the gauge table, select "18 Gauge" from the drop down list. The thickness will be automatically filled from the gauge table. Then select 0.050" from the bend radius drop down list to define the bend radius for our part. If needed, the material's thickness and bend radius can be overridden by selecting the corresponding checkboxes.

* *"Standard"* sheet metal gauge tables vary by country, material, standard, etc.

In order to convert the solid into a sheet metal part, first we have to select a face that will be the fixed face in the sheet metal, the edge(s) of the part that will be converted into bend(s), and the edge(s) that will be ripped (open). Select the left face of the part to be the fixed face (you can use the "Select other" command to select the hidden face.) After selecting the fixed face, make sure the thickness of the sheet metal is added *'inside'* as we want to maintain the dimensions given as external dimensions. You can zoom into a corner to see if the material is added inside or out, and check (or uncheck) the "Reverse Thickness" option if needed.

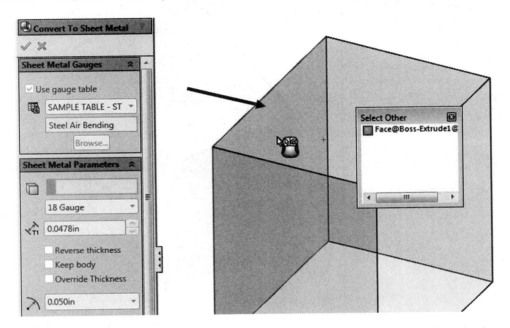

239. – Select the 3 vertical edges (labeled "Radius") that will become bends. The remaining vertical edge (labeled "Gap") is automatically selected as a "Rip Edge." Set the default gap for rips to 0.01″ and the "Auto Relief" type to Obround with a 0.5 ratio, a K-Factor of 0.5 and click OK to convert the part to sheet metal.

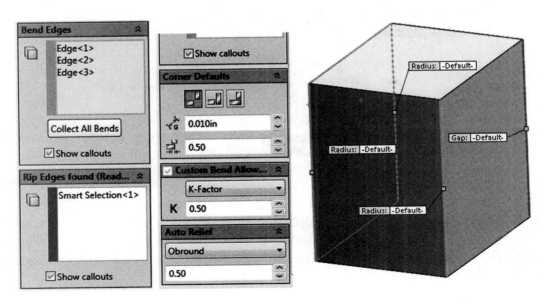

In the fly-out labels we can override the default settings for bend radius and gap distance if needed by clicking in them.

After converting the part to sheet metal, the top and bottom faces have been automatically removed and the sheet metal features added to the Feature Manager.

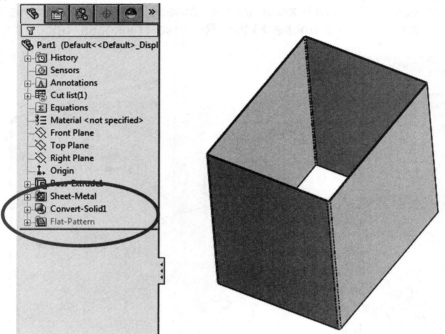

240. – Now we need to add miter flanges at the top and bottom, but instead of using the "**Miter Flange**", we'll use the "**Edge Flange**" command to show a different approach to accomplish the same result. Using the "**Edge Flange**" command, select all the edges as indicated.

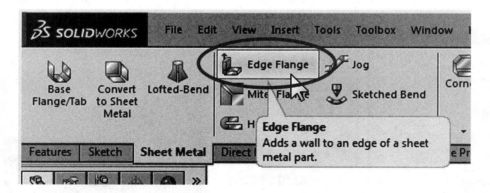

Using this technique the miter cuts will be automatically added to the corners. Make the flange 0.5″ long, with the option "Material Inside" and set the rest of the options as shown. The arrow in a flange can be used to reverse the direction if needed. Another advantage of using this approach is that we can change the flange's profile if needed.

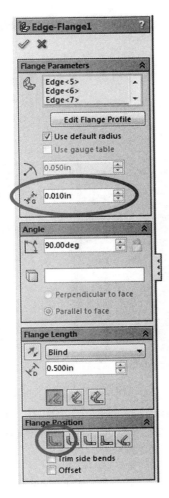

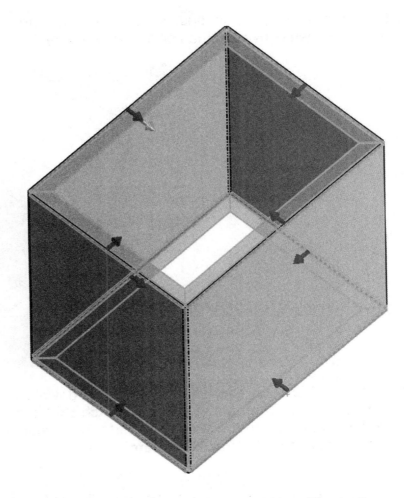

 In this case there is no difference between using a **"Miter Flange"** or an **"Edge Flange,"** but remember, the "Miter Flange" runs the full length of the edge and can change the profile, and an "Edge Flange" can change the length and shape of the flange but not the profile.

Miter Flange
Different profile, full length

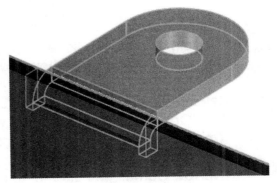

Edge Flange
Different shape, not full length

241. – After adding the flanges, we need to close the gap in the corner that was ripped when we converted the solid to sheet metal.

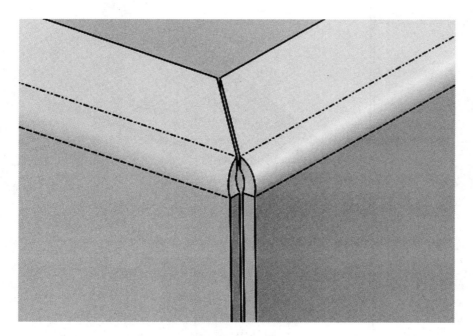

Select the "**Corners, Closed Corner**" drop down command. This feature will extend one side of the sheet metal to match the other using a butt, overlap, or underlap extension. In the "**Faces to Extend**" selection box, select the flat face on one side, and in the "**Faces to Match**" select the other flat face.

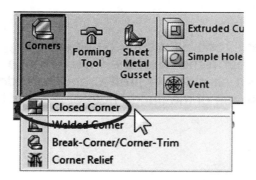

 Closing the corner is an option when converting to Sheet Metal, but instead we decided to show how to use this command separately.

- Select "Overlap" to make the first face extend over the second face.
- The "Open bend region" option will close the bend region too if selected.
- "Coplanar faces" will automatically select faces that are coplanar to the one selected.
- "Narrow Corner" will attempt to close the gap when using large bend radii.
- In the "Gap Distance" we can make the gap in the corner as small as needed. Use 0.005″ for our example.

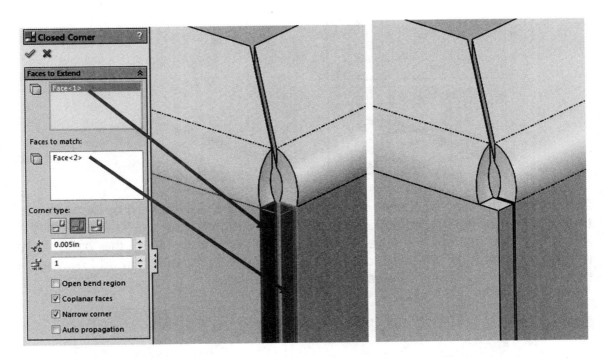

242. – After reviewing the other corners made when we added the edge flange we can see an extra piece of metal in the bend area. To eliminate it, select the *"Edge-Flange1"* made before, edit it and turn on the option "Trim side bends" in "Flange Position." Click OK to finish. Now all of the corners made with this Edge Flange command are trimmed. This option was not turned on before, because this way the difference of not activating it is better illustrated.

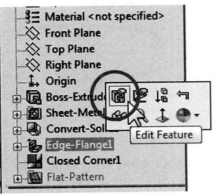

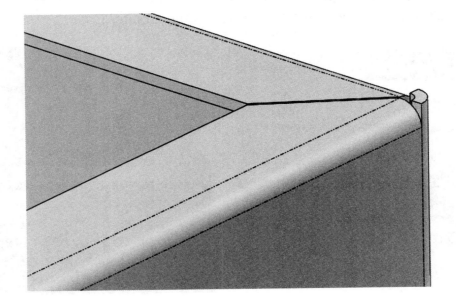

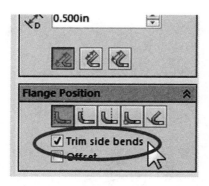

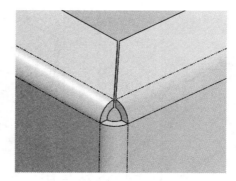

243. – Save the part as *'Locker Base'* and add it to a new assembly locating it at the origin, making it symmetric about the assembly's planes. The rest of the locker will be designed in the context of the assembly using the base as a reference; therefore, we have to save the assembly before we can add new parts to it. Save the new assembly as *'Full Locker Assembly.'*

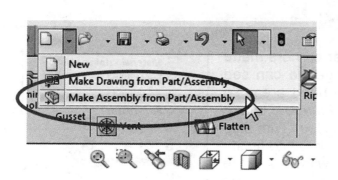

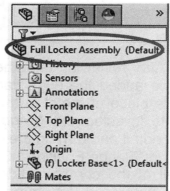

244. – Our next step is to add a new part to the assembly; it will be the locker's body. From the Assembly toolbar select the "**Insert Components**" drop-down menu and select "**New Part**." When asked to select a face or plane to add the new part select the face indicated at the top of the *'Locker Base.'*

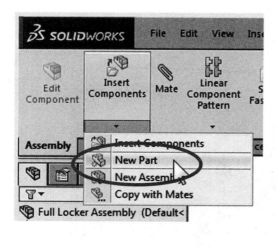

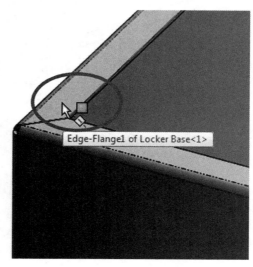

Immediately after selecting the Top face, a new part is added and, as we've seen before, it's listed in blue in the FeatureManager, the "**Edit Component**" icon is active, the *'Locker Base'* becomes transparent, and a new sketch is added.

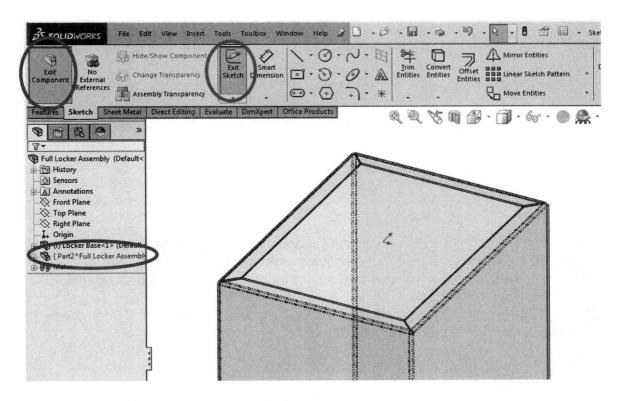

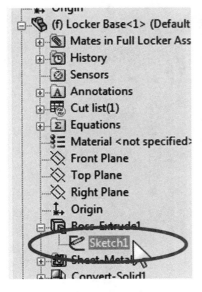

245. – Since we want to make the locker's body the same size as the *'Locker Base'*, we'll use the "**Convert Entities**" command. In the FeatureManager expand the *"Locker Base, Boss-Extrude1"* feature and select *"Sketch1."* After selecting it click in "**Convert Entities**" from the Sketch toolbar, projecting the first sketch of the 'Locker Base' to the sketch of the new part.

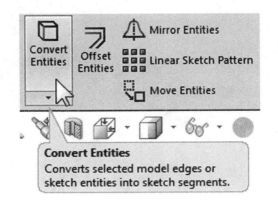

When working with sheet metal parts, it's usually easier to distinguish between sketch entities and model edges by using "Shaded" view mode.

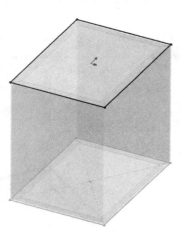

 By selecting it in the FeatureManager, the entire sketch is converted to new sketch entities. To selectively convert sketch entities we can make the sketch visible (Show) and select individual lines to convert them.

246. – After converting the sketch edges, extrude it up 16″. Switch back to "Shaded with Edges" mode for visibility.

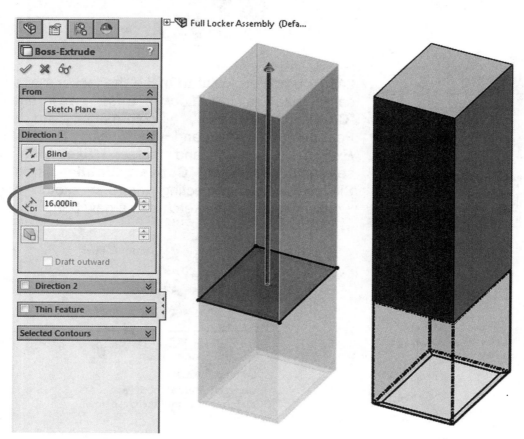

247. – When adding features that are not related to other components, we can either continue working in the assembly or open them in their own window. To some extent it's really a personal preference; however, when adding references to other component's geometry we have to do it in the context of the assembly. Open the locker's body (the new part) in its own window to convert it into a sheet metal component; even though it's an internal component in the assembly at this time, we can open it by itself.

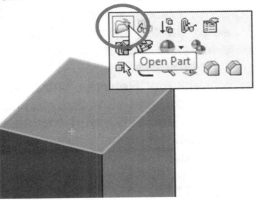

248. – After opening the part, use the Shell command with a 0.05″ thickness and remove the top and bottom faces, which will become the front and back of the locker's body. For this technique to work the shell's thickness has to be thicker than the sheet metal part's thickness we plan to make.

249. – We can convert this model to sheet metal without having to use a shell first like we did before, but we are doing it this way to show a different approach. From the Sheet Metal toolbar select "**Convert to Sheet Metal**." Select the outside face on the left as the fixed face and use the same settings we used for the 'Locker Base'. Activate the "Use gauge table" option, select the sample table for steel, select 18 gauge thickness and a bend radius of

0.05″. You *may* have to check the "**Reverse Thickness**" checkbox if the material is added outside, we have to make sure the material is added to the side we want. Remember, the dimensions in our part are outside dimensions. Zoom in a corner to verify this. Essentially the sheet metal preview will overlap the shell's thickness. Now we have to select the edges to bend or rip.

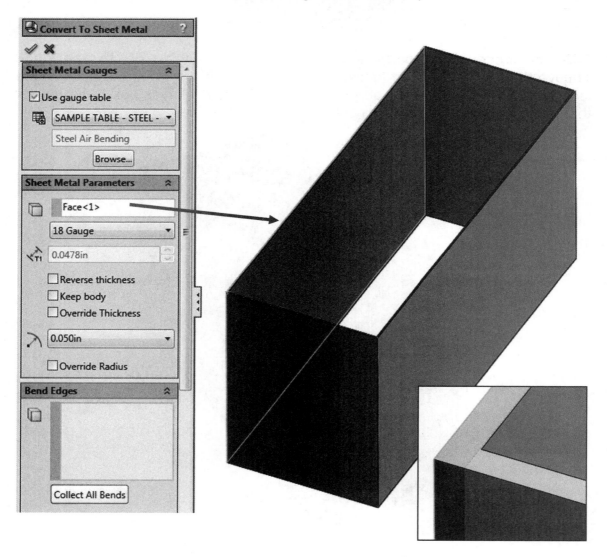

When we convert a shelled part to sheet metal, at the time of selecting the **"Bend Edges"** we have to select the edges connected to the face we selected. For example, if we picked the inside face of the shell, we have to select the inside edges; if the outside face was selected, the outside edges have to be selected. If the part is big compared to the shell's thickness, you may have to zoom in to the corners to select the correct edges. If the incorrect edge is selected you'll see the following message:

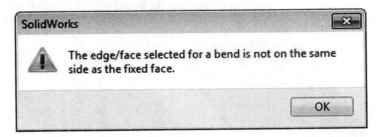

250. – Select three edges to convert to bends and the fourth edge will be automatically added to the "Rip Edges found" selection box. Set the default rips gap to 0.01", use a K-Factor ratio of 0.5, and set the default relief type to "Obround" with a 0.5 ratio. Click OK to complete and convert the shelled part to sheet metal.

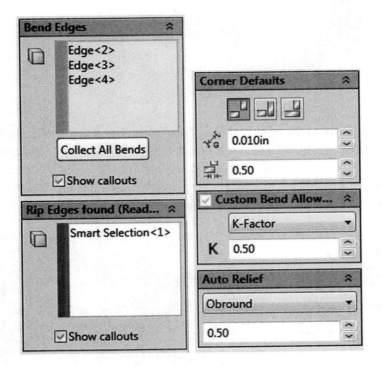

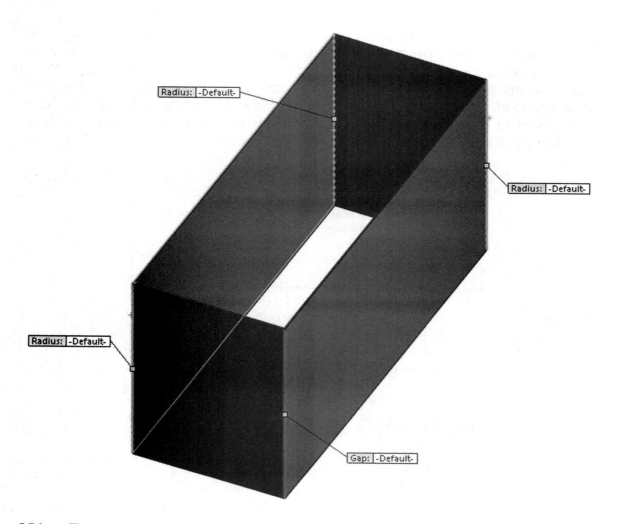

251. – The next step is to add a miter flange in the front and back of the locker. Zoom in the open corner with the gap and select the "**Miter Flange**" icon from the Sheet Metal toolbar. Immediately after selecting it, SolidWorks will ask us to either select a face/plane to sketch the flange's profile, or a pre-existing sketch with the profile. Use the magnifying glass if needed (Shortcut "G").

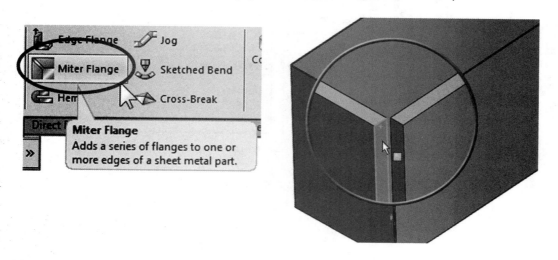

252. – Select either the flat face to add a sketch to it, or the top edge (either edge works) to add a plane perpendicular to it and a new sketch in it.

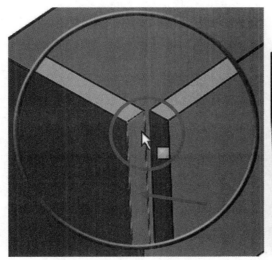

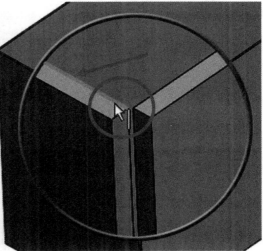

Face selected   Edge selected

Selecting faces is easier using a "**Selection Filter**." Selection filters allow us to select only the type of geometry or element we need. In this case press "**X**" on the keyboard (default filter shortcut for faces) to turn it on (press "**X**" again to turn it off when done.) We know the filter is active when the mouse pointer has a small funnel added next to it. The Selection Filters toolbar can be enabled selecting the menu "**View, Toolbars, Selection Filter**", or a right mouse click in any toolbar and selecting "**Selection Filter**", or using the default keyboard shortcut "**F5**." In this toolbar we can turn on/off any combination of filters as needed.

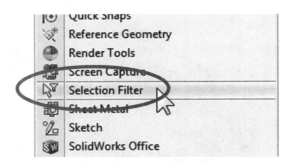

253. – After making the selection draw the profile for the miter flange as shown. It's a single 0.375″ long line.

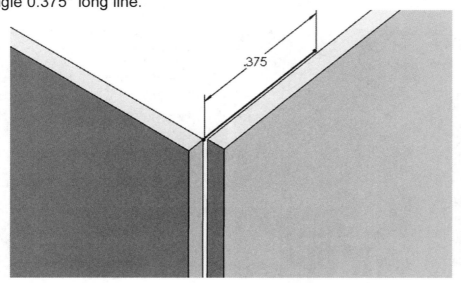

254. – After we select "**Exit Sketch**" or the "**Miter Flange**" icon the preview is activated. Make the flange with the option "**Material Inside**" and a 0.01″ gap distance for the corner openings. Individually select the edges on top or click in the "**Propagate**" icon to automatically select the three tangent edges connected and click OK to finish.

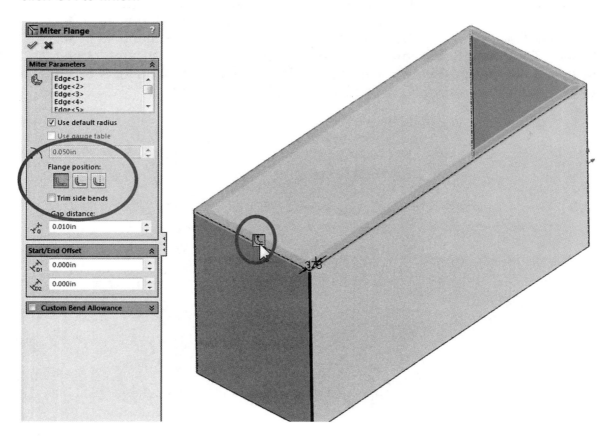

255. – Repeat the same "**Miter Flange**" command on the other side using the same options. Your part should now have miter flanges in the front and back and should look like this:

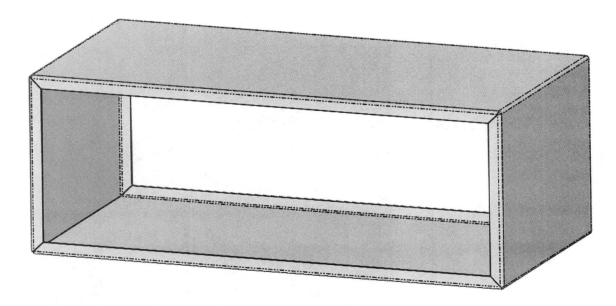

256. – Switch back to the assembly (if working in the part's window); notice we are still editing the part. Now we need to add a forming tool feature to support shelves inside the locker. Expand the "**Design Library**" and scroll down to "**forming tools, lances**" and find the "**90 degree lance**" form tool. Drag-and-drop the lance in the right side of the locker; before we release the mouse button we can press the "Tab" key to flip the lance's direction to go 'into' the locker or 'out' of it (We can also flip its direction it in the Property Manager.)

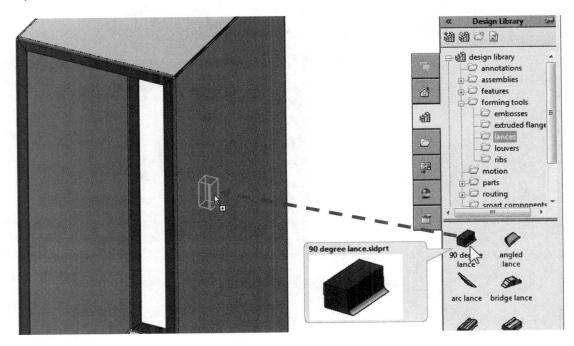

257. – After dropping the lance into the side of the locker we need to add the lance going 'into' the locker, and the bend is on top. Use the "Flip Tool" button and rotation angle options if needed. After making sure the rotation is correct. Select the "Position" tab and dimension its center 2″ from the front and 9″ from the bottom of the locker's body. After dimensioning click OK to complete it.

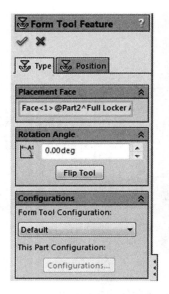

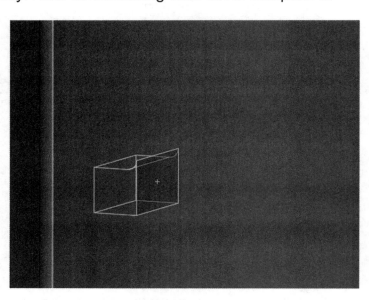

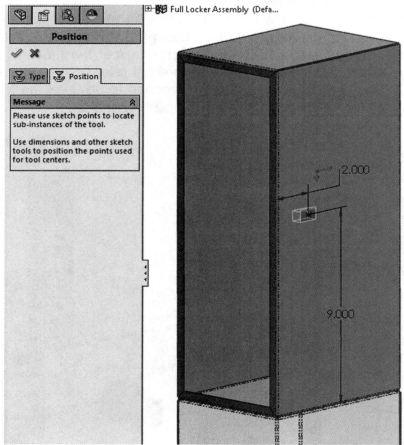

The lance should look like this from inside the locker when finished.

258. – Make a horizontal linear pattern with 3 copies spaced 2″ apart...

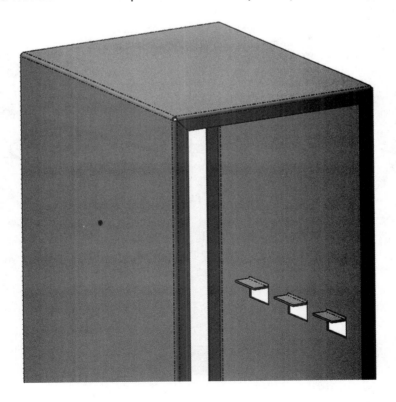

259. – … and then add a vertical linear pattern to copy all 3 instances 3″ up. It will be explained later why we added two patterns instead of making the first pattern in two directions. After adding the vertical pattern, make a mirror of all lance cuts about the *"Right Plane"* to add them on the left side as shown.

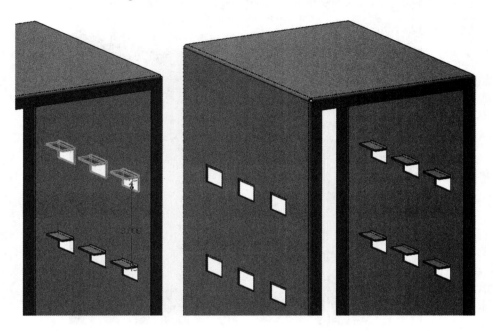

260. – Rename the internal part in the assembly as *'Locker Body'* and save to an external file.

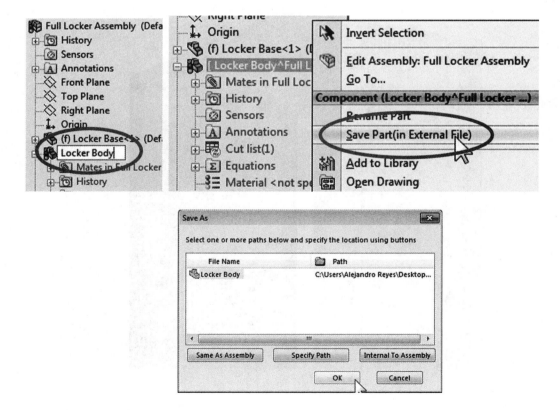

261. – Now that we are finished with the locker's body we can turn off "**Edit Component**" and go back to editing the assembly to add the next components: a cover in the back and a door.

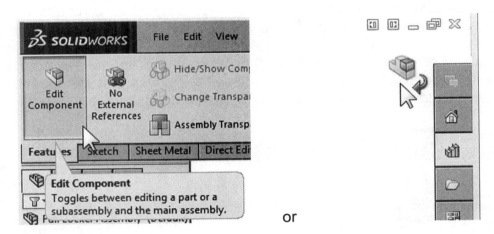

or

262. – The back is a simple sheet metal cover that will be welded or riveted in place. The cover needs to be the same size as the locker's body, and it will also be made in the context of the assembly. Select the "**Insert Components, New Part**" drop down icon and select the inside face of the rear miter flange to locate the back cover.

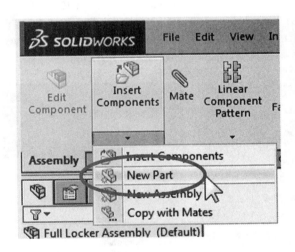

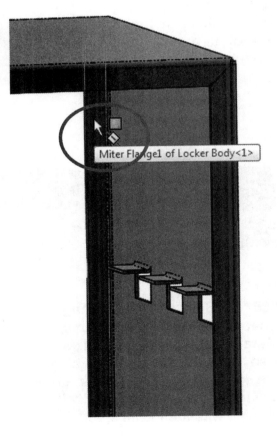

263. – After adding the new part, the locker's base and body turn transparent; we are editing the new part and a new sketch is ready for us to work on it. In this step it is easier to work in an opaque assembly, as it is difficult and/or confusing to reference other components when they are transparent. To change the other component's transparency select the "**Assembly Transparency**" icon in the CommandManager and select "Opaque." If the other components don't change to opaque, we may have to rebuild the assembly to update the component's display. If we rebuild the assembly we will exit the sketch. If that happens, add a new sketch to the Front plane of the new part to continue.

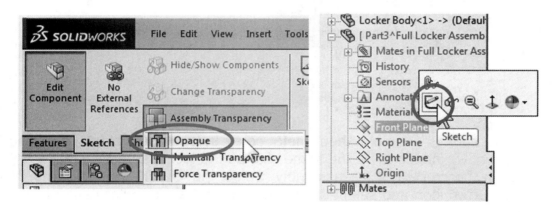

264. – Draw a rectangle starting in one of the lower corners (inside the locker) and finishing in the opposite upper corner. Be sure to capture a Coincident relation to a miter flange vertex. By adding these relations to the miter flange we are fully defining the sketch and are now ready to make the base flange.

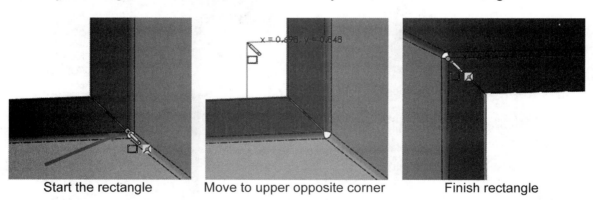

Start the rectangle   Move to upper opposite corner   Finish rectangle

265. – Select "**Base Flange/Tab**" from the Sheet Metal toolbar or the menu "**Insert, Sheet Metal, Base Flange**." Make the base flange also from 18 gauge steel sheet metal as the previous parts. Make sure the material is added towards the inside of the locker and does not overlap with the locker's body.

This is a view from above and behind the locker, showing the new part through the open corner to make sure the material is added inside the locker.

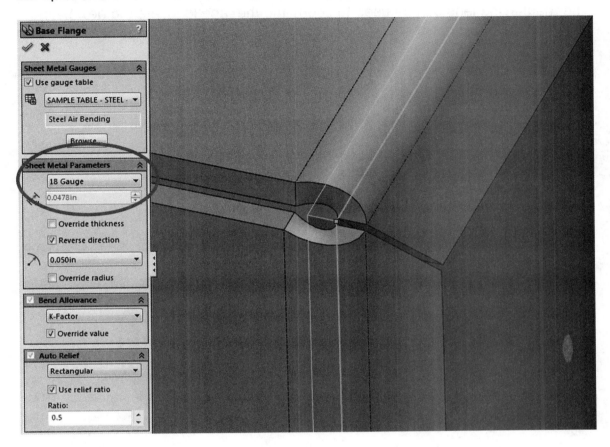

266. – Rename the part **"Locker Back"** in the FeatureManager, save it to an external file, and turn off **"Edit Component"** to continue editing the assembly.

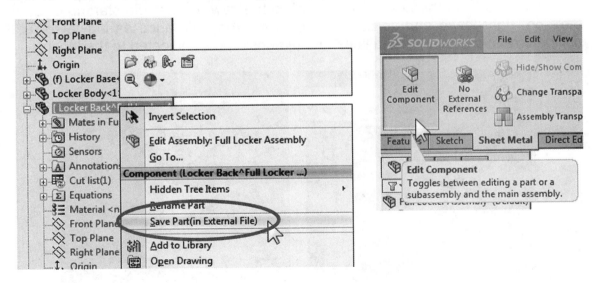

267. – The next component will be a shelf. Select "**Insert Components, New part**" and select the top face of one of the lower lances we just made to locate it. The magnifying lens can be used to zoom into the face to locate the new part.

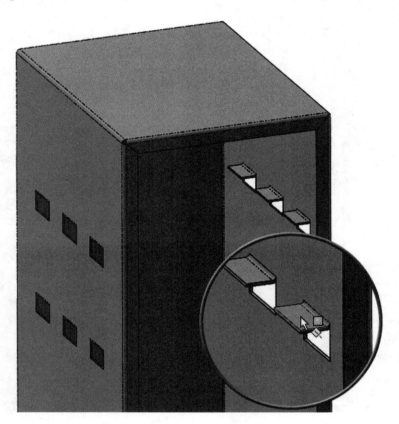

268. – Since we changed the transparency setting to "Opaque" in the previous part, the other components will not become transparent when adding this one; we are now editing the new part's first sketch. Draw a rectangle starting close to the front right lance and finishing close to the rear left lance as indicated, add collinear relations to the bend lines and dimensions as indicated to the lances.

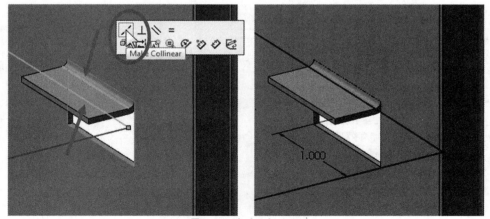

Front right lance

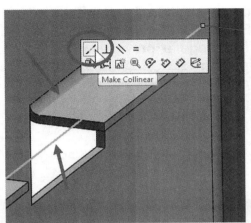
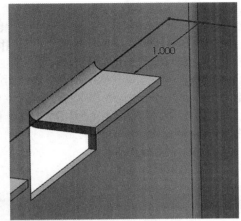

Rear left lance

269. – Click in the "**Base Flange/Tab**" command from the Sheet Metal toolbar and make the shelf using 16 gauge steel, as it needs to be thicker because this is the part that will be carrying the load. Be sure to add the material going up. Use a bend radius = 0.075", and Rectangular auto relief with a 0.5 ratio.

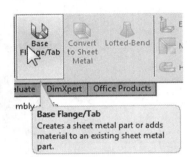

270. – Since a flat shelf like this is not strong enough, a flange will be added to all edges for reinforcement. Click in the **"Edge Flange"** icon and select all four edges. Immediately after selecting the first edge a preview will be displayed. Click towards the bottom to define the direction of the flange, and select the other three edges. Set the flange length to 0.625″; leave the flange gap to the default value of 0.01″ and the flange position to "Material Inside."

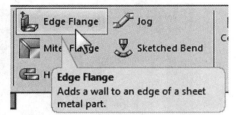

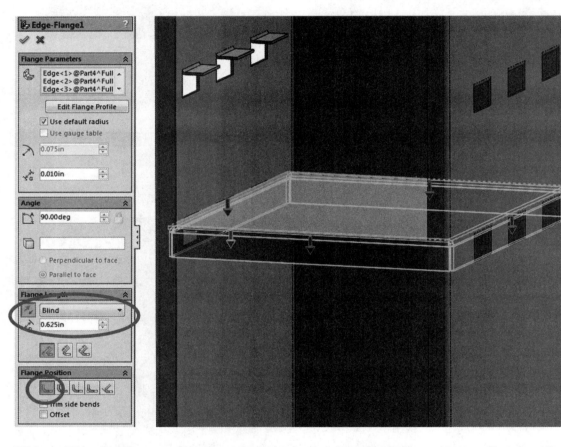

271. – The flanges we just made interfere with the lances in the *'Locker Body.'* To eliminate the interferences we'll unfold the side flanges, add cuts to clear the lances and then re-fold them. Select the **"Unfold"** command from the Sheet Metal toolbar; click in the top face of the shelf part to select it as the fixed face.

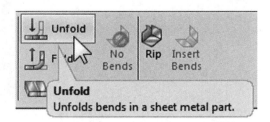

After selecting the fixed face, select the two side bends (the ones interfering with the lances) to unfold them. SolidWorks will automatically add a bends selection filter; click OK when done.

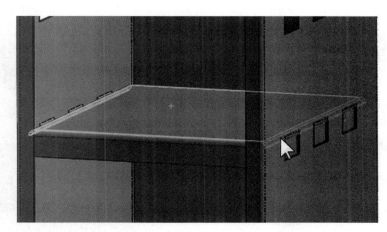

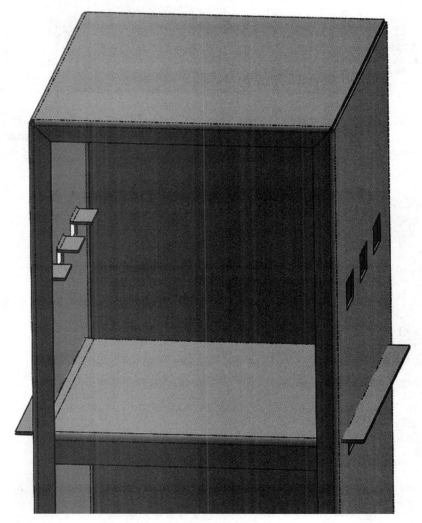

Unfolded bends

272. – To better visualize the next operation, we'll tilt the locker to view it from the bottom up and make the cuts needed to clear the lances in the *'Locker Body.'* The reason is that from this view angle we can see the lances we need to clear. Rotate your locker assembly until you can see the lances from below in one side.

273. – Create a new sketch in the bottom face. Draw a rectangle; add dimensions and relations to clear the lances as shown. Make one side of the rectangle coincident to the outside edge of the unfolded flange and the other side collinear to the bend line as indicated.

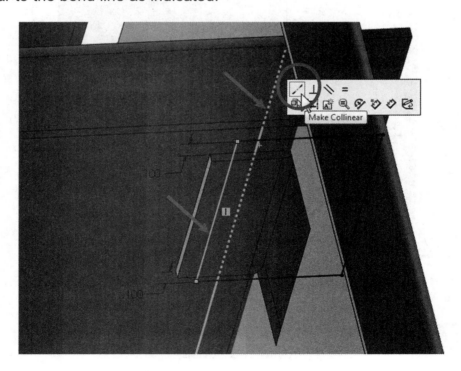

274. – With the sketch fully defined, our next step is to add a "**Sketch Pattern**." Not only can we make patterns of features, we can and also make linear and circular patterns of sketch elements *in the sketch*. Select the menu "**Tools, Sketch Tools, Linear Pattern**" (the icon *may* not be in the Sketch toolbar by default).

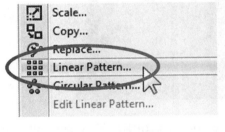

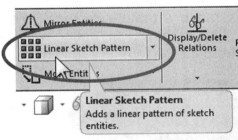

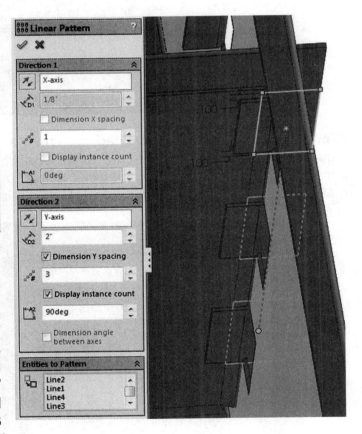

The sketch "**Linear Pattern**" behaves similar to a feature pattern. By default the "**X-axis**" direction is the sketch's horizontal direction (short red arrow) and the "**Y-axis**" direction the vertical (long red arrow.) Our pattern will have 3 copies vertically spaced 2″ apart.

 The direction of a "Sketch Pattern" can be changed by selecting either a model edge or a sketch line.

Turn on the option "**Dimension Y spacing**" to dimension the distance between instances, otherwise the spacing can only be changed by selecting an instance and selecting the menu "**Tools, Sketch Tools, Edit Linear Pattern**."

Fill in the values for the Y-axis (*or* select an edge for direction in the "Direction 1" options box). Add the rectangle's lines in the "Entities to Pattern" selection box and click OK to finish.

275. – If we look closely, the sketch is not fully defined. To fully define it, select the construction line added between the rectangle we drew and the first copy and add a vertical relation.

276. – With the sketch fully defined, we can make a sketch mirror to make all six cuts at the same time, OR make a cut *and then* make a feature mirror. We'll leave the option to you. Just remember to use the "Link to thickness" option when making the cut. Our part should now look like this from the bottom (still unfolded):

277. – To finish the part select the "**Fold**" command from the sheet metal toolbar. Either select both unfolded bends (the selection filter will only select bends) or click in the "**Collect All Bends**" button to add all un-folded bends to the selection list. Click OK when done. The refolded part is shown using the "**Force Transparency**" option.

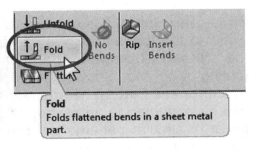

Fold
Folds flattened bends in a sheet metal part.

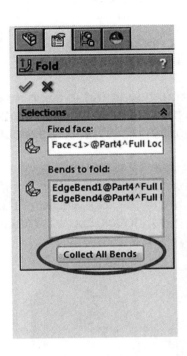

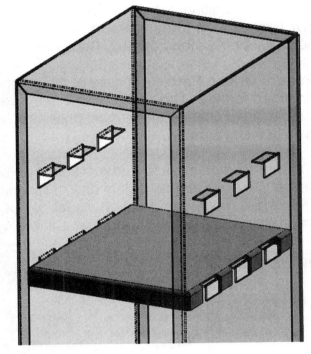

278. – The shelf part is complete; finish editing it by clicking on the "**Edit Component**" icon or the confirmation corner. Rename it *'Locker Shelf'* and save as an external part. After saving the part return to editing the assembly by turning off the "**Edit Component**" command.

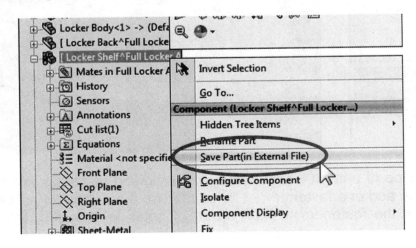

279. – The next step is to add another *'Locker Shelf'* in the top lances. We can add it using the **"Insert Component"** command and mating it to the *'Locker Body'* or using a **"Component Pattern."** Parts and assemblies can be patterned in an assembly using either linear or circular patterns just like features in a part. In this case, a better option to maintain design intent is to use a **"Pattern Driven Pattern."** While editing the assembly select

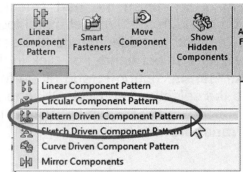

"Pattern Driven Component Pattern" from the **"Linear Component Pattern"** drop down list or the menu **"Insert, Component Patten, Pattern Driven."**

In a **"Pattern Driven Pattern"** a copy of the component is added in each instance of a patterned feature. This is the reason why we patterned the lances vertically; for example, if we change the vertical pattern of lances to 3 copies, we would get 3 shelves in the assembly.

280. – In the "Components to Pattern" selection box select the *'Locker Shelf'* part and for the "Driving Feature" select one of the lances in the top position. Turn on the "Propagate component level visual properties" checkbox to make the patterned·copies look like the seed part. Click OK to finish.

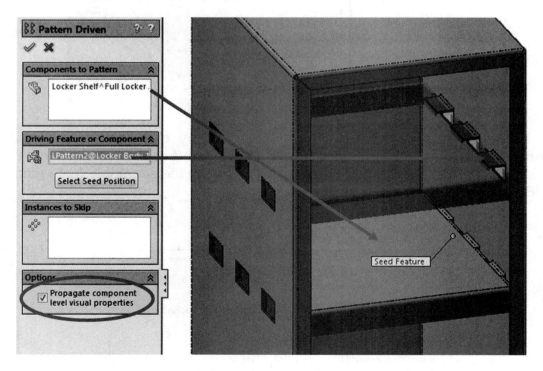

 This type of pattern is very handy when adding fasteners to an assembly; we can add one fastener to a hole, and use the pattern of holes to add the rest of the fasteners. If the pattern of holes is changed, the number of fasteners is also updated.

224

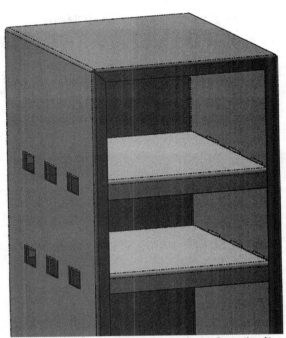

'Locker Shelf' color changed for clarity

281. – After adding the top shelf, we want to make a change to our design and make the top space smaller. Expand the *'Locker Body'* FeatureManager and find the *"LPattern2"* feature (the vertical pattern). Double click in it to display its dimensions and change the spacing from 3″ high to 4″. Rebuild the model when done to finish. Notice the pattern of shelves updates as expected.

282. – The next thing we need to do is to add a "**Hem**" to the *'Locker Shelf'* front flange. Hems are used to eliminate sharp edges and to reinforce a sheet metal part by folding or rolling the edge. The first thing we need to do is to edit the *'Locker Shelf'* in the assembly. Select it in the graphics area or FeatureManager and click in the "**Edit Part**" icon. If the rest of the components don't become transparent when editing the *'Locker Shelf'*, set the Assembly Transparency setting to "Force Transparency."

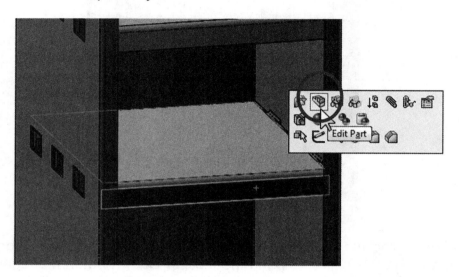

283. – Select the "**Hem**" command from the Sheet Metal toolbar. In the "Edges" selection box, pick the front edge of the *'Locker Shelf.'* A preview will show the hem's direction. Make sure the hem is inside the locker; click on "Reverse Direction" if needed. Besides editing the hem's width, the "Hem" command has the following options:

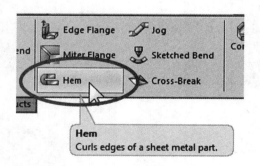

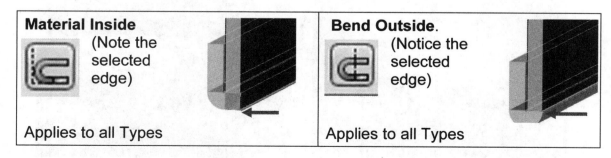

Material Inside		**Bend Outside**.	
	(Note the selected edge)		(Notice the selected edge)
Applies to all Types		Applies to all Types	

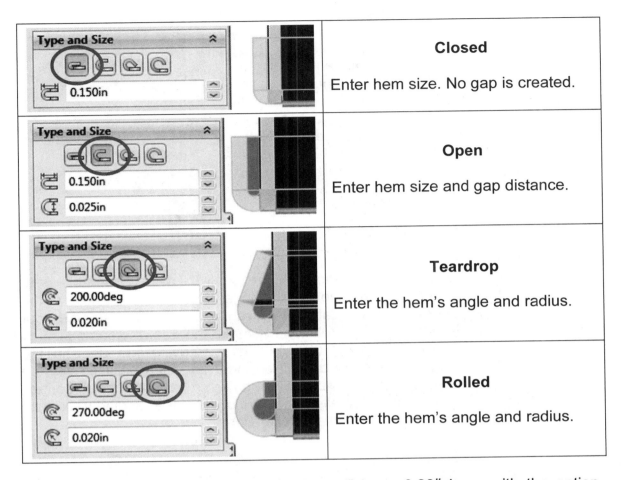

(Closed icon) 0.150in	(image)	**Closed** Enter hem size. No gap is created.
(Open icon) 0.150in / 0.025in	(image)	**Open** Enter hem size and gap distance.
(Teardrop icon) 200.00deg / 0.020in	(image)	**Teardrop** Enter the hem's angle and radius.
(Rolled icon) 270.00deg / 0.020in	(image)	**Rolled** Enter the hem's angle and radius.

284. – For the *'Locker Shelf'* use a "Closed" hem, 0.20″ long, with the option "Material inside." After finishing the hem notice that the top copy of the *'Locker Shelf'* also shows the hem just added.

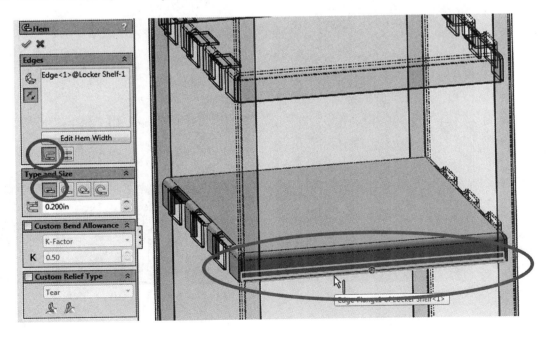

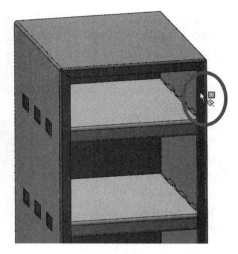

285. – Turn off "**Edit Component**" and return to editing the assembly. The last component to add to our assembly will be the locker's door, and it will also be designed in the context of the assembly. Select "**Insert Components, New Part**" and click in the front face of the *'Locker Body'* to locate the door.

286. – The purpose of adding relations in the context of the assembly is to propagate changes when other components change; however, in the case of the door, we have a slightly different situation. We want to be able to open the door and move it; however, if we capture relations to make the door the same size as the *'Locker Body'*, when we open the door later (moving the door in the assembly), it will be rebuilt and its size will be incorrect, fail to rebuild with an error, and/or produce unexpected results when trying to maintain the in-context relations. Notice how the door is smaller when we open it after adding in-context relations to make it the same size of the *'Locker Body.'*

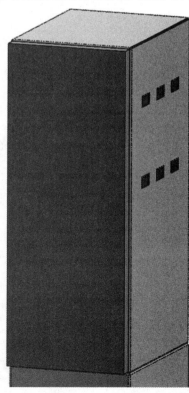
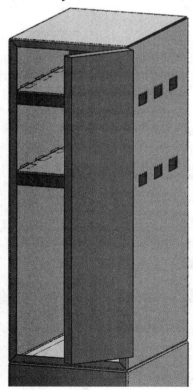

287. – One thing we can do to avoid capturing external relations is to activate the "**No External References**" option. While this command is active no external references are captured to other components, either automatic or manual references. Activate "**No External References**" and draw a rectangle for the door as indicated. Dimension it to match the locker's size and then add a Midpoint relation between the rectangle's lower line and the part's origin to center it about the origin (almost always a good idea). This relation will fully define the sketch. Select the origin in the FeatureManaeger if needed. Make the assembly opaque for clarity.

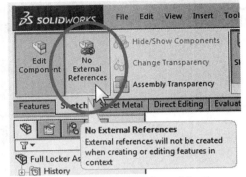

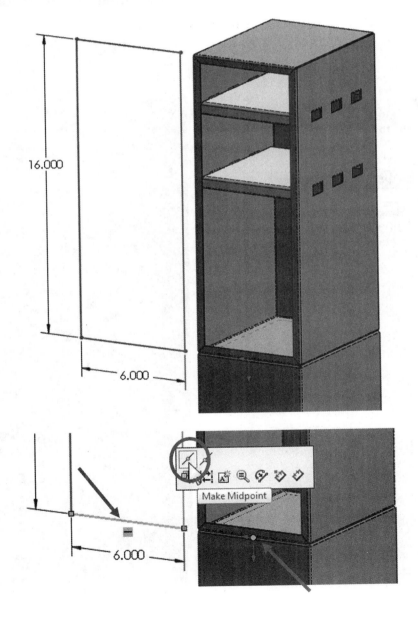

288. – Extrude the rectangle 0.625″ outward. After making the first extrusion we can see in the FeatureManager that no external references have been added to the extrusion or its sketch.

289. –The next step is to convert the solid door extrusion into a sheet metal part. From the Sheet Metal toolbar, select "**Convert to Sheet Metal**" to continue.

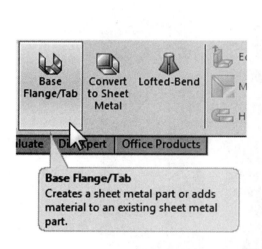

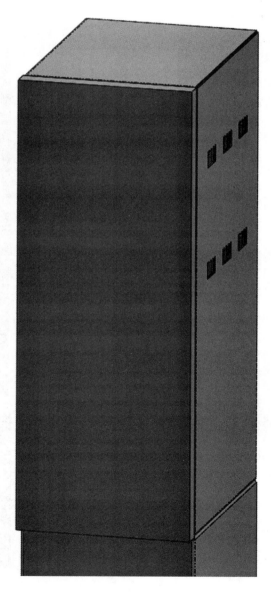

290. – To convert the part to sheet metal, select the front face of the door to be the fixed face; add the four edges around it in the "Bend Edges" selection box. Once the bend edges are selected, the small corner edges will be automatically added to the "Rip Edges found" selection box. Using the same sheet metal gauge table as before, make the door 18 gauge, default bend radius of 0.05", default gap for rips of 0.02", "K-Factor" of 0.5, and the "Auto Relief" option to "Tear." Make sure the material thickness is added inside, making our part dimensions external. Click OK to finish.

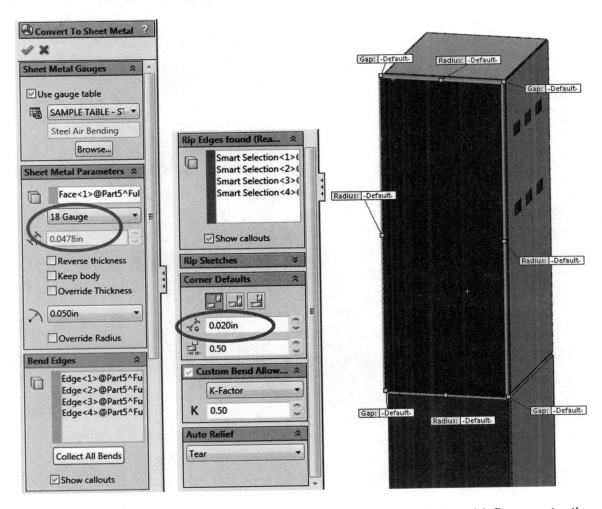

291. – After converting the part to sheet metal, we need to add flanges to the inside of the door. Open the part in its own window and turn it over to see the back (inside).

292. – Select the "**Edge Flange**" command and add all four edges in the back to the "Edge" selection box. Make the flange 0.625″ with a gap distance of 0.01″ and "Material Inside" option.

293. – Go back to the assembly; rename the new part as '*Locker Door*' and save it to an external file.

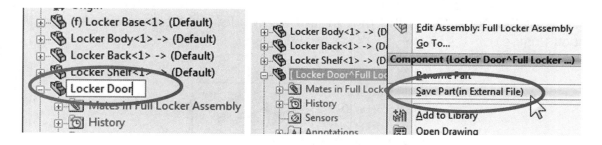

294. – Our next step is to add a set of louvers to the door to help ventilate the locker. Since the louver forming tool available in the default SolidWorks Design Library is too small for our design, we'll need to make a new one.

To create a "**Forming Tool**" library, the first thing we need to do is to build the shape of the tool, define a *'stopping face'*, which defines how deep the tool will go into the sheet metal part, and optionally the face(s) that will be removed (cut) as openings in the part.

Make a new part (by itself outside of the assembly), add the following sketch in the *"Top Plane"* and extrude it 0.375″ up. This extrusion will serve as the base to build the library on top of it.

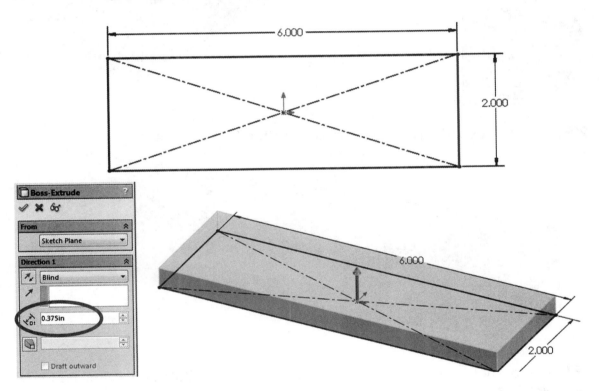

295. – Add the following sketch on top and extrude 0.375″, then round the two corners indicated with a 0.625″ radius.

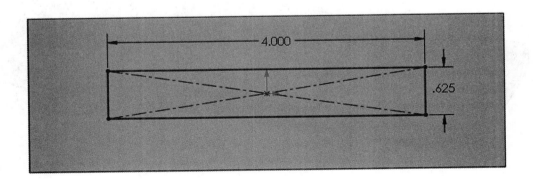

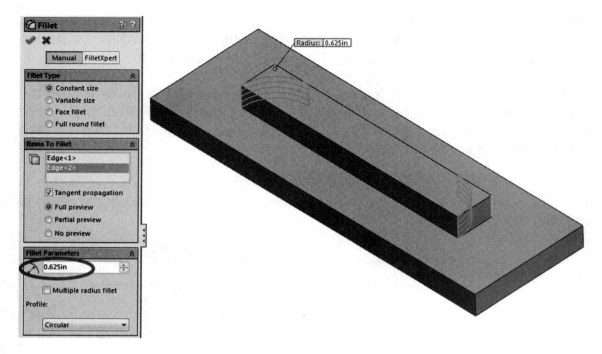

296. – Add a second Fillet to the top edge with a 0.25″ radius.

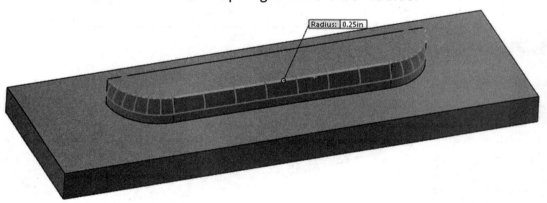

297. – Add another fillet to the bottom edge with a 0.375″ radius; our part now looks like this:

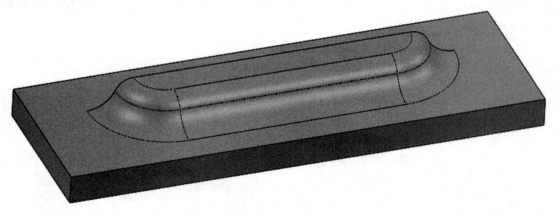

298. – Now we need to remove the base of the part (the first feature) to complete the forming tool. Add a sketch on the top face of the first feature, use "**Convert Entities**" to project the edges (or draw a square coincident to the base corners, your choice) and make a cut using the "Through All" end condition to remove the entire base. This cut could have been made in any face of the base, as long as the entire first feature is removed.

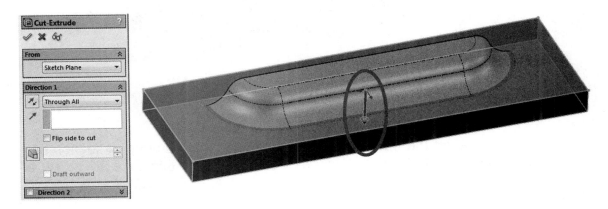

Our form tool now looks like this:

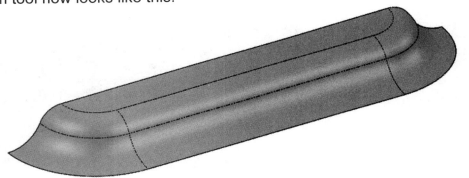

 Cutting the base is not required to make a form tool; in our case we had to do it to add the fillet at the bottom for our form tool.

299. – After the part is complete, select the "**Forming Tool**" icon in the Sheet Metal toolbar or the menu "**Insert, Sheet Metal, Forming Tool**."

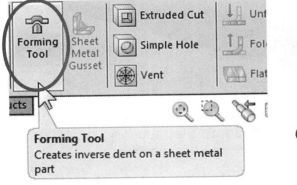

For the "Stopping Face" selection use the bottom face as indicated. This is the face that will define how deep the tool will go *into* the sheet metal part.

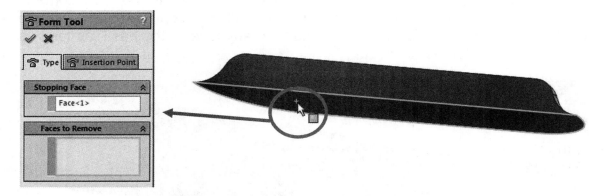

300. – And then select the front flat face in the "Faces to Remove" selection box. This is an optional selection, but in our case we want this face to make an opening in the sheet metal part when inserted. Click OK when done.

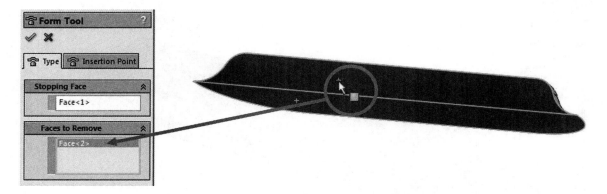

301. – After finishing the form tool, SolidWorks automatically changes the face's colors to indicate which are the stopping, forming, and open faces.

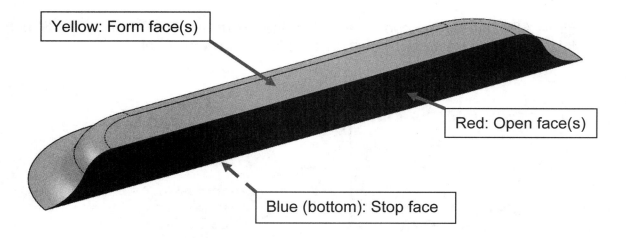

302. – Now we have to save it in our library. Save the part as '*4in Louver*.' Right-mouse-click at the top of the FeatureManager and select "**Add to Library**."

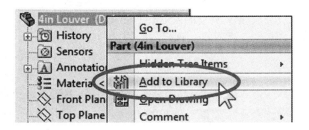

In the "File name:" field enter '*4in Louver*' and select the "louvers" folder. In the "File type:" drop down selection box leave the option as "Part (*.sldprt)."

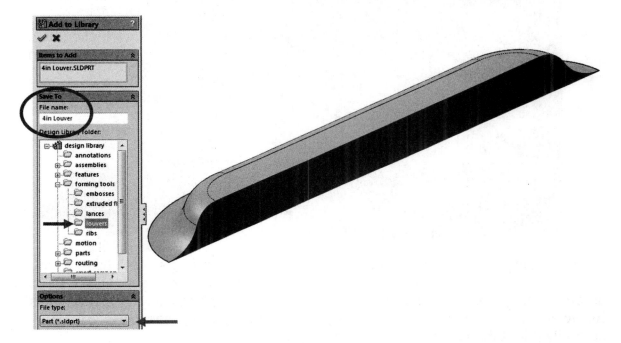

 The form tool feature automatically adds a sketch called "*Orientation Sketch*." This is the sketch that will be used to locate the form tool when it is inserted into a sheet metal part.

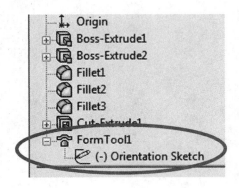

303. – Close the *'4in Louver'* library part and go back to the assembly. Make sure we are still editing the *'Locker Door'* in context of the assembly to add the library to it. After opening the Design Library we can see the new *'4in Louver'* tool, select it and drag it to the front of the locker door to insert it.

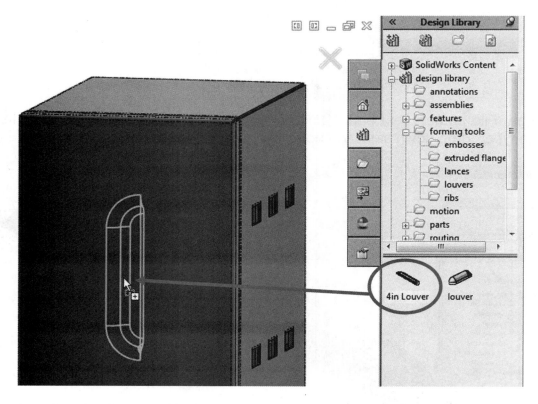

After dropping the form tool in the door use the "Rotation Angle" options to rotate the form tool, and if needed use the "Flip Tool" to reverse the direction to make it go out, getting the louver correctly oriented as in the next image.

Be aware that the values entered *may* be different in your case; the goal is to correctly orient the louver to match the preview.

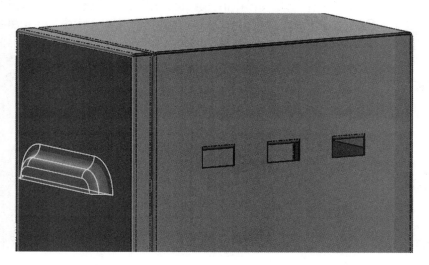

304. – To locate the louver select the "Position" tab at the top of the Property Manager. Add a 2" dimension from the louver's center point to the top edge of the door, and a "Vertical" geometric relation between the louver's center and the door's origin to center it. It may be useful to turn on the display of origins to select the door's origin for the vertical relation. Click OK to finish when done.

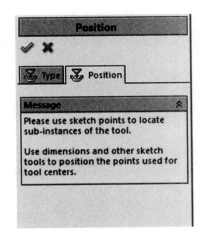
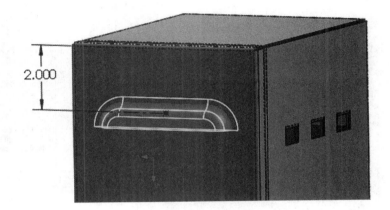

305. – When the louver is done make a linear pattern of 4 louvers spaced 1.25". Finish editing the part (turn off "**Edit Component**") and save the assembly and all components at the same time.

306. – After reviewing the assembly in a Left view, we notice that the door protrudes beyond the locker's base. Measuring the difference we can see the *'Locker Body'* needs to be 0.625″ smaller (the thickness of the door) to make the door flush with the base. To correct this we need to edit the first sketch in the *'Locker Body'* to change its size.

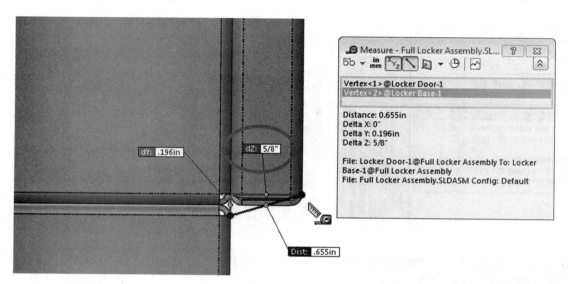

307. – Expand the *'Locker Body'* FeatureManager and edit the *"Boss-Extrude1"* sketch. When we edit a part's sketch in an assembly, the **"Edit Component"** mode is automatically activated.

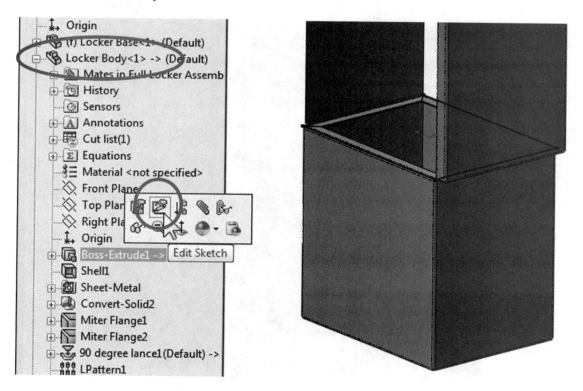

308. – Select the front sketch line (close to the door) and delete the "On Edge" relation (created when we used the **"Convert Entities"** command).

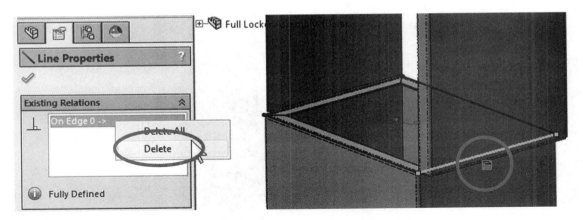

309. – After deleting this external relation (we know it's external because it is followed by the "**->**" symbol), drag the front line a little to the back and dimension it. (Be sure to reference the edge of the base and not the door to locate it.)

 If the **"No External References"** command is active you will NOT be able to add dimensions (or relations) to other components. Turn it off if it's active.

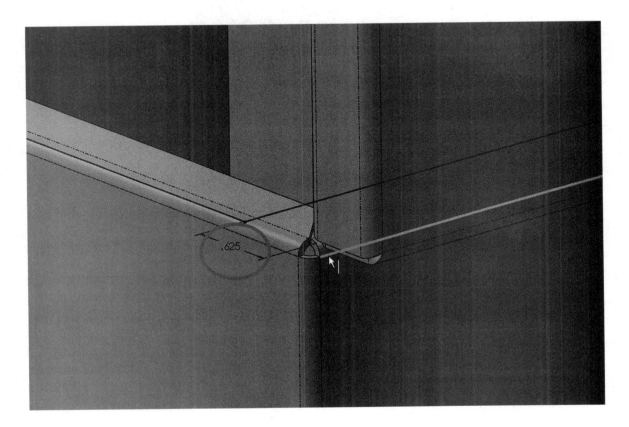

310. – Close the sketch and turn off "**Edit Component**." All components will rebuild correctly. Your assembly now looks like this:

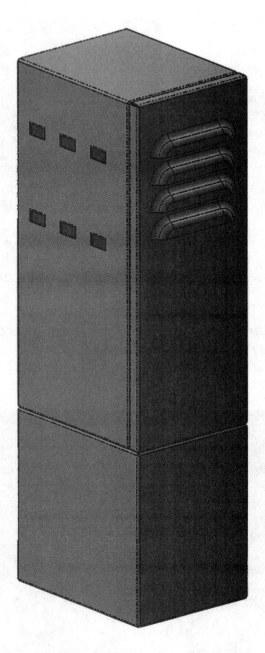

311. – After reviewing the assembly, the *'Locker Shelf'* part is protruding through the back of the locker. When we dimensioned the shelf, it was referenced to the lances in the body, and after the lances were moved when the *'Locker Body'* was resized, the shelf moved too causing this problem. To correct it we need to edit the *'Locker Shelf'* sketch and change the dimension to reference the back of the *'Locker Body'* instead of the lances. Expand the *'Locker Shelf'* FeatureManager; edit the *"Base-Flange1"* sketch and delete the dimension in the back. Hide the *'Locker Back'* part for better visibility.

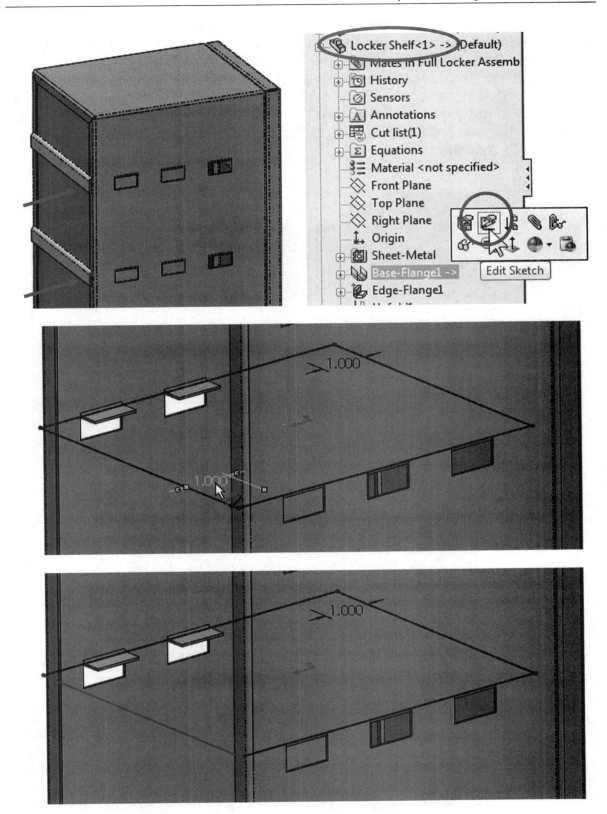

312. – Move the back line to be inside the locker and dimension it to the edge of the *'Locker Body'* as indicated; this way the shelf will be referenced to the *'Locker Body'* and not the lance's position. Be aware that we need to reference a horizontal edge in the *'Locker Body'* to dimension the sketch. We can either go to a top view to get the correct reference, or select the top or bottom miter edge to reference the dimension. Exit the sketch and finish editing the part in the assembly to continue.

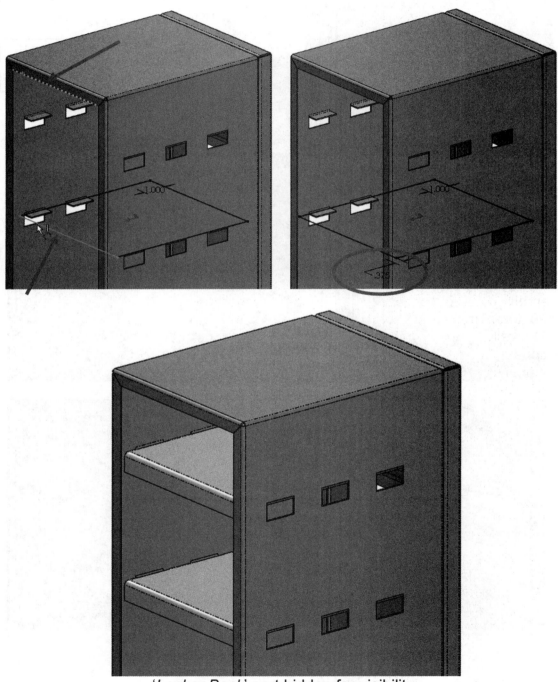

'Locker Back' part hidden for visibility

313. – After correcting the shelf the last step is to add hinges to the locker to open the door. While editing the assembly (not a part) if we try to move the door we get the following message:

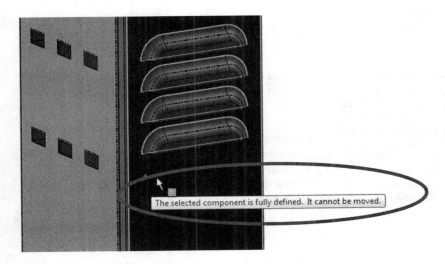

The selected component is fully defined. It cannot be moved.

When we add components in the context of the assembly, an **InPlace** mate is automatically created when we select the face where we want to locate a new part. The InPlace mate completely immobilizes a part, and in order to be able to move it, we have to delete (or suppress) this mate. Expand the FeatureManager for the *'Locker Door'* part and scroll down to the folder *"Mates in Full Locker Assembly."* This folder contains all the mates referencing the door in this assembly. If you remember, we did not make the door the same size as the body with external relations, but with dimensions, the reason was so that we could later be able to move the door using assembly motion. If we had added external relations in the sketch, after moving the door, its size would have changed and most likely produced undesired results. Now delete the **InPlace** mate and drag the door to move it away from the locker. The hinges will be added in the next step.

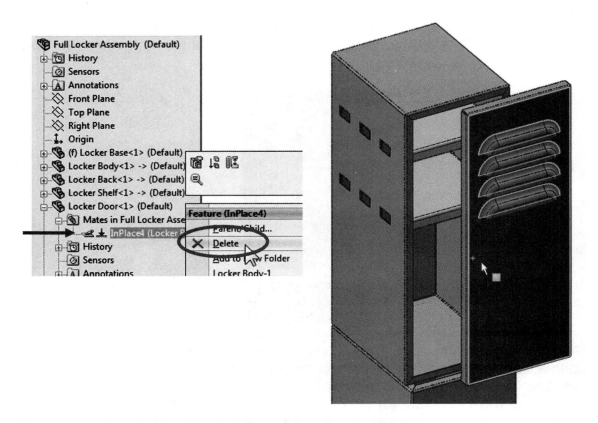

314. – From the accompanying files open the '*Hinge*' assembly and add it to the locker assembly.

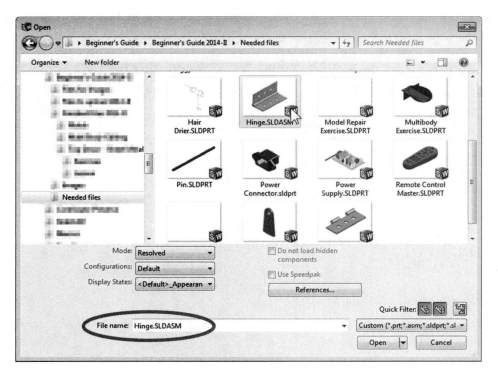

315. – Add a hinge close to the right side of the locker and move the door a little to the left. Mate the *'Hinge'* and the front flange of the *'Locker Body'* using a Coincident mate.

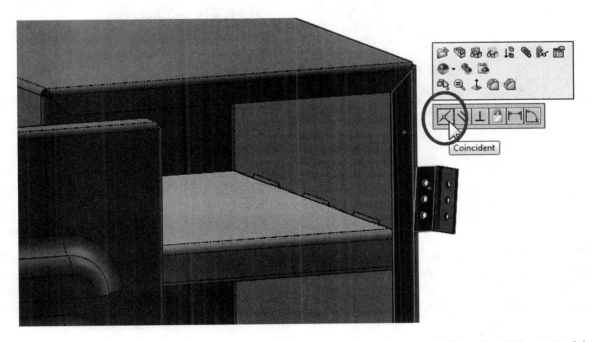

316. – Add a second mate to make the top of the *'Hinge'* sub-assembly Coincident to the top face of the upper shelf. Select an edge or a face of the hinge; either one works the same.

 To select the smaller faces of the hinge we can use a selection. The default shortcut to filter the faces is **'X.'** if you use a selection filter, don't forget to turn it off when you are done using it.

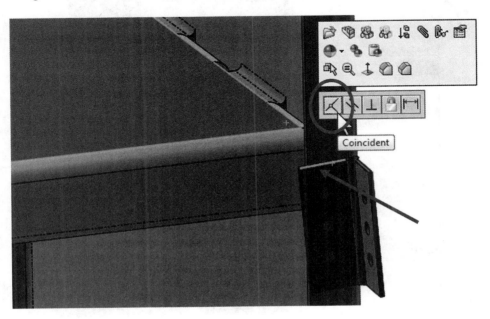

317. – And finally make the indicated edge of the *'Hinge'* coincident to the outside face of the *'Locker Body'*. (View looking from the back.)

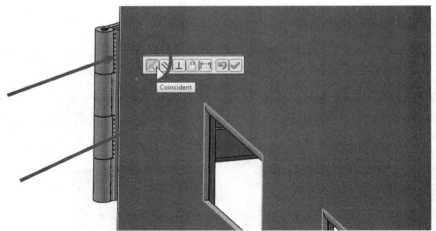

318. – With the hinge located in place, make a linear pattern and add two more *'Hinges'* spaced 5 inches apart. Select the menu "**Insert, Component Pattern, Linear Pattern**" or the "**Linear Component Pattern**" from the Assembly toolbar. Use the *'Hinge'* subassembly as the component to pattern and any vertical edge for the direction.

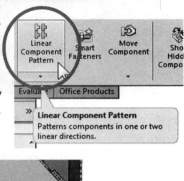

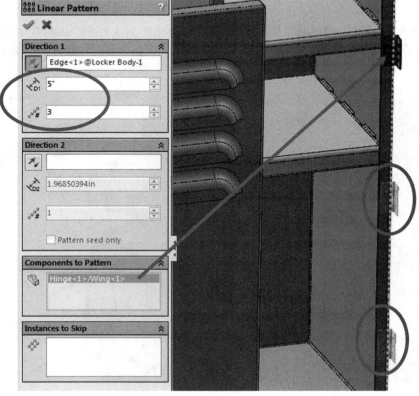

A component pattern feature is added to the assembly's FeatureManager with the two additional hinges.

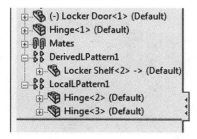

319. – Now we have to add mates to locate the door. Rotate the door a little to mate the face of the hinge and the door's miter flange as indicated.

320. – Rotate the assembly and make the side of the door Coincident to the edge of the hinge (similar to how we mated the hinge to the locker's body).

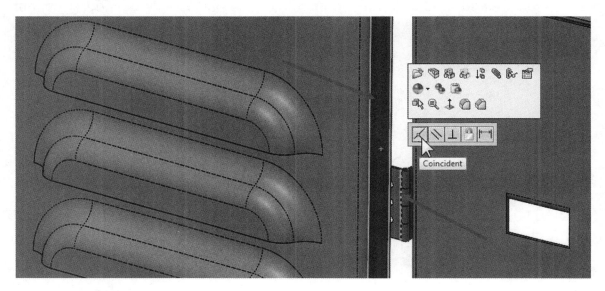

321. – Finally, add a coincident mate between the top face of the door and the top face of the locker's body.

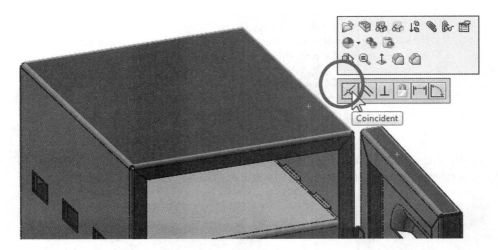

322. – After the door is completely mated, if we try to move it, we get a message letting us know that the component is fully defined. To illustrate how a subassembly behaves in SolidWorks, open the *'Hinge'* assembly.

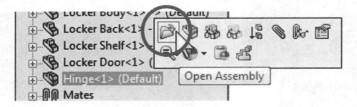

323. – When we add mates to assemble components together, we are removing degrees of freedom (DOF). By default, a component has six degrees of freedom. Three translations along the X, Y and Z axes, and three rotations about the X, Y and Z axes. When a component is under defined, it means that *at least* one DOF is not constrained, and it can move or rotate. If a part is fully defined, it cannot move or rotate. This is the basis for assembly motion in SolidWorks.

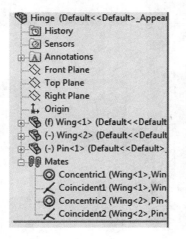
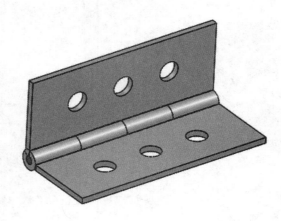

In our *'Hinge'* assembly one *'Wing'* and the *'Pin'* parts have a (-) sign before their name. This means that these parts can move or rotate about *at least* one axis. In this case, both can rotate about the *'Pin'* axis. Since the first *'Wing'* is automatically fixed in place (because it was the first part added to the assembly) the other two rotate about it. If we click-and-drag the under defined *'Wing'* part, it will move as a hinge (as expected) because it was restrained (mated) with one Concentric and one Coincident mate.

324. – If we keep moving the *'Wing'*, it will go past through the fixed *'Wing.'* This is the normal behavior. We can optionally activate the "**Collision Detection**" to make parts stop when they touch each other. Move the hinge to a position where there is no interference and click in the "**Move Component**" icon in the Assembly toolbar; under the "Options" section select "Collision Detection." By default "All components" is selected; as we move a component, SolidWorks will calculate collisions between all components in the assembly.

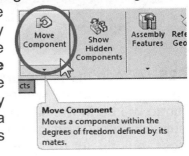

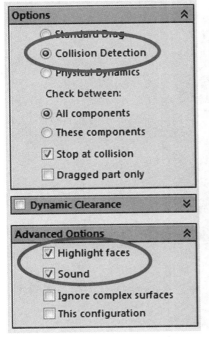

If the assembly has a small number of components, this option is fine (*"small"* varies, from a few to tens or even hundreds depending on the computer's capacity); but if we have a large number of components (and *"large"* also varies...) we can limit the scope of detection by selecting the option "These components" and select only the ones we are interested in, greatly enhancing the speed and accuracy of the calculation. With only three parts using the "All components" option should be fast enough in just about any computer. Turn on the option "Stop at collision" to prevent a part from going through another, "Highlight faces" to visually identify the component faces that are colliding, and "Sound" to hear when a collision is detected.

Drag the moving wing in the hinge towards the other wing part and see how it stops when the faces touch each other.

If we start a **"Collision Detection"** with interference between the components being analyzed, the "Stop at Collision" option will be ignored, and the part's motion will continue. To enable collision detection again the assembly must be set to a non-colliding position, or select non-interfering components in the "These components" selection box. Then re-activate the "Stop at Collision" option.

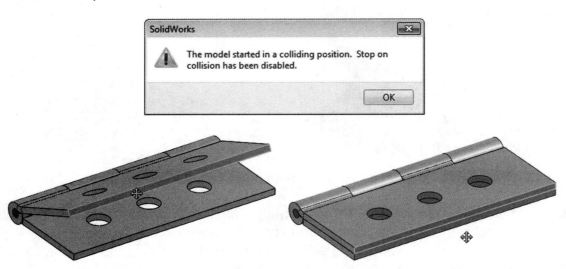

325. – Open the hinge approximately 90 degrees and switch back to the locker's assembly window, where we'll be asked to rebuild it. After selecting "Yes" the hinge will be in the same position we had it in its own window. Moving the '*Hinge*' sub-assembly and going back to the main assembly is one way to move the door, but it's not really the easiest way, or the behavior we want.

326. – What we need to do is to propagate the subassembly's Degrees of Freedom (motion) to the main assembly. By default, subassemblies are calculated as rigid bodies without motion for performance reasons. Select the *'Hinge'* subassembly in the FeatureManager, click in the **"Component Properties"** icon from the pop-up menu and activate the "Solve as Flexible" option. Click OK when done.

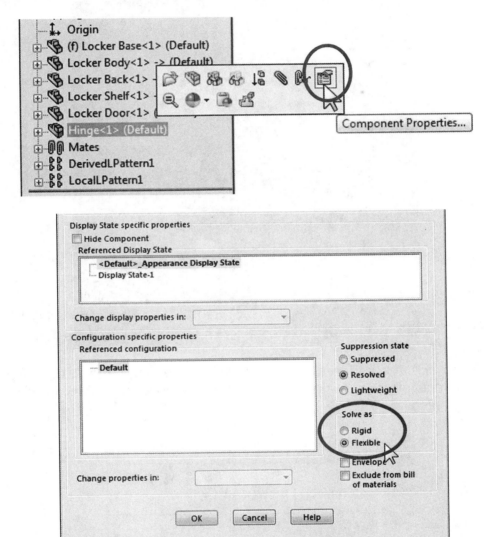

Note the different assembly icons when a subassembly is made flexible.

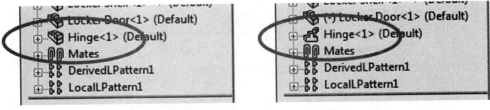

Rigid sub-assembly   **Flexible sub-assembly**

327. – By making the *'Hinge'* subassembly "Flexible" we can click and drag the door to move it freely. Notice that the lower two patterned hinges are not flexible and when we move the door they remain in the previous position.

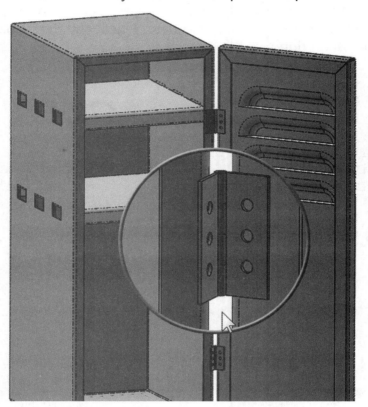

328. – To make the other two hinges flexible, expand the *"Local Pattern1"* feature, and activate the "Make Subassembly Flexible" checkbox in the pop-up menu in both hinges.

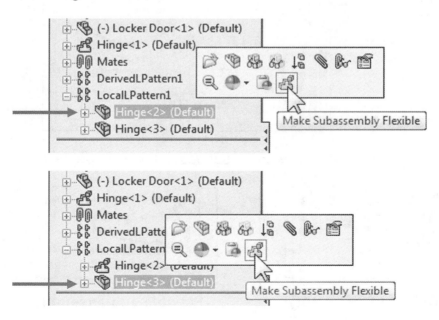

After making the subassemblies flexible, mate them to the door. To avoid over defining the assembly add a parallel mate between the hinges and the door, or between hinges. The assembly may have to be rebuilt to restore assembly motion after making the sub- assemblies flexible.

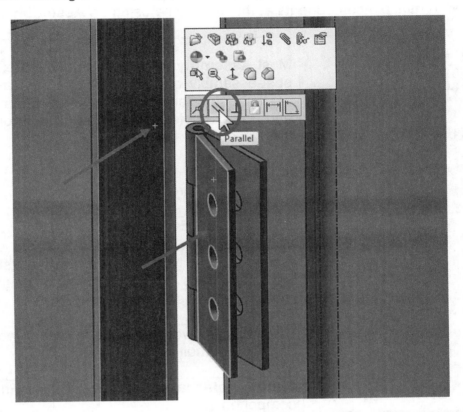

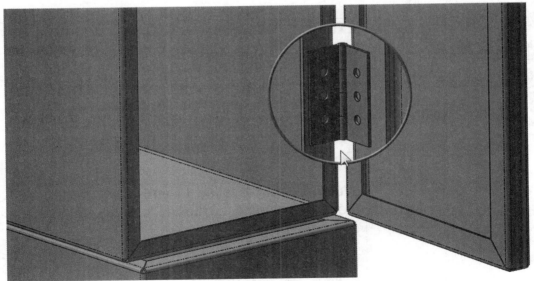

329. – The last thing we need to do is to add a series of holes to hold the *'Locker Back'* with rivets or screws. We can add the holes to one part, and then make the holes in the other part using in-context relations, but that would take too long. In this case, the holes will be added while editing the assembly, not a part, and will be added to multiple components at the same time using an **Assembly Feature**.

	Most **Assembly Features** remove material, like a post-assembly operation, and in our case would be the equivalent of holding both parts together and, for example, drilling through both of them at the same time. Assembly features include:
	Hole Series can make holes using the Hole Wizard through several parts with the option to make different types of holes in the first, middle and last parts like counterbore, countersunk, clearance threaded, etc.
	Hole Wizard works the same way it works in a part, but for multiple components.
	Simple Hole makes single round holes in multiple components.
	Extruded Cut, Revolved Cut and **Swept Cut** work like a cut feature in a part, but in the assembly through multiple components.
	Fillet and **Chamfer** work exactly as in a part, but in the assembly they are used mostly in the weldments environment to prepare mated parts for welding.
	Welded Bead is used to add welds in structural steel, including manufacturing annotations (Covered later.)
	Belt/Chain is used to add Belt mates between pulleys or sprockets to maintain their relative motion and calculate belt/chain length, or to define the belt/chain length and locate components.

330. – Select the "**Hole Wizard**" option from the "**Assembly Features**" drop-down icon. We'll use ⅛″ rivets to secure the *'Locker Back'* to the *'Locker Body.'* Turn the locker view to see the upper corner from the back. Using the simple hole type, select a ⅛″ size. Making the hole's end condition "Blind" with a depth of 0.25″ will be deep enough to go through both parts.

Notice the additional option box at the bottom titled "Feature Scope." This is where we can define which parts will be affected by this feature and if we want the feature to be propagated to the individual parts. If we leave the "Propagate feature to parts" option unchecked, the holes will only exist in the assembly and not in the parts. This is useful when making post-assembly operations, like drilling. If the option is checked, the holes will be added at the part level, will have an external reference, and will be in the assembly, like any other feature.

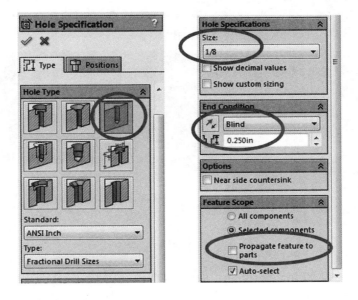

331. – Select the "Positions" tab, select the back face of the miter flange in the *'Locker Body'* to add the hole. Immediately the "**Sketch Point**" tool is activated. Add a point and locate a single hole as indicated. We'll use a linear pattern to add the rest of the holes. Click OK when done to finish the hole.

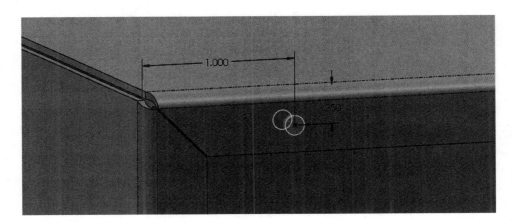

A hole is added to both parts in the assembly. Since the feature was not propagated to the parts because the option "Propagate feature to part" was not checked in the Hole Wizard, when we open either part the hole will not exist.

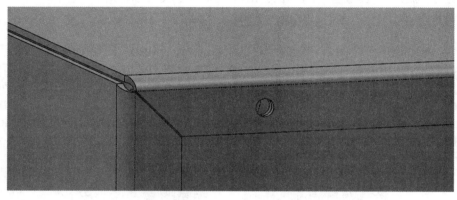

Assembly

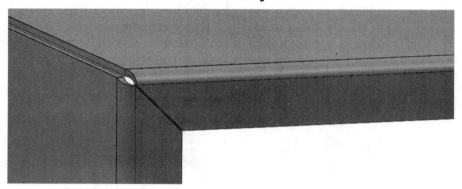

Part

332. – When assembly features are added, the "**Assembly Features**" drop-down icon will be expanded to include linear, circular, table and sketch driven patterns.

Using the "**Assembly Features, Linear Pattern**" make a horizontal pattern to add 5 holes spaced 1".

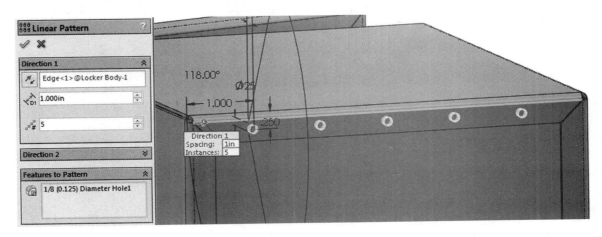

333. – Make similar assembly features and patterns in the other three flanges to complete the assembly and save when done.

 Edit the Hole Wizard in the top to add a second hole in the bottom flange to pattern both holes together; use a vertical pattern in the sides using a linear pattern of 8 holes spaced 2" apart. Save the assembly when done.

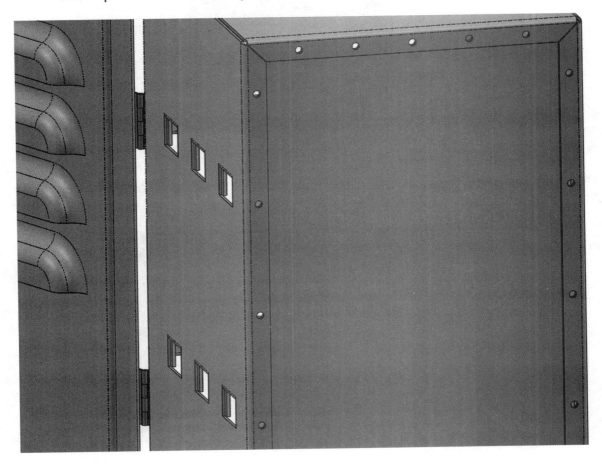

CHALLENGE EXERCISES:

1: Change the *'Locker Body'* to be the same size as the base with a collinear relation to the *'Locker Base'* as originally designed.

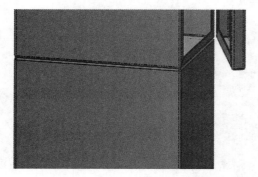

2: Change the *'Locker Base'* dimension to 8″ wide x 10″ deep x 10″ high.

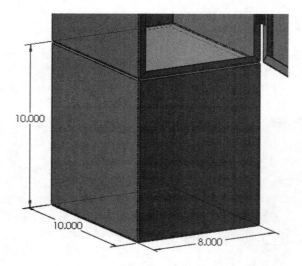

3: Add assembly features using the Hole Wizard to attach the *'Locker Body'* to the *'Locker Base'*. Choose the location and spacing.

4: Add equations in the assembly to make the *'Locker Door's* width ⅞" narrower than the *'Locker Base'* and the *'Locker Door's* height ⅞" smaller than the *'Locker Body.'* Equations in an assembly work just like in parts, except the dimension's name include the component's name. Turn on each component dimension's display in the assembly and rename the dimensions as needed.

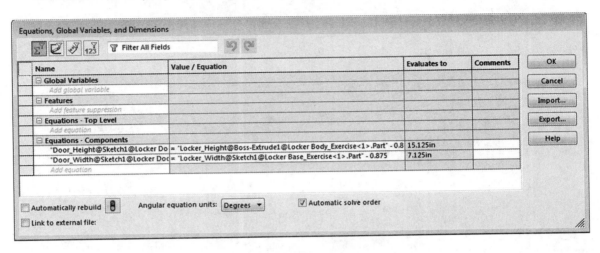

5: Modify the miter flange in the front of the *'Locker Body'* by editing its sketch as shown to accommodate the *'Locker Door'* inside and make it flush with the front of the assembly.

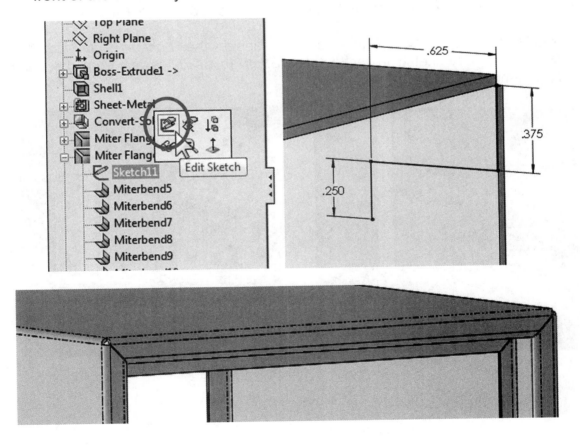

 When the miter flange is edited, if its edge was used for a linear pattern's direction we may get an error in an assembly. If this happens edit the pattern and re-select the direction's edge if needed to clear the error.

6: Modify the *'Locker Shelf'* to fit correctly inside the *'Locker Body.'* Edit the *"Base-Flange1"* sketch in the *'Locker Shelf'* to be ¼" away from the miter flange modified in the previous step.

7: Add and delete mates in the assembly as needed to make the *'Locker Door'* flush with the front face of the *'Locker Body'* and mate the hinges in the front of the locker assembly. Add a "Width" mate to center the door vertically between the miter flanges. Use "**Collision Detection**" to adjust the mates in the hinges and make sure the door opens and closes as intended without interferences.

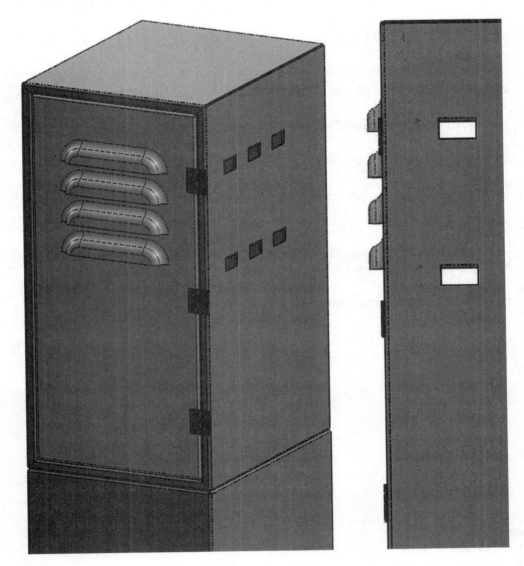

8: Make detail drawings of all parts including a flat pattern and an assembly drawing with an exploded view and bill of materials.

Extra Credit: Design a suitable lock for the door and incorporate it into the assembly.

Review and Questions:

a) Name the different options available in SolidWorks to calculate sheet metal bend allowances.

b) How is the Relief Ratio calculated (Rectangular and Obround reliefs)?

c) What is and how is the K-factor calculated?

d) Which external reference state will propagate changes immediately in an assembly (In-Context, Out of Context, Broken, Locked)?

e) How can external references be temporarily "frozen"?

f) What does the option "Merge faces" do in a sheet metal component?

g) What is the "Link to thickness" option and what does it do?

h) What is the extension used for library features?

i) What is the option "Flexible" in a subassembly used for?

j) Explain the difference between a Miter Flange and Edge Flange?

Answers:
a.- Bend allowance, Bend Deduction, K-Factor
b.- It's the relation between the sheet metal thickness and the width of the relief.
c.- It's the relation between the distance from the inside bend radius and the neutral plane to the thickness.
d.- In Context.
e.- Locking them using under "List External References."
f.- It eliminates all the bend lines and bend regions in the flat pattern making it a single surface.
g.- It's an option available when adding extrusions and cuts to sheet metal parts to make the extrusion (or cut) the same depth as the sheet metal's thickness.
h.- *.SLDLFP
i.- If a subassembly has mates that permit motion, the motion will be available at the top level assembly.
j.- In an Edge Flange, the width and shape of the flange can be changed but not its profile; in a Miter Flange, only the profile can be edited.

3D Sketch and Weldments

A 3D sketch, as its name implies, exists in 3D space and is not limited to a 2D plane. 3D sketches are typically used for guide curves and paths for sweeps and lofts, as "centerlines" for structural profiles (weldments), as reference to create auxiliary geometry, etc.

A Weldment part in general refers to a multi body part that is made from extruded (or swept) profiles along a 2D or a 3D sketch. It is used for designing structures, welded components, tubular parts, etc. and includes special features for welding processes and manufacturing annotations like weld types, calculating cut lists of each type and size of material used (very much like a Bill of Materials) and more trade specific tools to fully design and document welded structures.

Notes:

3D Sketch

To learn how to make a 3D sketch, we'll build a tubular frame for a chair. We'll need to make two sweeps, each using a 3D sketch as a path.

334. – Before we start the 3D sketch, we'll make a new part; add a 2D sketch, and an auxiliary plane to help us draw the 3D sketch. Draw a 14″ construction line in the *"Right Plane"* as shown, and exit the sketch.

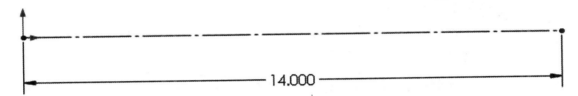

14.000

335. – Switch to an Isometric view and select the **"Reference Geometry, Plane"** command. Create an auxiliary plane at 20 degrees from the *"Right Plane."* Select the *"Right Plane"* in the "First Reference" selection box and the sketch line we just drew in the "Second Reference" selection box. Make the plane 20 degrees to the left (you may have to "Flip" the direction). Click OK to finish. This plane will be used later for reference in the 3D Sketch.

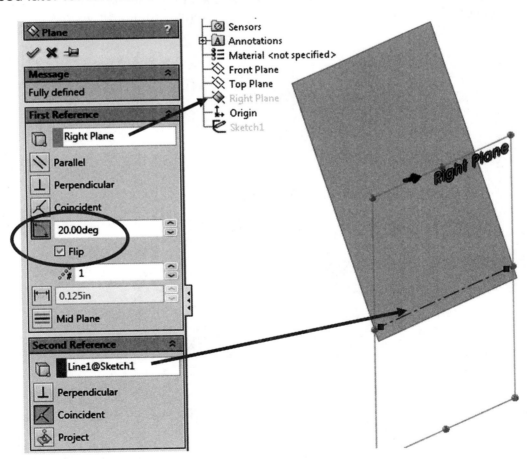

336. –The chair will be made from two bent tubular parts, each made with a sweep feature using a 3D sketch path. To make the first path for the tubular frame select the menu "**Insert, 3D Sketch,**" or the "**3D Sketch**" icon from the drop-down "**Sketch**" command. When we start a 3D sketch, notice we are not asked to select a plane and the only difference from a regular sketch is that we get *"3DSketch1"* in the FeatureManager, status bar and window title.

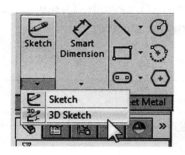
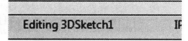

337. – In a 3D sketch we have most of the tools available in a 2D sketch. After selecting the "**Line**" tool, we immediately see a bigger than usual sketch origin, aligned parallel to the *"Front Plane"* and an "XY" label is displayed next to the mouse pointer. This is to let us know that, when we draw a line, we'll be working in a 'plane' parallel to the *"Front Plane"* (XY-plane within the 3D Sketch). Click in the origin to start drawing a line; move towards the right and align it with the 'X' direction. The yellow inference icon is somewhat small, but visible.

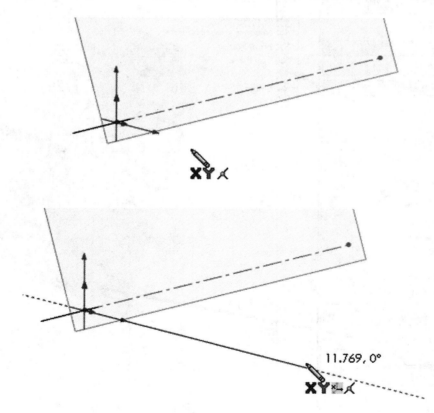

338. – For the second line, go down and right, but be aware we are still in the XY-plane. Notice the origin is now located at the start of the new line to alert us of the plane's orientation we are working on. Repeat the same two lines starting at the other end of the construction sketch; the line tool will snap to the end of the line but will remain in a plane parallel to the XY-plane until we change it.

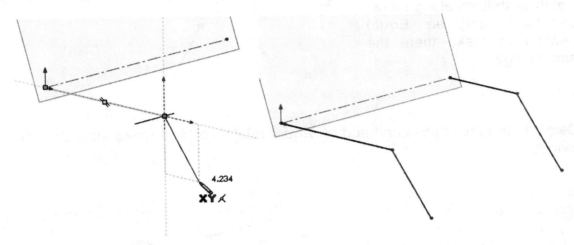

339. – When working in a 3D sketch, besides the geometric relations we are used to in a 2D sketch, we have a few more options to help us define geometry in a 3D sketch:

Relation:	Between entities:	Result:
↕X Along X	Line or two points	Line or endpoints aligned along the global X axis.
Y↕ Along Y		Line or endpoints aligned along the global Y axis.
↕Z Along Z		Line or endpoints aligned along the global Z axis.
Normal	Line and Plane or flat face	Makes the Line perpendicular to the plane or face.
On Plane	Plane or flat face and Line or point	Makes the line or point Coincident to the plane.
ParallelYZ	Plane or flat face and Line	Makes the selected line parallel to the YZ plane.
ParallelZX		Makes the selected line parallel to the ZX plane.

269

340. – First we need to add the relations we know we need up to this point.

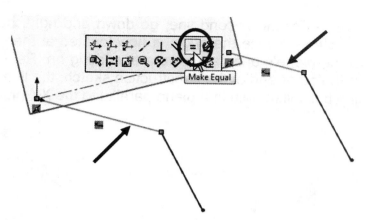

Select both lines along the X-axis and add an **Equal** relation to make them the same length.

Select both angled lines and add an **Equal** relation to also make them the same length.

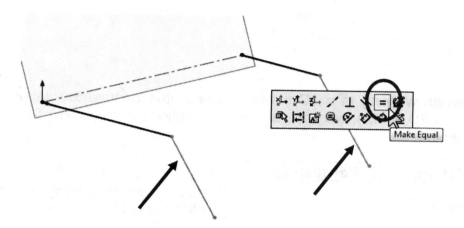

341. – Add an "**On Plane**" relation between the first diagonal line and the *"Front Plane."* By adding this relation, the line will not 'move' along the Z-axis.

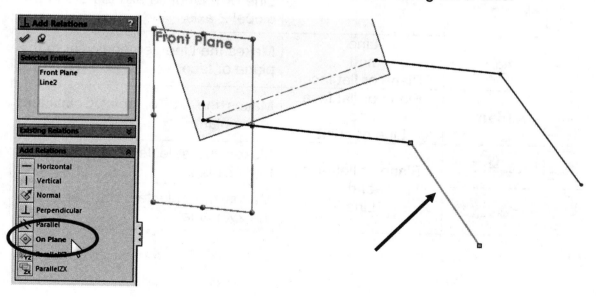

342. – Add a Parallel relation between the two diagonal lines.

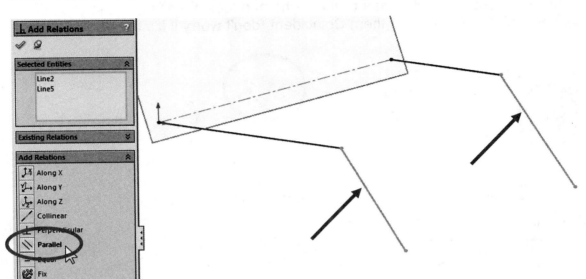

343. – Now add dimensions to the 3D Sketch. The sketch should be fully defined after dimensioning it.

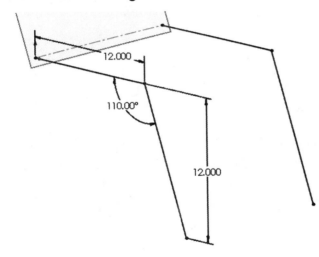

- Dimension the horizontal line 12″ long.

- Add the angular 110° dimension between the two lines.

- Add the vertical dimension between the horizontal line and the endpoint at the bottom. If you select the lower line only, you'll dimension its length, not the distance from the horizontal.

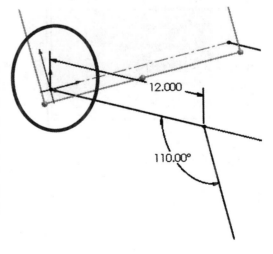

344. – These four lines will be part of the seat and front chair legs. The next step is to add the lines for the backrest of the chair. The plane we added at the beginning will help us align the 3D sketch's X, Y and Z axis directions about it. Pre-select *"Plane1"* and then turn on the "**Line**" tool. The sketch origin and the local directions will now be temporarily aligned to *"Plane1."*

345. – With the new plane alignment, draw a new line starting at the part's origin going 'up' along *"Plane1,"* then to the 'right' and 'down' ending at the start of the second set of lines to make them Coincident (don't worry if it's not parallel).

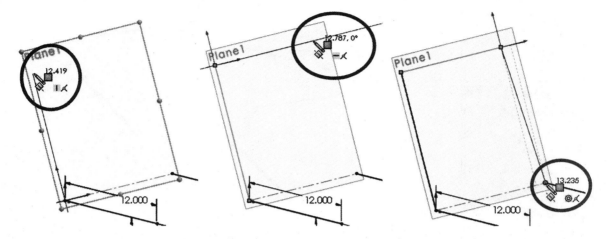

346. – Turn off the "**Line**" tool and add a Parallel relation between the two lines aligned with *"Plane1."*

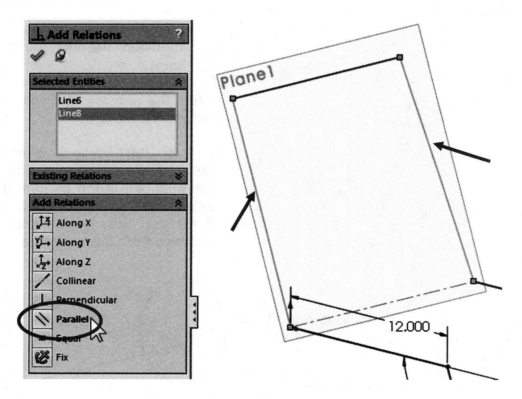

347. – And dimension the length of the backrest 18 inches (its length, not height.)

348. – To finish the 3D sketch, select the **"Sketch Fillet"** tool and round all corners with a 2″ radius. When adding the fillet we are warned about segments with equal relations; keep adding fillets until done. Since we are rounding all segments, they will still be equal at the end. Click "Yes" to continue when asked.

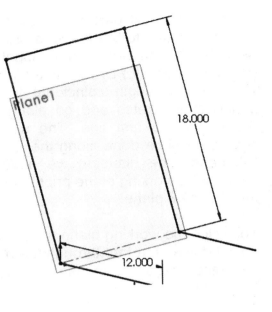

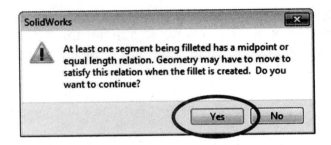

Click OK to finish the sketch fillets and exit the 3D sketch.

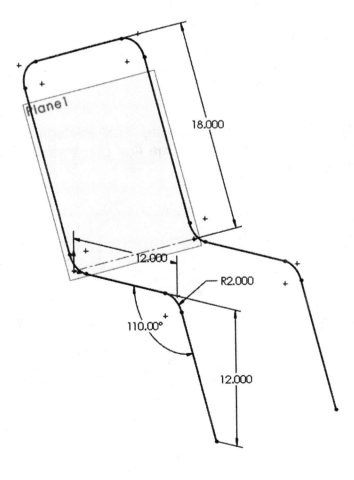

349. – The next part of the chair will be done similar to the first one. Start a new 3D Sketch (rotate the model as needed for clarity.) Draw a line starting close to the origin coincident to the construction sketch and go along the X-axis for the first line. The second segment will be done along the Z-axis; to change the direction we need to change the working plane orientation to the YZ or XZ plane.

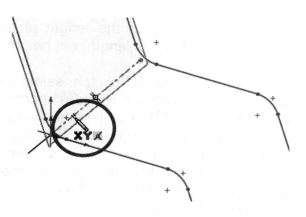

To switch the working plane on-the-fly, press the "Tab" key (just once; no need to hold it down) to rotate through the other work planes until we see either the YZ or XZ plane next to the mouse pointer. In the second picture we are working in the YZ plane.

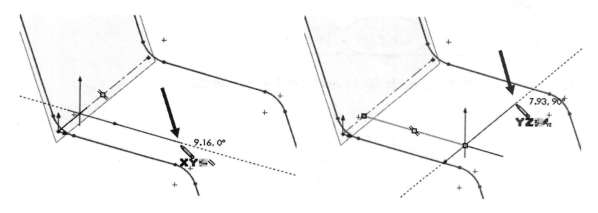

350. – After adding the second line press the "Tab" key again to switch to the XY plane and draw the third line parallel to the first one making it coincident to the construction sketch line as shown.

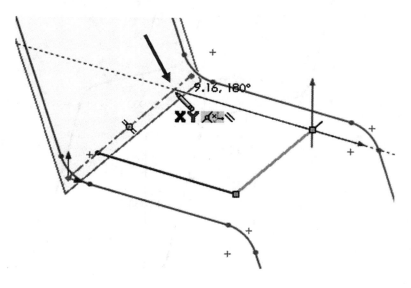

351. – The last line will go 'down' at an angle for the rear leg of the chair, also along the XY plane.

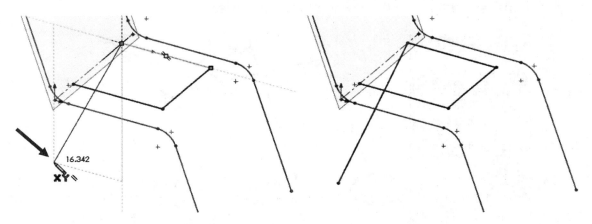

352. – Add a similar line to the other side where we started the second 3D sketch.

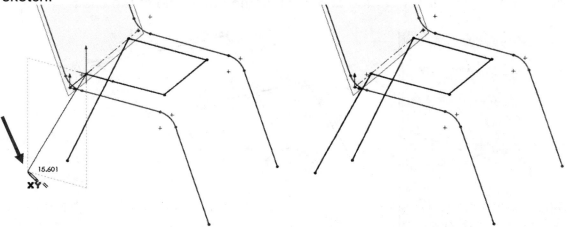

353. – Dimension the horizontal lines 0.75" from the previous 3D Sketch, and the angle between the two lines and the length of the horizontal as shown.

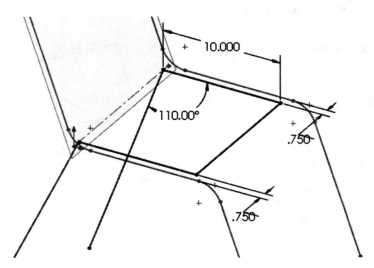

354. – Now we need to add some geometric relations. Select the endpoints at the bottom of the legs on one side of the chair and add an "**Along X**" relation; this way we'll make them lie on the same plane and axis.

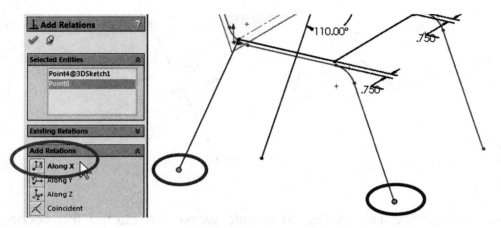

Repeat the same relation for the end lines on the other side of the chair.

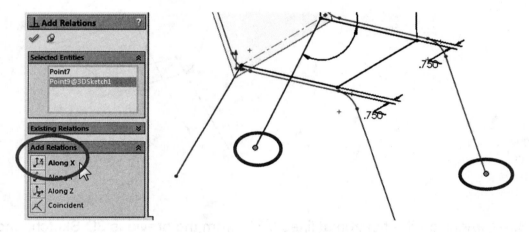

355. – And finally, add an "**Along Z**" relation between the endpoints of the two lines in the back to fully define the sketch.

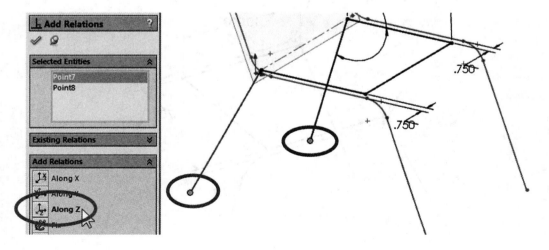

356. – To finish the second 3D Sketch, add a 2″ sketch fillet to all corners as we did in the first one. Exit the sketch to finish.

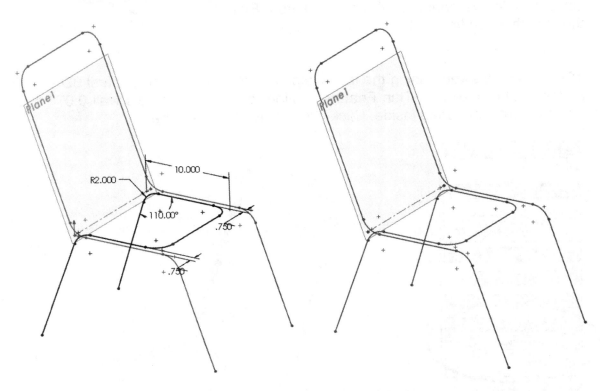

357. – To make the tubes of the chair, we need a plane at the endpoint of a leg to locate the sweep's profile. Add an auxiliary plane Parallel to the *"Top Plane"* ("First Reference") and Coincident to the endpoint of the chair's leg ("Second Reference").

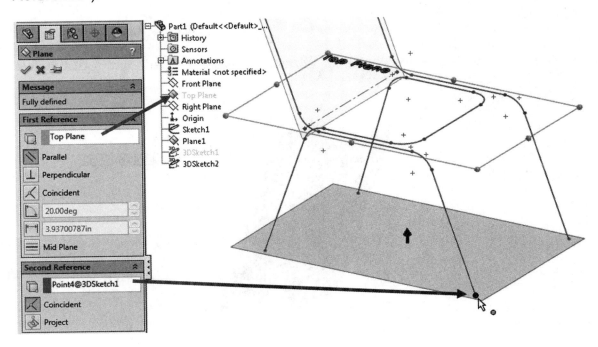

358. – When the new plane is finished, add a new 2D sketch in it. Be sure to make the circle coincident to the endpoint of the 3D sketch. Exit the sketch when finished.

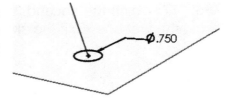

359. – Make a sweep using the last sketch as the "Profile" and the first 3D sketch as the "Path." Use the "Thin Feature" option to make the tube's wall 0.07″ thick with the material added inside. Click OK to finish the first tube.

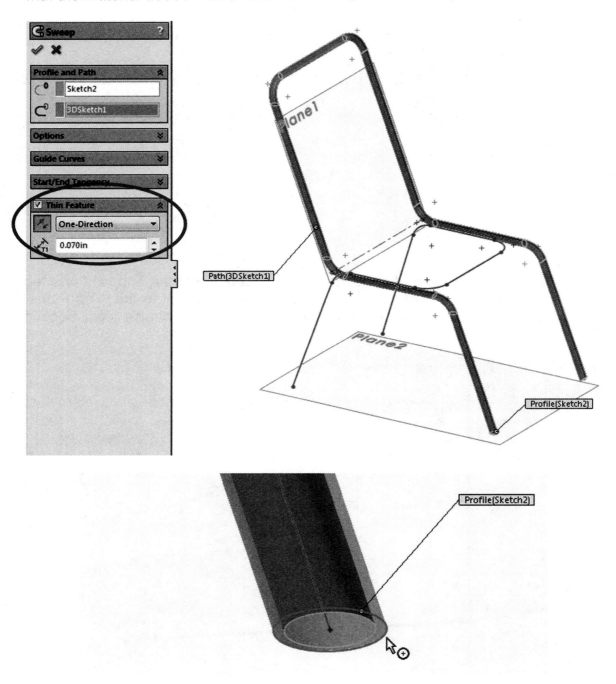

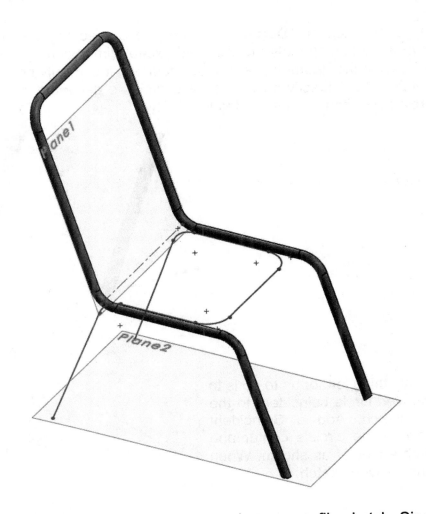

360. – To make the second sweep we need a new profile sketch. Since we want both profiles to be the same, we'll use a function called "**Derived Sketch**" to make the second profile exactly the same as the first one. Using this approach, if the first profile sketch changes, the second one updates, too. To add a derived sketch <u>we have to pre-select</u> the sketch we want to derive *and* the plane/face where we want to put it. Expand the *"Sweep-Thin1"* feature, select the profile sketch, hold down the "Ctrl" key and select *"Plane2"* in the FeatureManager or the graphics area. Now, select the menu "**Insert, Derived Sketch**". (If we don't pre-select them, this command is disabled.)

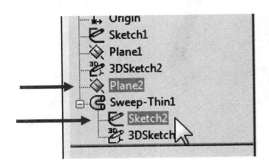
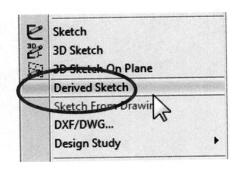

Immediately after selecting "**Derived Sketch**" we are editing the new sketch. Notice that most sketching tools are disabled except dimensioning and relations. In a derived sketch the geometry cannot be modified, it can only be located. The derived sketch will be exactly on top of the sketch we derived it from; click and drag the circle away from the chair's leg for clarity.

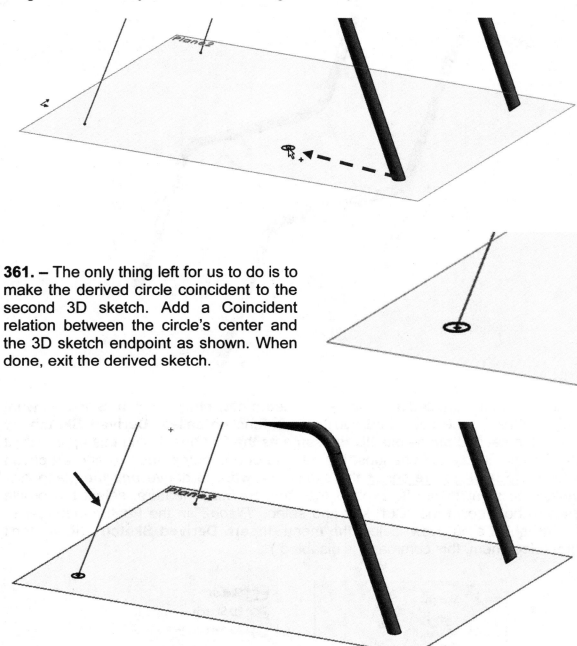

361. – The only thing left for us to do is to make the derived circle coincident to the second 3D sketch. Add a Coincident relation between the circle's center and the 3D sketch endpoint as shown. When done, exit the derived sketch.

362. – Make the second thin sweep with the same options as the first one; after finishing it hide all planes and sketches.

 A derived sketch's link to its parent can be broken, or 'underived' from the original by selecting the derived sketch with the right mouse button and selecting "**Underive**" from the pop-up menu, then we can modify the sketch as needed. Un-deriving a sketch cannot be undone.

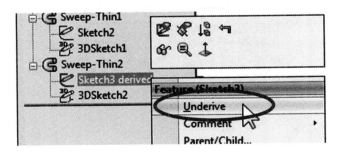

363. – After making the sweeps we need to finish their ends. Change to a Front view and zooming into the leg ends we can see the chair legs are uneven. To correct this we'll have to extend the legs just enough to go past the 'floor' (*"Plane2"*) and then trim them to size.

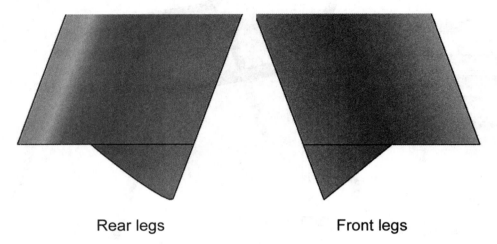

Rear legs   Front legs

364. – Select the flat bottom face at the end of the first sweep (the uneven leg), add a new sketch in it and project both circles using "**Convert Entities.**"

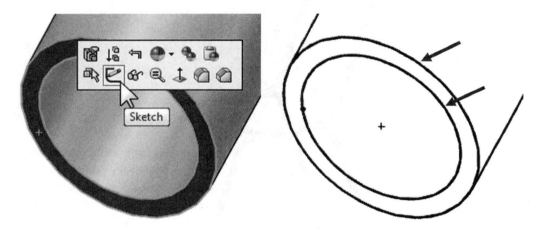

365. – Expand the *"Sweep-Thin1"*, make *"3DSketch1"* visible and select the "**Boss Extrude**" command. If the option to globally show/hide sketches is set to hide, turn it back on (menu "**View, Sketches**").

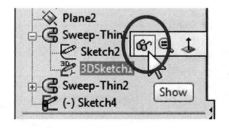

When we add an extrusion to a part, it's always perpendicular to the sketch plane, and after we select the "**Boss Extrude**" command we can see the preview of the extrusion (Back view shown here) is perpendicular to the sketch plane giving us an undesired result. To correct it, select the "Direction of Extrusion" selection box, and select the (now visible) end of "*3DSketch1*" to make the extrusion follow its direction. Make the extrusion long enough to cross "*Plane2*" and click OK to finish. Repeat the previous steps at the end of the second sweep.

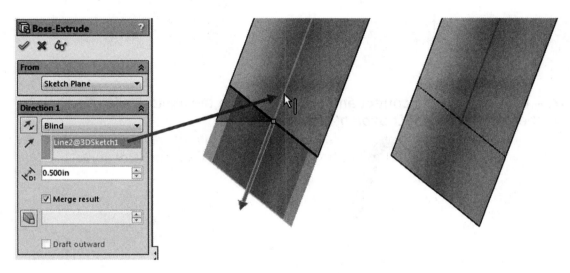

366. – After finishing the extrusions, we need to cut the excess material we added to level the legs. We could draw a sketch and make a "Cut Extrude" but we are going to use a different command instead. Select the menu "**Insert, Cut, With Surface**." We can use any plane or surface and use it as a cutting tool, including cutting multiple bodies at the same time. Select "*Plane2*" for the surface, and flip the direction if necessary. "**Cut With Surface**" will cut everything on the side of the plane/surface the arrow is pointing in. Click OK to finish.

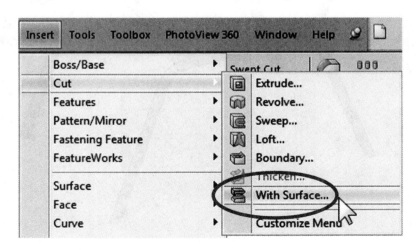

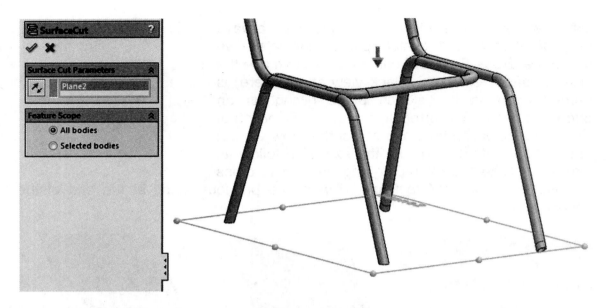

367. – Now our chair is correct and complete. Hide the sketches for a better look, save the part as *'Chair 3D Sketch'* and close it.

368. – A different way to make the path to sweep the frame of the chair is by using a **"Projected Curve."** By drawing the desired curve's projections in the *"Front Plane"* and *"Top Plane"*, we can combine them into a 3D curve and use it as a sweep path. Make a new part and add the following sketch in the *"Front Plane."* The sketch's origin is at the end of the 18" line. Exit the sketch when done.

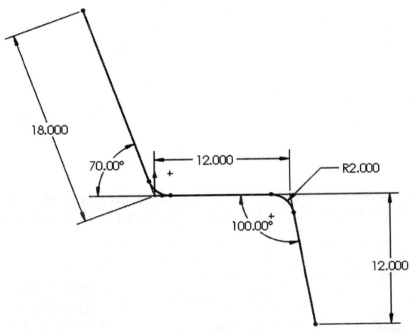

369. – Make a new sketch in the *"Top Plane"* as shown. Add geometric relations to the previous sketch endpoints to make it exactly the same size. (You may have to rotate the model to select the endpoints.) Exit the sketch when done.

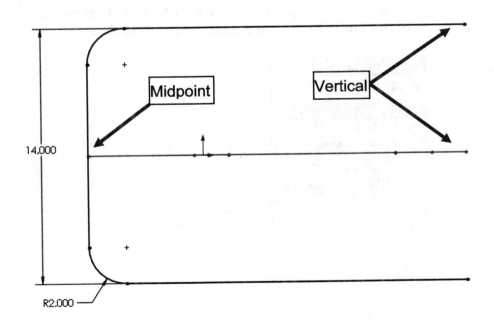

This is what the two sketches look like in an isometric view:

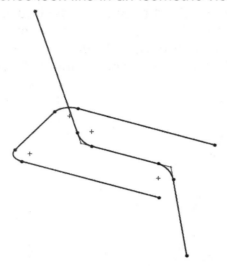

370. – Now, select the "**Projected Curve**" command from the "**Curves**" drop down menu in the Features toolbar or the menu "**Insert, Curve, Projected.**"

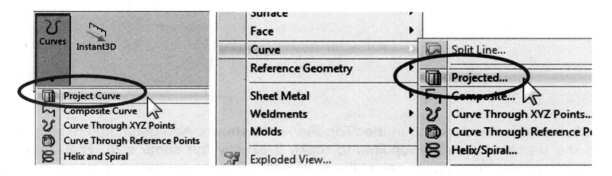

Select the "Sketch on Sketch" option and pick both sketches previously made; note the preview of the curve to be generated. Click OK to finish.

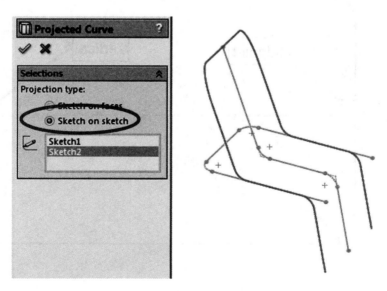

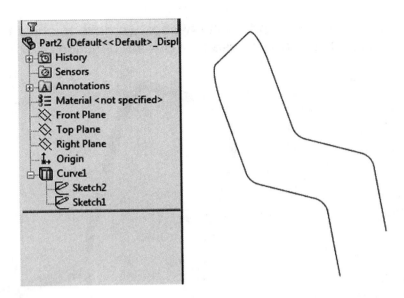

Now we can use the projected curve as a path for the sweep operation.

When projecting curves onto each other the radius may not be correctly projected depending on the angle, as is the case with the curves in the backrest. In this case building the part with a 3D sketch produces a better result. Making this part with a projected curve is a good option to show how to make a projected curve, as well as some of its limitations.

CHALLENGE EXERCISE: Create the second curve for the chair's frame using a projected curve and add the two sweep features.

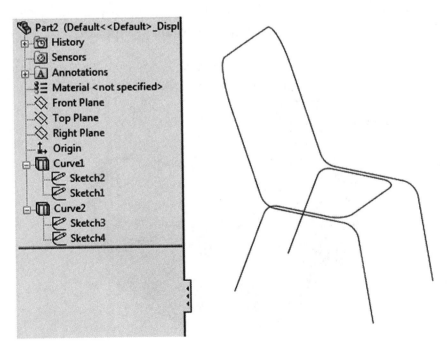

287

To make the extrusions at the end of the sweeps we have to show the sketches used before projecting the curves to use them for the direction of extrusion.

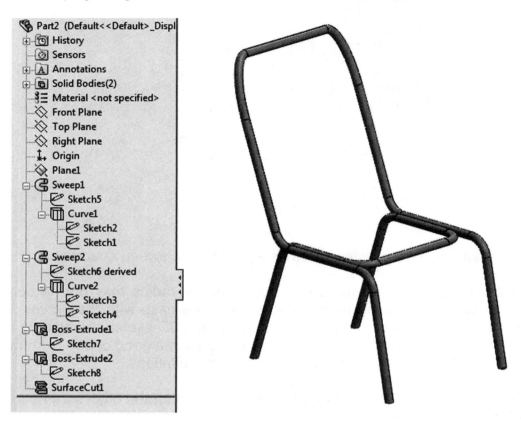

 Open the "Projected Curve" tutorial from the accompanying files for more information and learn how to make the following model.

Weldments

In SolidWorks, a welded structure can be defined as a part whose elements are long and thin and can be made with standard structural members. The weldments environment in SolidWorks allows us to design welded, modular or similar structures made from constant cross section elements (i.e., round or square tubing) easily, including weld beads, gussets and end caps in a multi body part. Detail drawings of weldments usually include material cut lists with length and cut angles ready for manufacturing.

Besides including a library of structural members in both metric and standard sizes, SolidWorks allows us to add our own structural member profiles to the built-in library.

When we work with weldments, the first thing we need to do is to make one or more 2D and/or 3D sketches that will serve as the structure's 'skeleton,' defining the structure's centerline, inside, or outside dimensions, just like in sheet metal. In a weldments part, each sketch line can be used to locate one element of the structure.

 A weldments part is probably one of the few instances where it's efficient to use a part file instead of an assembly to represent multiple 'parts.'

To learn how to use the weldments environment, we'll make the following steel structure using 2″ x 2″ welded steel.

371. – The first thing we need to do is to make a 3D sketch (or a combination of multiple 2D sketches) that will be the *skeleton* for our welded structure. Just as with sheet metal, we need to decide if the sketch dimensions will represent inside, outside or centerline dimensions of the welded structure. In our case, the conveyor's leg dimensions will be at the centerline, and for the top of the structure the dimensions will be outside dimensions; in other words, the width and length of the conveyor are not as important, but the height is. Open a new part and start a 3D sketch (menu "**Insert, 3D Sketch**"). Select the "**Line**" tool and press the "Tab" key to switch to the "**YZ**" plane. By default, 3D sketches always start in the "**XY**" plane.

TAB …   TAB…

 To make the 3D sketch instructions easier, the reader will be directed to go along the X, Y or Z axes in the positive or negative direction.

Start the sketch line in the origin using the YZ plane; first go up in **Y**, then negative **Z**, finally negative **Y** and stop. Don't worry about the size; It will be dimensioned after adding some relations.

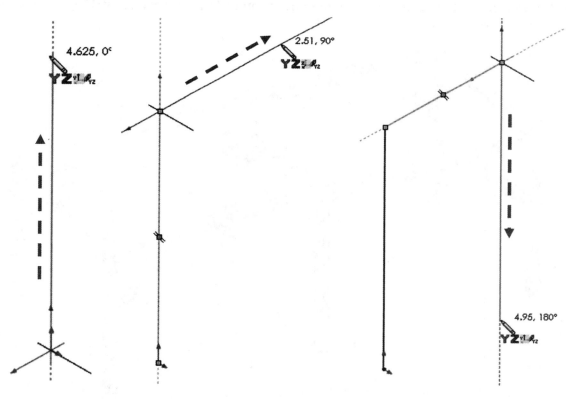

372. – Still with the "**Line**" tool active, start in the upper left and right corners and draw two lines going in the **X** direction. Press the "Tab" key if needed to switch to the **XY** or **XZ** plane; either one works in this case.

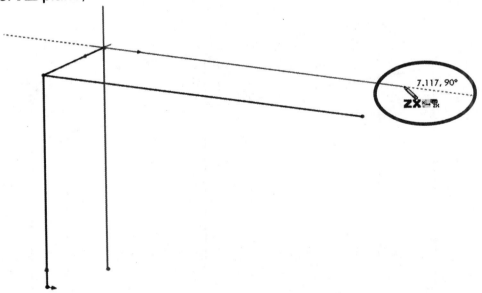

373. – An advantage of working in a 3D sketch is that we can 'snap' to other geometric elements even if they are not in the same plane we are working on. Draw the next four legs of the structure by capturing Midpoint and Coincident relations to the two long lines as shown. Switch planes if needed by pressing the "Tab" key.

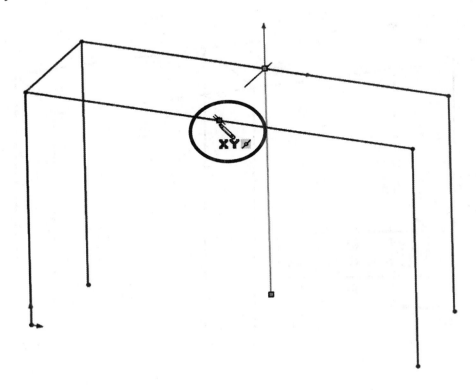

374. – Draw the next two lines along the **Z** direction to complete the geometry. Snap to the endpoints, even if the lines are not exactly along the **Z** direction. We'll correct them in the next steps.

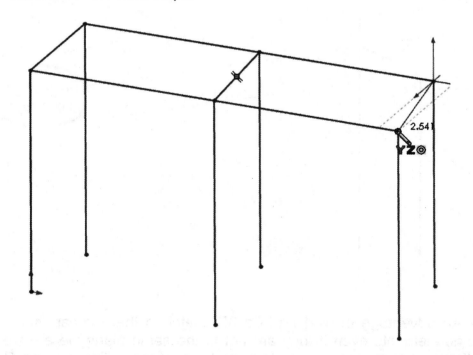

375. – Add **Along X** relations to the three endpoints of the legs on each side, one side at a time, *or* make all six legs Equal.

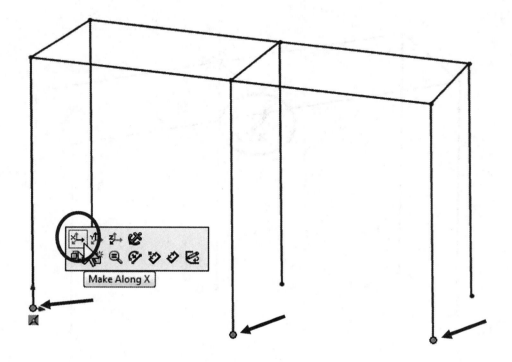

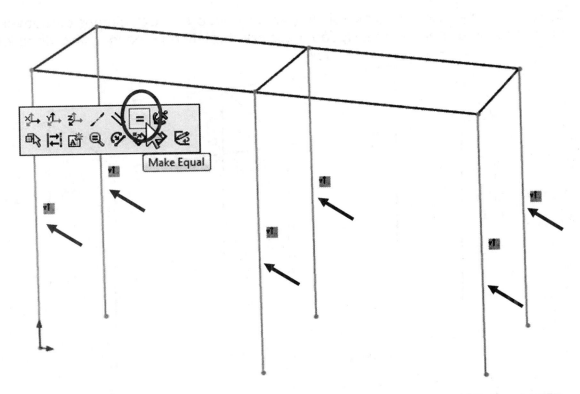

376. – Since we 'snapped' two of the transversal elements to existing endpoints, they may not be along the **Z** direction. We need to add a relation to make them either **Along Z** *or* **Parallel** to the first transversal line.

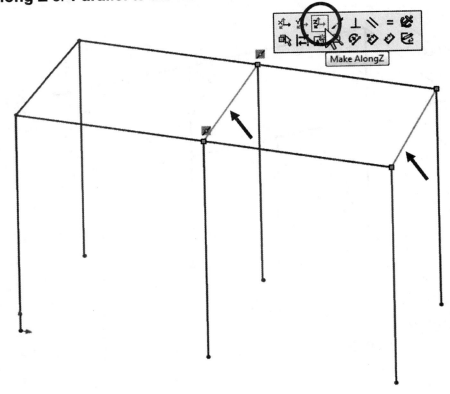

377. – Add the following dimensions (we only need one 24″ dimension between the vertical lines because we added a **Midpoint** relation to the long line along X):

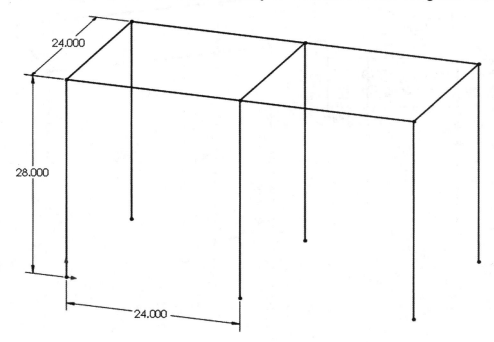

378. – Add the corner reinforcements and dimension one corner as indicated. Draw the lines by 'snapping' to the existing geometry (in this case it doesn't matter which plane we are working on.) Make all the reinforcement lines **Equal**.

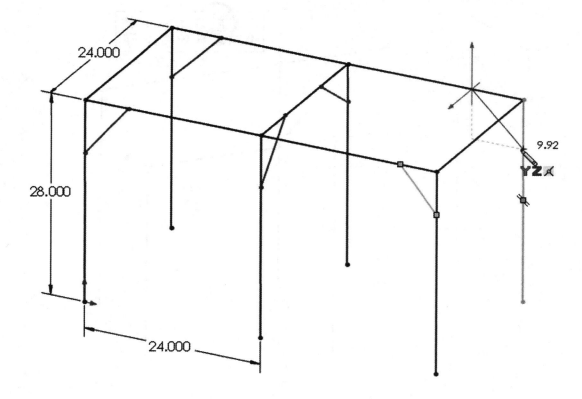

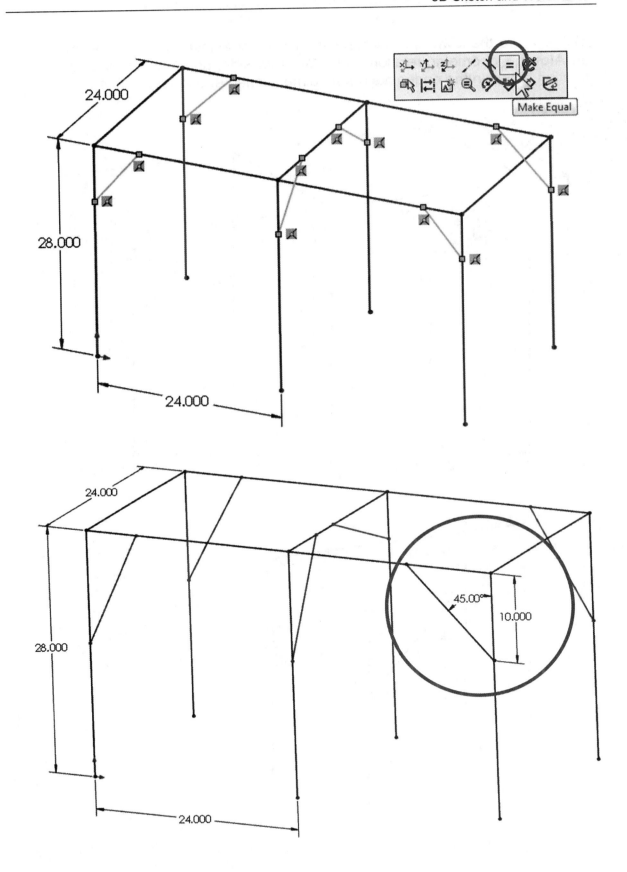

379. – Make the lower three endpoints at the same level on one side by adding an **Along X** geometric relation, then the other side, and finally add an **Along Z** relation between the endpoints on the dimensioned end to fully define the sketch.

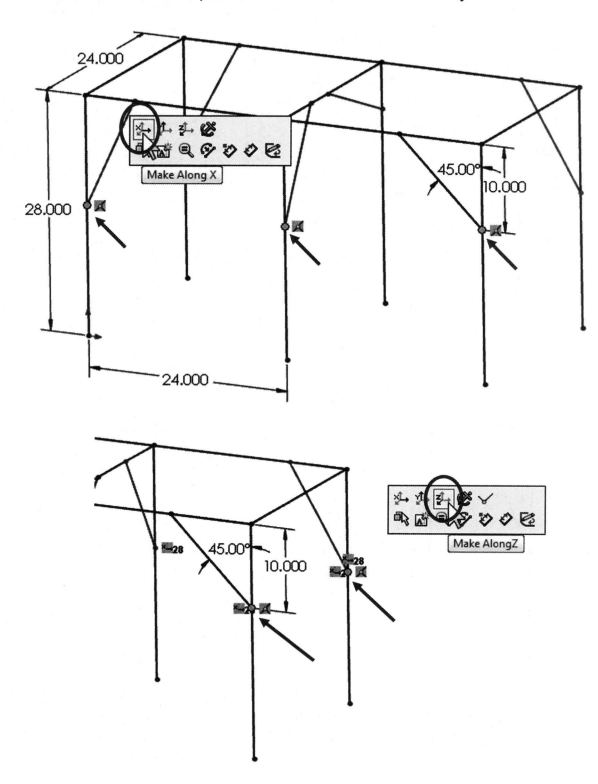

380. – The fully defined 3D sketch now looks like this. Exit the sketch to continue.

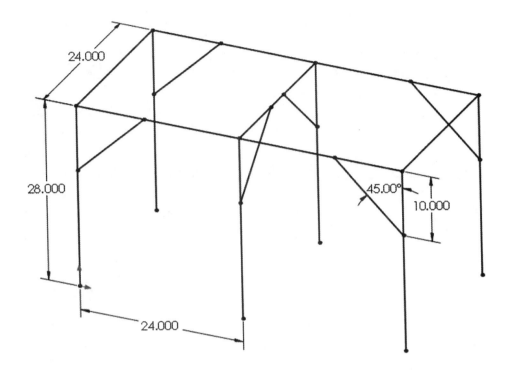

381. – Weldments work similarly to sheet metal in the sense that a special feature is added to the FeatureManager that enables the weldments environment, which in this case is a "**Weldment**" feature and a "**Cut list**" folder that takes the place of "**Solid Bodies**"; the reason for this is because a welded part is essentially a multi body part, where each body represents a piece of the structure. To continue make sure the weldments toolbar is enabled. Right mouse click in the CommandManager's tabs and activate the "**Weldments**" tab, and/or right mouse click in a toolbar and turn on the Weldments toolbar.

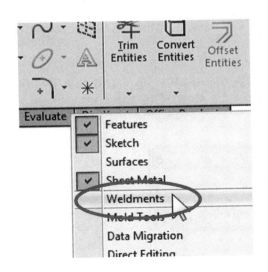

382. – In the Weldments toolbar is a "**Weldment**" icon; selecting it enables the weldments environment in the part but, just as with sheet metal, it is also automatically added when we make the first "**Structural Member**" feature.

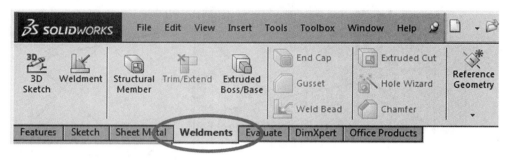

To start our welded part, select the "**Structural Member**" command. From the "Selections" options, select "ansi inch" in "Standard," "square tube" in "Type" and "2x2x0.25" in "Size."

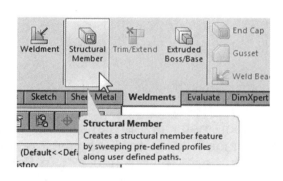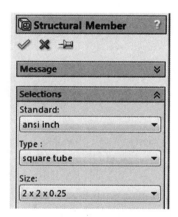

383. – When we add structural members, they are added in groups oriented in the same direction or laying in the same plane. The first group to be added will be the structure's 'legs.' As we start selecting them in the screen, we see a preview of each one with the selected profile centered in each line (we'll talk about locating the profile later). Click OK when finished selecting the legs to create the first group.

 Since we'll be adding more members and the legs are just the first group, click in the "Keep Visible" option and then OK to create them. By doing this we can continue adding more groups without having to re-open the command.

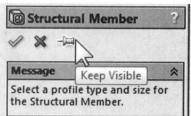

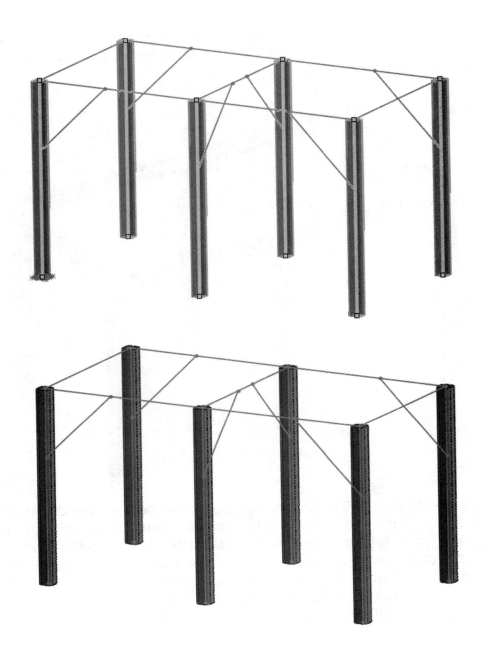

When selecting sketch lines to add Structural Members to a group, remember that the lines in a group can be:

a) Discontinuous <u>parallel</u> segments in the same or different planes, or
b) Continuous segments connected by endpoints.

 The idea of having groups is to be able to modify its properties in a single operation, like profile to use, orientation, location, etc.

384. – The second group will be the four lines that make the outside perimeter at the top. We won't be able to select the cross member in the middle because it's not connected to an endpoint.

385. – Before clicking OK to add the structural members we need to make a few modifications to how this group will be located about the sketch, and how the corners will be trimmed. Zoom in a corner (the one with the sketch of the member's profile is a good option, but any one works). For this group we'll use the "End Miter" corner treatment. By using this option, all the connecting elements <u>in this group</u> will get this treatment. Be sure to turn off the option "Merge miter trimmed bodies," otherwise the four elements will be merged into a single body.

The different options for corner treatments available are:

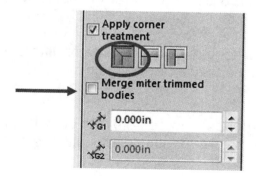

The command's preview is translucent; these images are the end result for clarity.

End Miter   End Butt 1   End Butt 2

 The first distance box after corner treatments is used to specify a gap between the connected segments of the same group, and the second to add a space between one group and the previous. This is usually done to add a space to weld the structural members together.

Also, if we use the "End Butt" corner treatments, we can have a simple cut or a coped cut end condition to match the other piece.

 Simple Cut    Coped cut

 If needed, we can change the type of corner treatment individually by clicking in the pink point at the corner of the preview and selecting a different option of corner treatment, merge bodies or weld gaps.

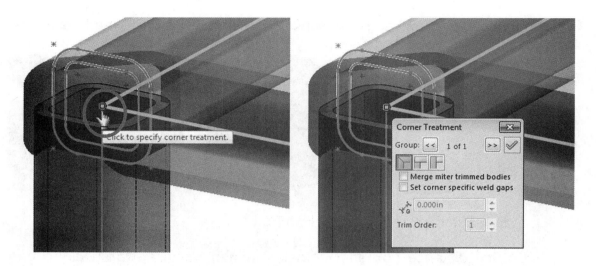

386. – By default, when adding new structural members, the sketch line pierces the profile's origin automatically. If we remember, the conveyor must have a height of 28″; if we don't change the profile's location, our conveyor will be higher than needed by one half of the profile height.

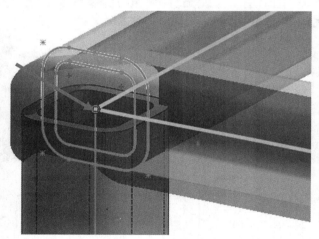

To change the profile's location we have to make the element's sketch line pierce the top of the profile instead of its center. At the bottom of the "Structural Members" options, click in the "Locate Profile" button. The part will zoom into the profile's sketch where we can select the point in the profile's sketch to be coincident to the sketch line. Any vertex or sketch endpoint in the profile can be selected as a pierce point.

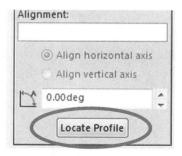

387. – Select the top middle point as indicated; the preview will update the profile's location to the desired position. Click OK to locate the profile, add the group, and continue. Do not worry about the members intersecting each other; we'll trim them as soon as we finish adding the last group.

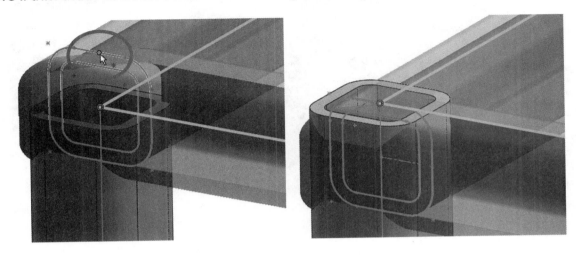

388. – For the next group, select the top middle line, and change the profile's location as we did in the previous step. Click OK to continue to the next group.

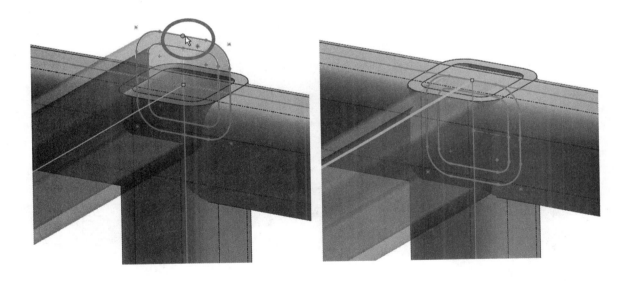

389. – For the next group, select two of the diagonal reinforcements in one side. (Since only two are parallel and oriented in the same direction, that's why we can only do two at a time.) After selecting the sketch lines to add the structural members we notice the profiles are not correctly aligned. They are located in the centerline, but rotated. To change the profile's rotation and align them correctly to the structure, scroll down to "Rotation Angle," and enter "45" degrees.

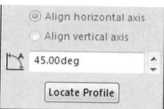

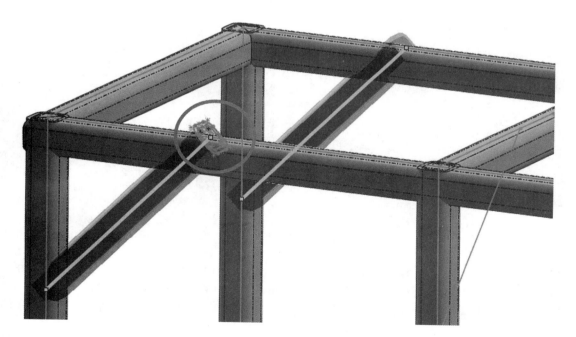

 Other options to orient the profile include selecting a line or edge in the "Alignment" selection box and aligning the horizontal or vertical axis. We can also mirror the profile horizontally or vertically if needed.

390. – Repeat the same process with the reinforcements on the other side (another group of two) and finally make a group for each of the reinforcements in the center; since they are not parallel to each other or any other member, we can only add one at a time. Close the "Structural Member" command when done. Our structure now looks like this:

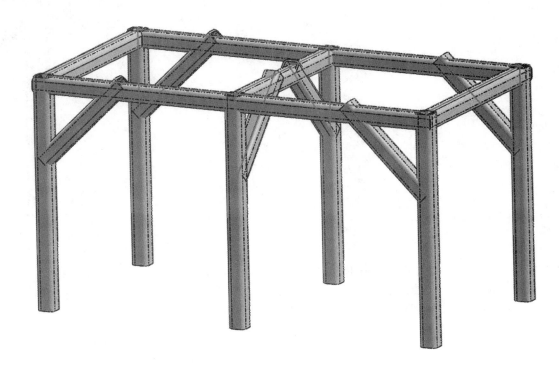

 Notice the new "*Weldment*" feature automatically added to the Feature Manager after adding the Structural Members.

391. – Since the structural members are intersecting each other, the next step is to trim the excess material. From the "Weldments" toolbar, select the "**Trim/Extend**" command. We have the option to trim elements with one or more faces/planes, or other bodies. In our example, we'll trim the structural members using a model face. Select the "End Trim" option to cut the bodies using a selected surface. In the "Bodies to be Trimmed" selection box, select two or four elements to trim at a time. (We can select all bodies to trim at the same time, but the screen will be cluttered with "keep" or "discard" labels.)

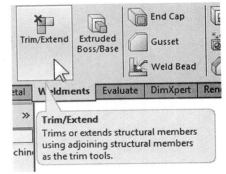

305

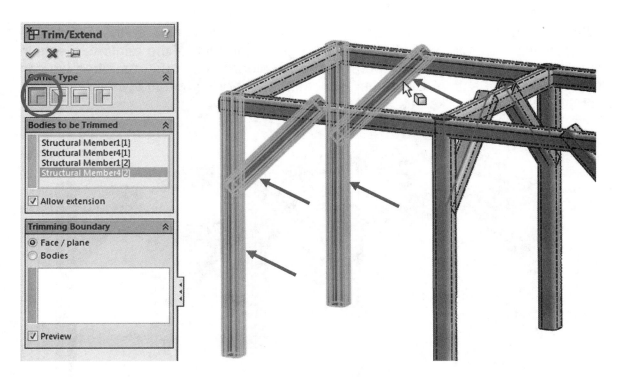

392. – In the "Trimming Boundary" selection box, select the "Face/Plane" option and pick the face under any of the top horizontal members to use it as the trim boundary. Rotate the model to select the correct face if needed.

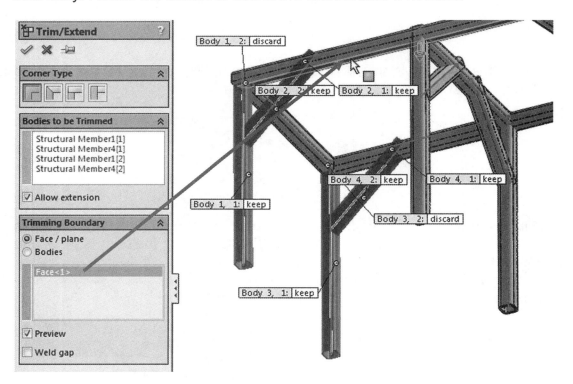

393. – After selecting the face, the bodies are 'cut' in the preview and we have to review which bodies we want to keep and which bodies to discard. Check each member being cut. Review the pop up labels of each segment to toggle the "keep" or "discard" message by clicking in it. The labels can be moved around for clarity. By default, smaller segments are *usually* marked correctly to be discarded. The discarded elements preview is lighter (almost invisible) making it easier to identify discarded elements. After reviewing the elements to be discarded, press the "Keep visible" push pin in the "Trim/Extend" command and click OK to continue trimming the rest of the structure. Trim the top part of the other elements using the same trimming face.

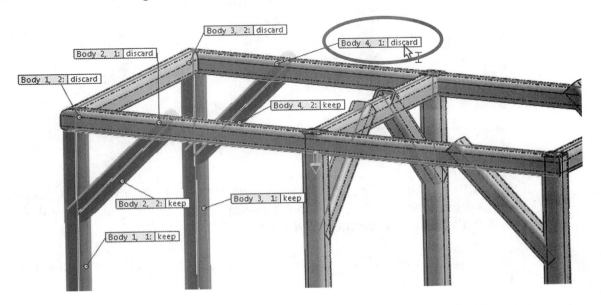

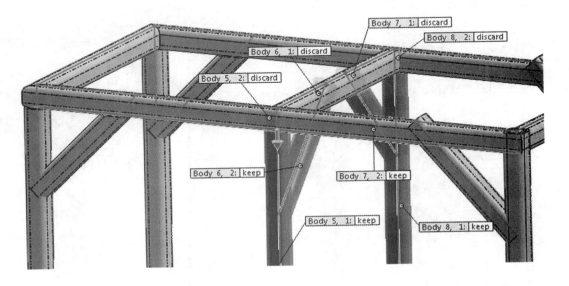

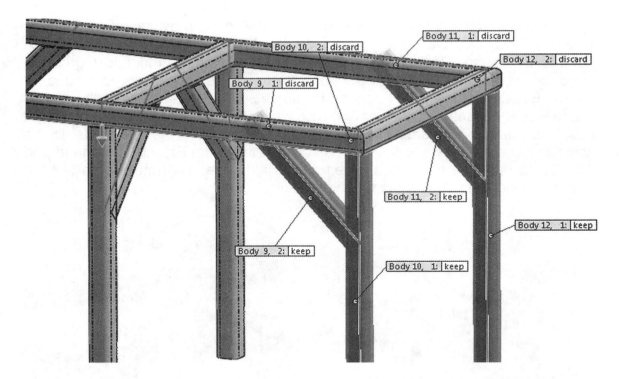

 If we had chosen to trim all the legs at the same time, the screen would have been cluttered with all of the "keep/discard" labels; that's why it was shown with just a few elements at a time.

394. – Trim the lower part of the corner diagonal reinforcing elements using the faces indicated.

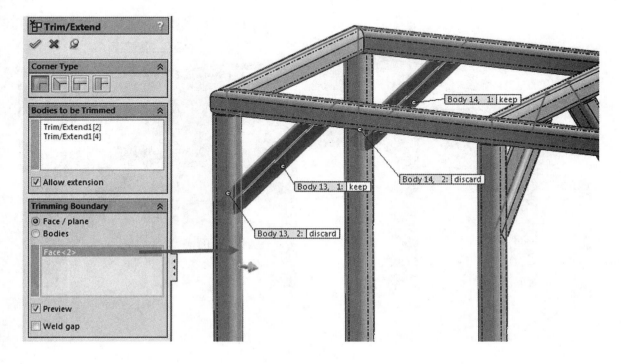

Trim the other side reinforcements.

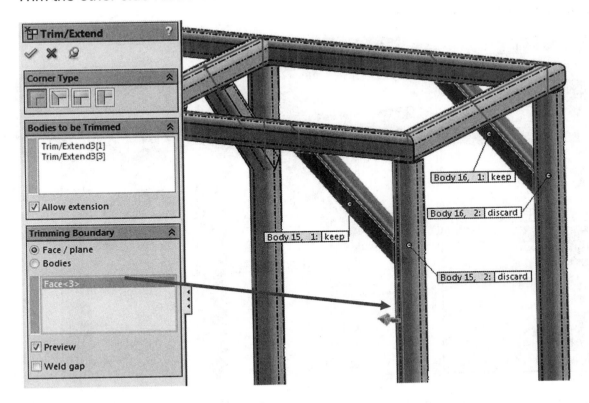

Trim the center bodies on one side,

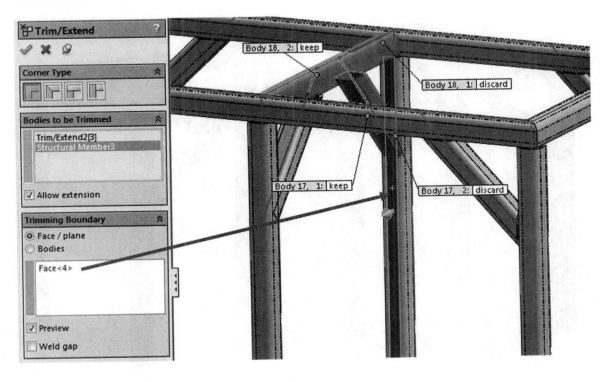

And then trim the other side of the center bodies. Close the **"Trim/Extend"** command when done to finish.

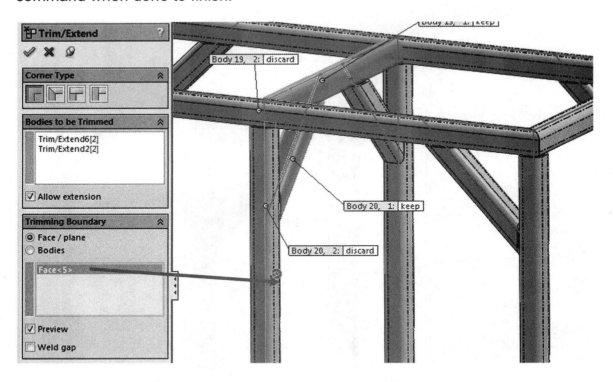

Our structure is now completely trimmed and no bodies intersect each other.

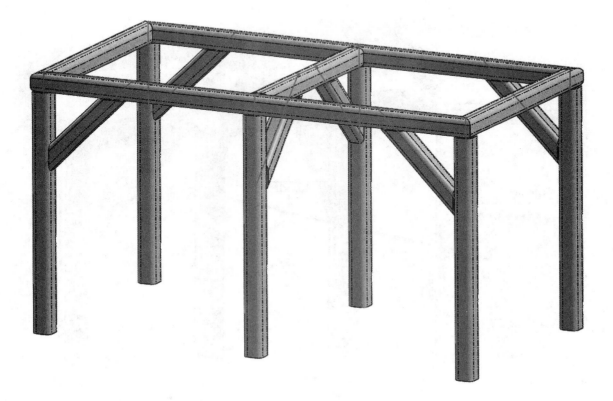

395. – The next step is to add gussets to our structure. Gussets are steel plates used to add strength and support at the intersections of structural components. For our example, we'll add gussets to the corners, and weld beads after that. To add a gusset we have to select the faces that will be connected, and then define the gusset's dimensions. Click in the

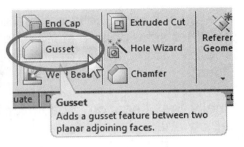

"**Gusset**" icon in the Weldments toolbar, select the faces indicated below in the "Supporting Faces" selection box and enter the dimensions and location shown. (It's easier if we zoom in a corner looking at it from the bottom.)

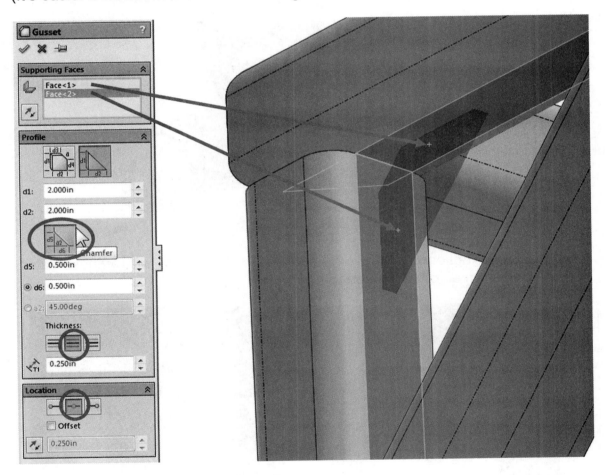

396. – We can make polygonal or triangular gussets. In our example we'll use the triangular option, and dimension it 2″ x 2″ (measured from the corner.) A chamfer is usually added to allow space for a weld bead in the corner and make room for the gusset to fit. Activate the "Chamfer" button and set the chamfer dimensions ("*d5*" and "*d6*") to 0.5″ x 0.5″. (Chamfers can also be defined by one dimension and an angle.)

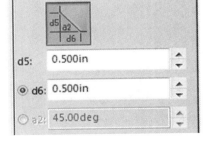

The gusset's thickness can be added on one side, the other side, or both sides. We'll make our gusset is 0.25″ thick using the midplane option ("Both Sides").

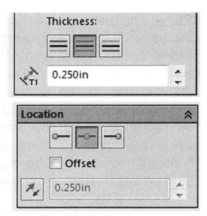

Finally the "Location" option for the gusset can be set on one side of the face, the center, the other side, or offset from any of them. We'll make ours in the center. Press the "Keep Visible" push pin to keep adding gussets at other corners. Press OK to continue.

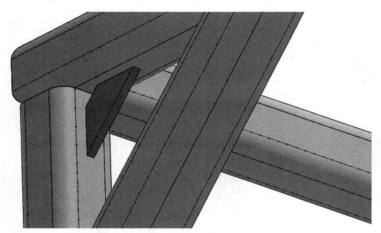

Gusset location options can have all possible combinations of thickness sides, location and even an offset if needed.

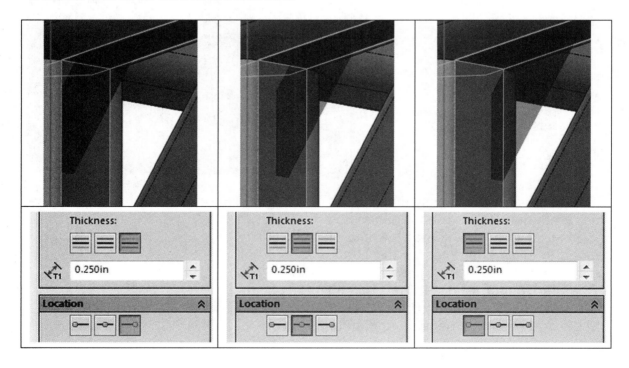

397. – Add the remaining gussets to the other inside corners using the same settings. After adding a gusset SolidWorks remembers the previous settings, and we only have to select a new set of faces and the new gusset is automatically positioned; then we only have to click OK to continue to the next one until we are done.

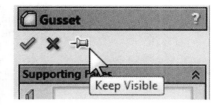

After adding all gussets, our conveyor looks like this (gussets highlighted):

398. – Another feature unique to weldments is the "**End Cap**." An end cap will add a cover to close the open end of a structural member. Rotate the part to look at it from below and zoom at the end of one leg. Click on the "**End Cap**" command in the Command Manager and select the end face of a leg. For our example the end cap's thickness will be set to 0.125″. The option **inward** in "Thickness Direction" reduces the structural element's length by the end cap's thickness maintaining the original length including the end cap.

The "Offset" option refers to the distance from the edge of the end cap to the edge of the structural member, essentially making it smaller than the structural member's profile, usually making room for a weld bead. This offset can be set using a ratio of the structural member's thickness or a specified distance turning off the "Use thickness ratio." Set the offset ratio to **0.5** (half the structural element's thickness), and activate the option "Chamfer corners" with a value of 0.125″.

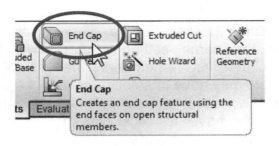

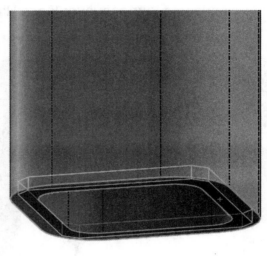

 A note about "Thickness direction:" If we make the end cap going "Outward", its thickness will be added to the length of the structural member. If we make it "Inward," the structural member will be shortened by the end cap's thickness, as in this case to maintain the original height.

Select the six legs' end faces to add all end caps with one command. (A selection filter for faces is automatically activated while using the command.)

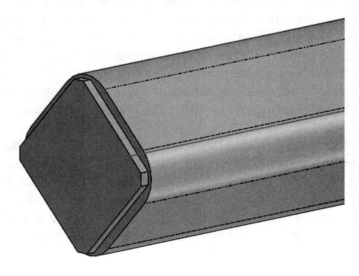

An important detail to keep in mind when we work with weldments is that the "Merge result" option in all features that add material is turned 'Off' by default. If we add an extrusion (any type) and want it to be merged to one or more bodies, we have to turn the "Merge result" option 'On' following multi body part rules.

399. – To complete our structure, we'll add weld beads and the corresponding welding annotations. Some advantages of adding weld beads to our model include automatic creation of welding annotations for drawings, welding symbols for manufacturing, and the possibility to add a user-defined welding path. Select the **"Weld Bead"** command from the Weldments toolbar or the menu **"Insert, Weldments, Weld Bead."**

400. – Zoom in a corner looking at it from the bottom. There are two ways to add a weld bead: selecting faces or using the Smart Weld Selection Tool. In the first case we can select the two faces indicated in the inside corner; notice the weld bead preview at the intersection of the faces to be welded. Set the weld bead size to 0.25″ and turn on the "Tangent Propagation" option. In this case we want the weld to go around the sides, and it only goes half way because the weld can only be made between two bodies; in this corner we have three bodies. Notice the preview goes around the tangency where both bodies touch. If we had merged the mitered bodies when the structural member was done, the weld bead would go all the way around.

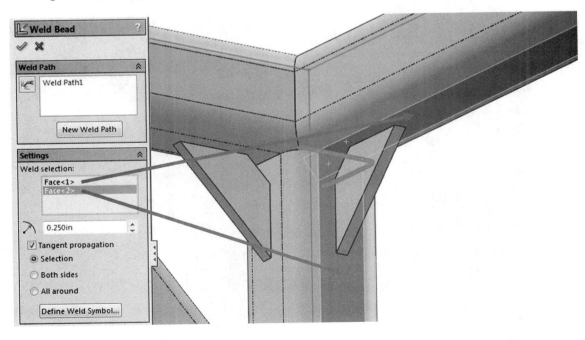

401. – To add the weld bead and proceed to the next one, select the "New Weld Path" button; select the next two faces to complete the weld around the leg. Notice the weld bead settings remain the same. If we continue adding weld beads that share the same settings, they will be grouped together. Press the "New Weld Path" again to add the next two welds using the same settings.

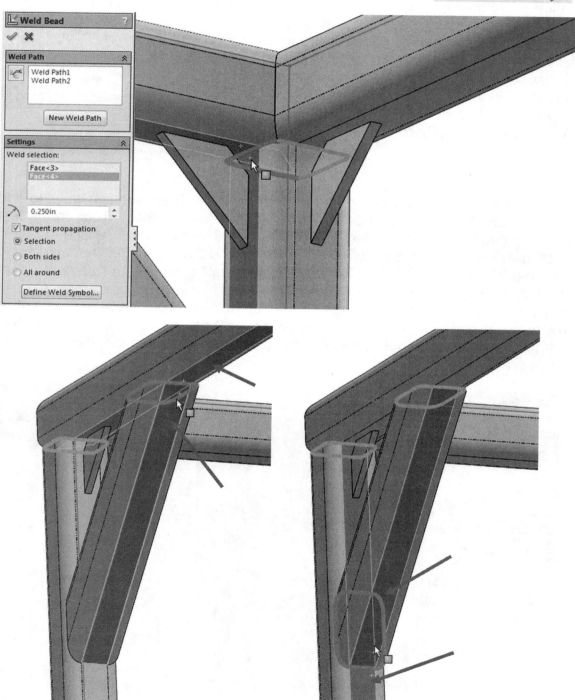

402. – We can keep adding the remaining weld beads selecting the faces, but now we are going to use the second approach. Click in the "New Weld Path" to continue adding the rest of the weld beads and zoom in the second leg of the structure. Instead of selecting faces individually, we'll use the **Smart Weld Selection Tool**.

After activating the Smart Weld Selection Tool the mouse pointer changes to a pencil icon.

To use this tool we need to click-and-drag across the intersection we wish to add a weld bead to, going from one face to the other. Notice that as we make a new selection, a new 'Weld Path' is automatically added to the list.

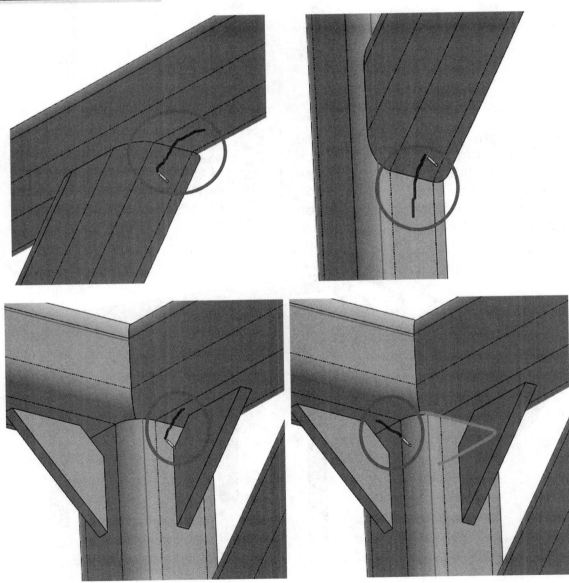

403. – Add the rest of the weld beads between all structural members including the corners and click OK to finish the weld beads. We'll add weld beads to the gussets next.

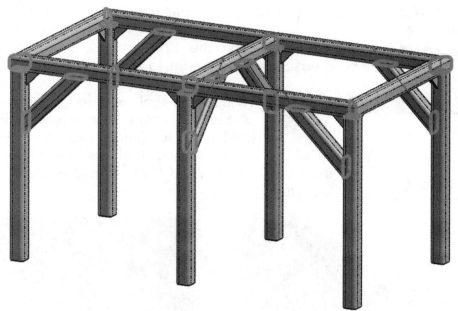

404. – Right after finishing the weld beads the corresponding weld beads and manufacturing annotations are added to each weld. Right-mouse-click in the "*Weld Folder,*" and turn on the option "Show Cosmetic Welds."

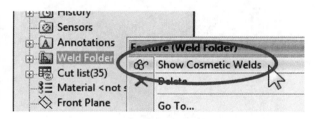

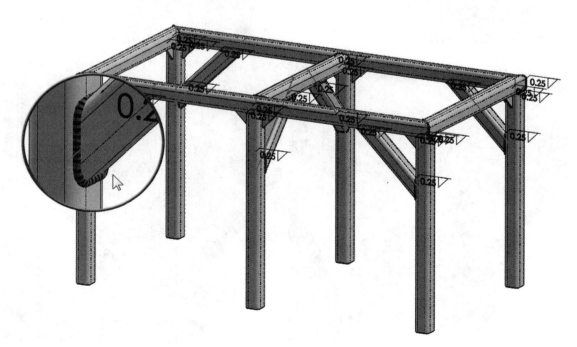

Under the *"Weld Folder"* we see the new *"0.25in Fillet Weld"* including all the weld beads added listing the length of each one, useful for cost estimating and manufacturing purposes. Depending on the selections made when the weld beads were added, you may have more or fewer welds.

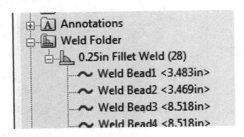

 To turn off the cosmetic weld beads right mouse click in the *"Weld Folder"* and select "Hide Cosmetic Welds." To hide the welding annotations, right mouse click in the *"Annotations"* folder and turn off "Display Annotations."

405. – To add the weld beads to the gussets, we'll use different options. Select the "**Weld Bead**" command from the Weldments toolbar and zoom in one of the gussets (annotations have been turned off for clarity). Select the two faces indicated to weld the gusset to the structural element using the "Smart Weld Selection Tool"; make the bead size 0.125" and in this case select the option "Both sides" under "Settings." Notice a new weld bead is added to the opposite side of the gusset, saving us time (similarly, if we select "All around," the entire loop is selected).

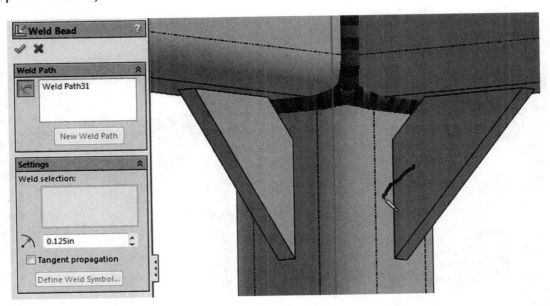

 When adding weld beads using the "Smart Weld Selection Tool," click and drag <u>starting in an unselected face</u>; otherwise it will get unselected and you'll have to re-select again to add a weld bead.

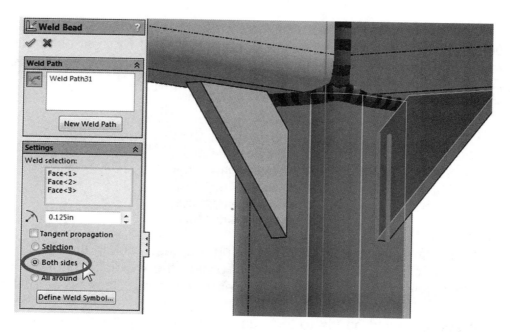

 In these images it's easier to see the reason to add the chamfer in the gusset, letting the weld bead to pass through the corner. Weld Bead options include defining a starting offset, bead length and an intermittent (optionally staggered) weld bead.

406. – Add weld beads to all gussets using the same settings. Click OK to finish.

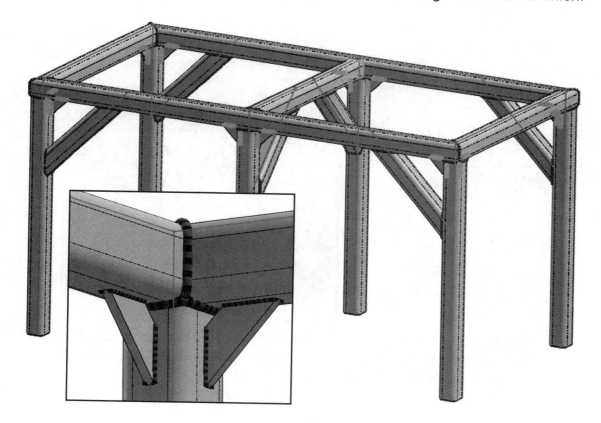

Every time weld beads are added, the part's annotations are displayed again. Turn them off to avoid a busy screen.

407. – Now that we have completed the welded structure, SolidWorks automates the creation of the 'cut list' by grouping structural members of the same type and size together. If we expand the "**Cut list**" folder in the FeatureManager, we'll see a long list of solid bodies. To group them make a right mouse click in the *"Cut list"* folder and select "**Update**" from the pop-up menu.

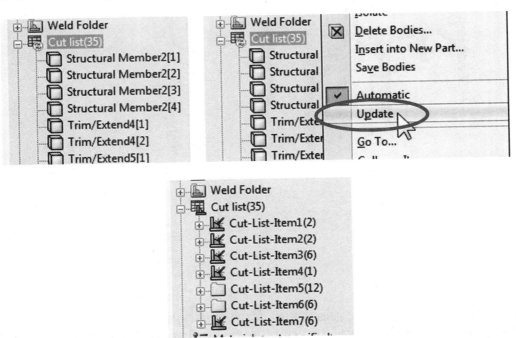

If desired, the "Cut-List-Item" folders can be reordered; the reason to do this is because this is the order in which they will be listed when we import the cut list into the drawing. A good idea is to reorder the elements by type and size, and leave the gussets and end caps at the end. To reorder the items, simply drag-and-drop them up or down as needed.

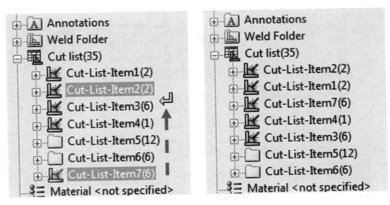

408. – Save the part as *'Weldments Table'* and make a new drawing with an isometric view; shaded with edges view mode and tangent edges removed will work fine for this example. Import the model dimensions to this view.

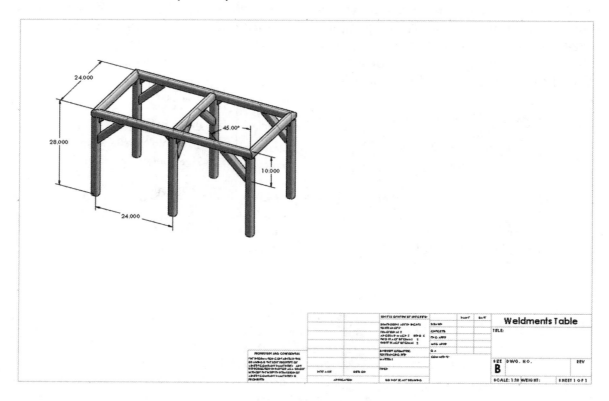

409. – Select the isometric view and from the right-mouse-click menu, select **"Tables, Weldment Cut List,"** or from the menu **"Insert, Tables, Weldment Cut List."** Accept the defaults for the table and click OK. Locate it in the upper right corner. Adjust the view's scale if needed.

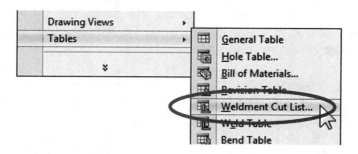

ITEM NO.	QTY.	DESCRIPTION	LENGTH
1	2	TUBE, SQUARE 2.00 X 2.00 X .25	50
2	2	TUBE, SQUARE 2.00 X 2.00 X .25	26
3	6	TUBE, SQUARE 2.00 X 2.00 X .25	25.875
4	1	TUBE, SQUARE 2.00 X 2.00 X .25	22
5	6	TUBE, SQUARE 2.00 X 2.00 X .25	11.899
6	12		
7	6		

410. – The rows without description in the list correspond to the gussets and end caps. For SolidWorks to fill in the correct information we have to go back to the part file; in the *"Cut list"* select the corresponding folder with a right-mouse-click and select "Properties." In the "Cut List Summary" tab we can see all the properties for each group of elements, or we can select the "Properties Summary" tab to see a list of properties with the

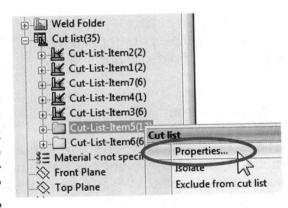

values of each group. Fill in the "Description" and "Length" for the groups missing one and click OK to finish. Go back to the weldment drawing to see the updated table.

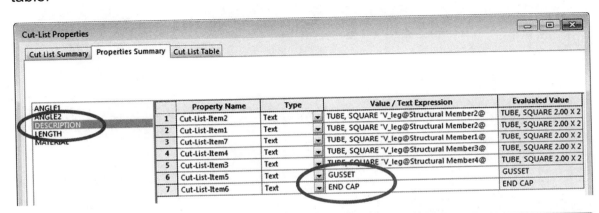

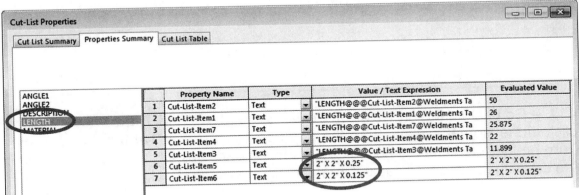

ITEM NO.	QTY.	DESCRIPTION	LENGTH
1	2	TUBE, SQUARE 2.00 X 2.00 X .25	50
2	2	TUBE, SQUARE 2.00 X 2.00 X .25	26
3	6	TUBE, SQUARE 2.00 X 2.00 X .25	25.875
4	1	TUBE, SQUARE 2.00 X 2.00 X .25	22
5	6	TUBE, SQUARE 2.00 X 2.00 X .25	11.899
6	12	GUSSET	2" X 2" X 0.25"
7	6	END CAP	2" X 2" X 0.125"

411. – Go back to the *'Weldments Table'* part; expand the *"Weld Folder"* and right-mouse-click in each of the *"Fillet Weld"* items and select "Properties" from the menu.

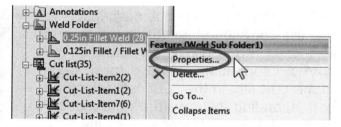

412. – In the Weld Bead Properties we can fill in additional properties like material, process, weight per unit length, cost, welding time per unit length and number of passes. We are also provided with the total number of welds, total weld length, mass, cost and time. This information is not required, but is useful if we add weld tables to our weldment drawing. Here are sample values for our example.

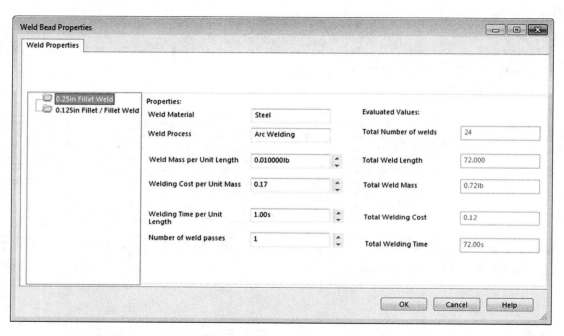

413. – Go back to the drawing, where we can add a "**Weld Table**." Right-mouse-click in the isometric view, from the pop-up menu, select "**Tables, Weld Table**." Turn on the "Combine same weld type" option and click OK to locate the table in the drawing. Change the table's font and size as needed.

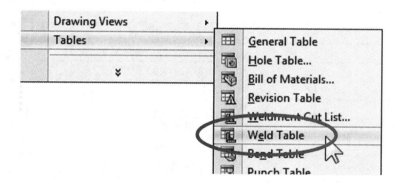

ITEM NO.	WELD SIZE	SYMBOL	WELD LENGTH	WELD MATERIAL	QTY.
1	0.25	△	192.56	Steel	1
2	0.125	△	144	Steel	1

414. – Add the identification balloons to identify each structural element using the "**Auto Balloon**" function. Select the "**Auto Balloon**" command from the Annotation toolbar, or the menu "**Insert, Annotations, Auto Balloons.**" In the "Balloon Layout" options select "Faces" to attach balloons to the structural members' faces. Under "Balloon Settings" select the "Circular Split Line" style, and pick "Quantity" in the "Lower text" selection list. Arrange the balloons as needed.

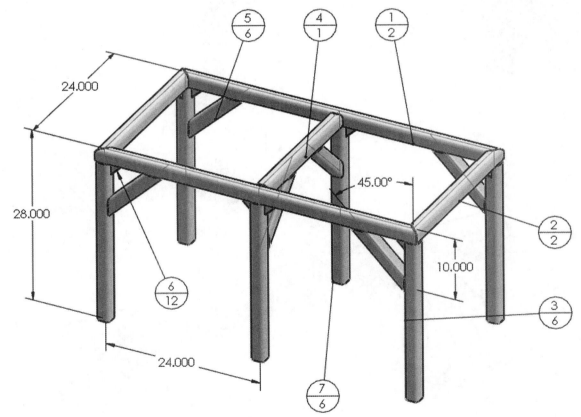

 In the "Balloon text" or "Lower text" selection lists we have the option to select "Custom Properties." If we do, a new selection list is revealed where we can select any property available including weldment cut list properties.

415. – To identify the welds, select the "**Weld Symbol**" command from the Annotation toolbar or the menu "**Insert, Annotations, Weld Symbol**."

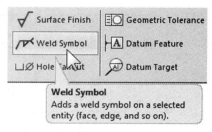

In the Weld Symbol properties window we can add all the necessary specifications for a welding instruction sheet, or we can click in the Weld Beads in the drawing, and the symbol will be automatically filled. Click to select a weld bead, locate the symbol in the sheet and repeat as needed.

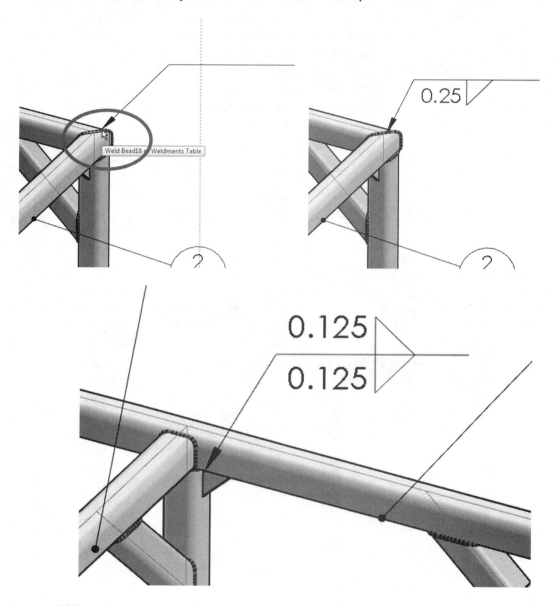

 When adding more Weld Symbols, the last type used is remembered. When a different weld bead is selected, the symbol is updated to reflect the currently selected bead. Close the Weld Symbol properties window when done adding welding annotations.

416. – Save the finished drawing and close.

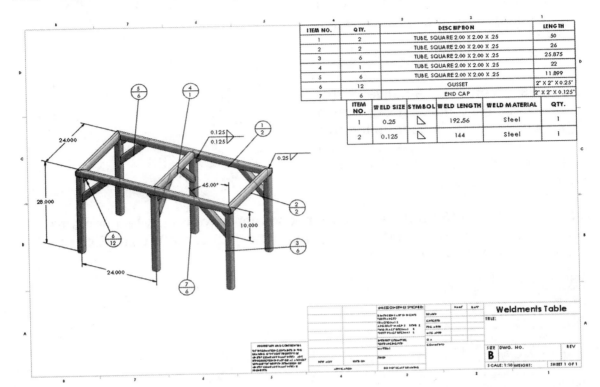

ITEM NO.	QTY.	DESCRIPTION	LENGTH
1	2	TUBE, SQUARE 2.00 X 2.00 X .25	50
2	2	TUBE, SQUARE 2.00 X 2.00 X .25	26
3	6	TUBE, SQUARE 2.00 X 2.00 X .25	25.875
4	1	TUBE, SQUARE 2.00 X 2.00 X .25	22
5	6	TUBE, SQUARE 2.00 X 2.00 X .25	11.899
6	12	GUSSET	2" X 2" X 0.25"
7	6	END CAP	2" X 2" X 0.125"

ITEM NO.	WELD SIZE	SYMBOL	WELD LENGTH	WELD MATERIAL	QTY.
1	0.25		192.56	Steel	1
2	0.125		144	Steel	1

417. – To review more weldments options, make the following (2D) sketch in the *"Top Plane."* The center of the arc is coincident to the origin; this will save us from making an auxiliary plane for the next step. When working with weldments we can only use circular arcs and straight lines. Exit the sketch when finished.

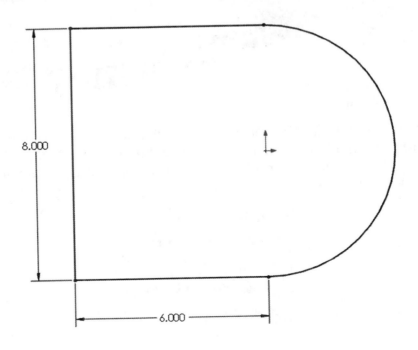

418. – Switch to an isometric view; add the following sketch in the *"Right Plane,"* and exit the sketch when done.

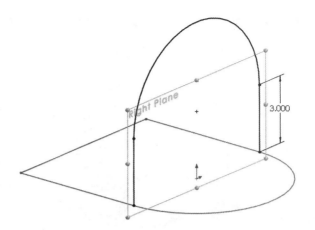

 As an alternative (or challenge), make a 3D sketch including the lines from the two sketches. When we work with weldments, it makes no difference if we use 2D or 3D sketches.

419. – Select the **"Structural Member"** command from the Weldments toolbar. Select the "ansi inch" standard, "pipe" type and "0.5 sch 40" size. Pick the lines indicated in the preview.

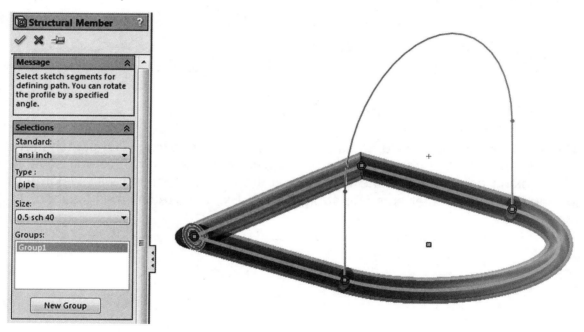

For "Corner treatment," instead of applying a global option, we'll manually select the treatment for each connection. Be sure the option "Merge arc segment bodies" and "Merge miter trimmed bodies" are turned off when adding structural members in this exercise.

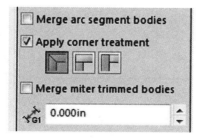

420. – Click on the point in the corner and select the following corner treatments:

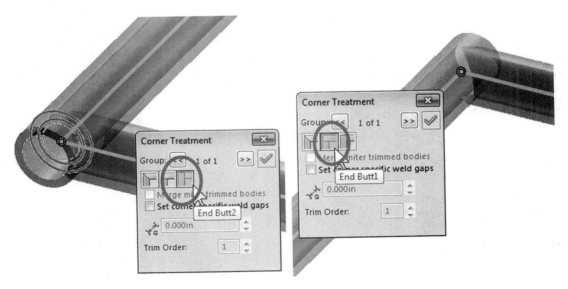

 The option "Set corner specific weld gaps" adds the specified gap between the connecting elements for welding.

421. – In the case of the joint between a straight line and an arc, changing the corner treatment gives us the option to merge the segments into a single element. In our example we'll leave this option unchecked.

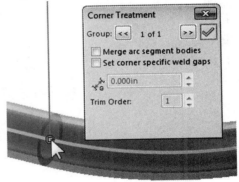

422. – Add the rest of the structural elements using the same structural member with the same settings. Click OK to complete it. Save the part as *'Welded Frame'*.

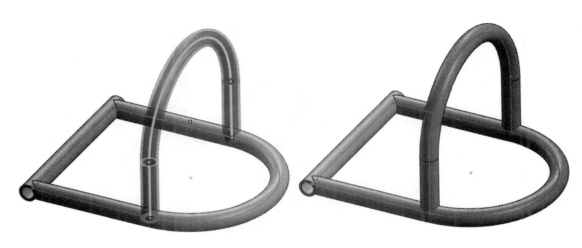

423. – Now select the "**Trim/Extend**" command to trim overlapping elements. Select the "End Trim" corner type option. Select the vertical elements in the "Bodies to be Trimmed" selection box. In "Trimming Boundary" use the option "Bodies" and select both of the bottom straight members and the curved one. Make sure the trimmed segments are correctly marked "keep" or "discard." Be sure to use the option "Simple cut between members" and click OK to finish.

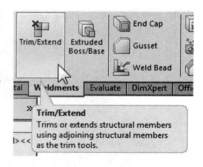

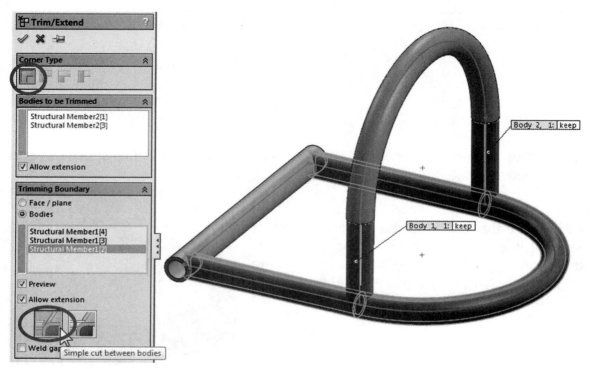

 After completing the structural members hide the sketch(s) used.

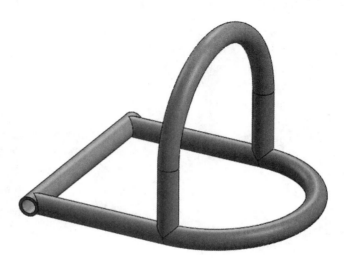

424. – Right mouse click in the *"Cut list"* folder and select "Update" from the menu to group elements together. Remember that we can preview the cut list by selecting "Properties" from the cut list item's right-mouse-button menu.

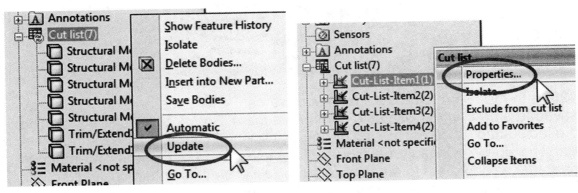

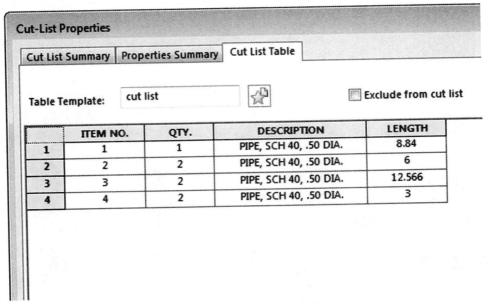

	ITEM NO.	QTY.	DESCRIPTION	LENGTH
1	1	1	PIPE, SCH 40, .50 DIA.	8.84
2	2	2	PIPE, SCH 40, .50 DIA.	6
3	3	2	PIPE, SCH 40, .50 DIA.	12.566
4	4	2	PIPE, SCH 40, .50 DIA.	3

425. – When working with multi body parts (like weldments), we can save each body to a new part file and/or make a new assembly with those parts at the same time. To save a single body to a new part we can select it from the "Cut-List-Item" folder or the graphics area with the right-mouse-button and

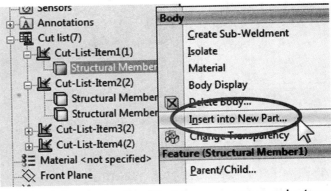

select "Insert into new part..." from the pop-up menu. But in our case, what we want to do is to save all of the bodies, each to a part, and at the same time make an assembly using those parts. Right-mouse-click in the *"Cut list"* folder and select the "**Save Bodies**" option.

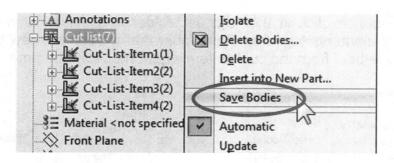

426. – In the "**Save Bodies**" command we can select one or more bodies to be saved to a part. In our example we'll name them 'Body1' to 'Body7' by clicking in the label attached to each body or double-clicking in the row for each body in the list. The option "Consume cut bodies" removes the bodies from the multi body part after saving them to an external file. We'll leave this option unchecked. Since we also want to make a new assembly, select the "Browse…" button in the "Create Assembly" option box and name it *'Welded Frame Assembly'* and Click OK to finish.

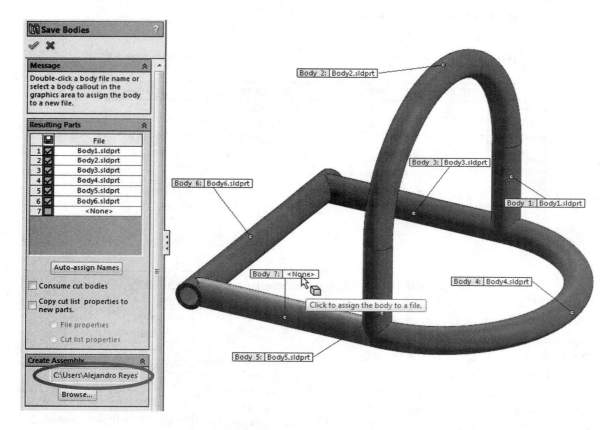

437. – After creating the assembly, a feature called *"Save Bodies1"* is added to the FeatureManager; a new part file is generated for each body saved as well as the new assembly which is already open. Switch to the new assembly window just made. An inconvenience of making an assembly from bodies using this technique is that all the parts will be automatically "fixed" in the assembly, and identical parts are given different names. We could skip duplicate parts when saving the bodies, but then we have to manually add and mate them to the assembly. It's up to the reader to pick one approach over the other, depending on what we need/want to accomplish.

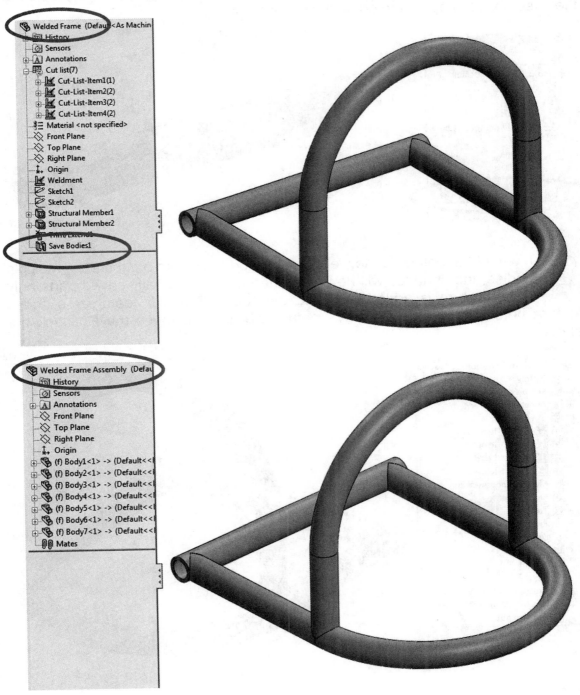

 The bodies saved to a part have an external reference to the original welded part. In this case, the reference status is "In context" because the '*Welded Frame*' part is open and loaded in memory.

438. – Opening one of the assembly parts shows the only feature is a saved body from a multi body part, listed as *"Stock-part-name."* From this point on, we can add more features to a part if needed, and these changes will be reflected in the assembly, but not in the original multi body part.

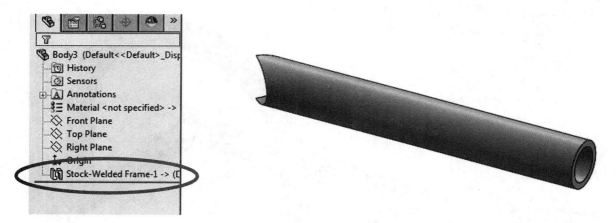

If we add a feature to the welded <u>multi body part</u>, like a cut across multiple bodies, the changes will be propagated to the externally saved parts only if the feature is added BEFORE the *"Save Bodies1"* feature. To add a feature before the *"Save Bodies1"* feature we have to move the Rollback bar above it.

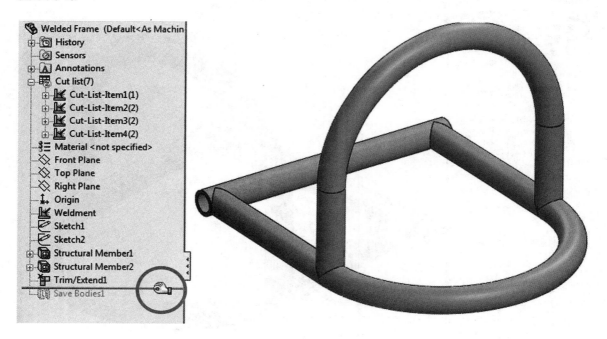

Structural Member Libraries

A word about structural profile members: SolidWorks includes a small number of profiles pre-loaded in the "**Structural Member**" library. To add custom profiles we have to make a new part, draw the desired sketch profile, and save the sketch as a library feature part in the folder designated for structural members.

439. – Make the following sketch in a new part. We can use any plane. Units are in millimeters (mm). Don't forget the sketch points in the middle of each side; the reason to add these points is to be able to use them as location points to position the profile in reference to the weldment sketch. Exit the sketch when done.

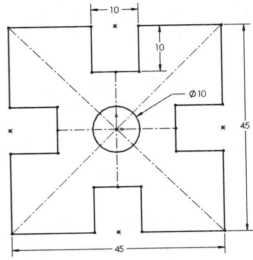

440. – To have a description in the cut list when using the library, we have to add a custom property (menu "**File, Properties**"). The Property Name is *"Description"* and the value *"Square Profile, 45mm x 45mm."* Click OK to continue.

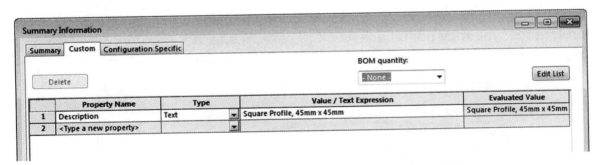

441. – To make a new weldments profile, <u>pre-select</u> the sketch in the FeatureManager and save the file as a Library Feature Part (Lib Feat Part *.sldlfp) in the weldment profiles folder. To find the folder where the weldments profiles are saved, go to the menu "**Tools, Options, System Options, File Locations**" and from the drop-down menu select "Weldment Profiles." To add weldment profiles to the default weldment profiles library, save the file to this folder. The default location is:

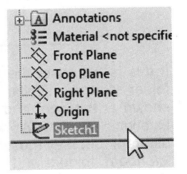

"<SW_Install_Folder>\lang\language\weldment profiles"

The folders in this location reflect the standards and types of libraries available. Here we can select a Standard and Type to save our library, or if we want, we can create our own library folder. For our exercise we'll save the library as "**45 x 45**" in the following folder:

"***<SW_Install_Folder>***\lang*language*\weldment profiles \iso\square tube"

If you can save to this folder this profile is now available for structural members.

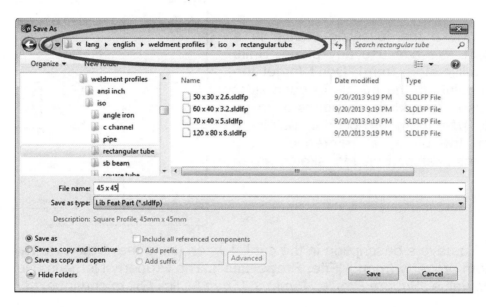

442. – Depending on your operating system and user permissions, OR personal preference, you may not be able/want to save files to the *'Program Files'* folder. If you get this warning we have to save it to a new location.

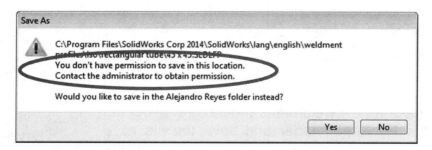

If this is the case, the new profile has to be saved to a folder structure similar to the original weldment profiles shown at the right and added to the "Weldment Profile" locations in SolidWorks. The folder names can be different, but they must have a *'home'* folder, one or more *'standard'* folders and one or more *'type'* folders. Using Windows Explorer, add a folder structure similar to this at a location of your choosing and save the library to the *'type'* folder.

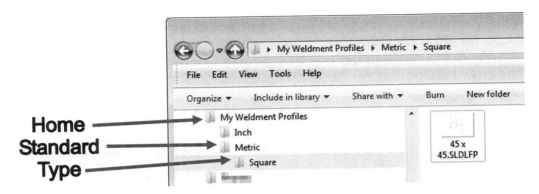

Home

Standard

Type

443. – To add the new weldment profile library to SolidWorks, go to the menu "**Tools, Options, System Options, File Locations**" and select "Weldment Profiles" from the drop down list. Click the "Add" button, browse to the *'home'* folder we made previously and select OK to finish.

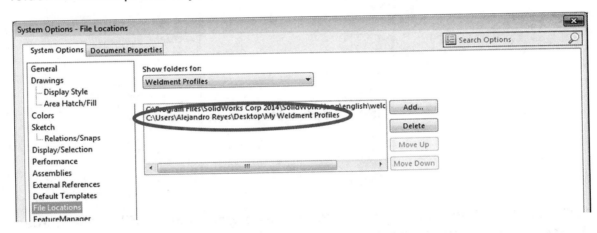

And now the new library is ready for use in the "Structural Member" command.

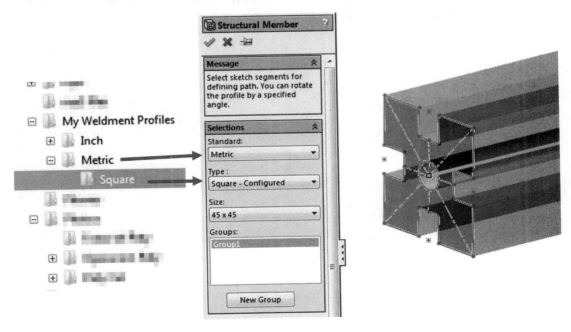

To increase the number of weldment profiles available to our designs, we can download a zip file with a more complete list of weldment profiles from the SolidWorks "Design Library, SolidWorks Content, Weldments" section. After downloading the zip file, extract its contents to the custom weldment profiles folder.

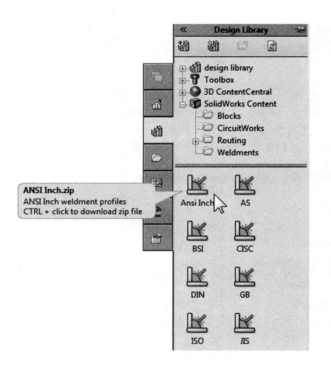
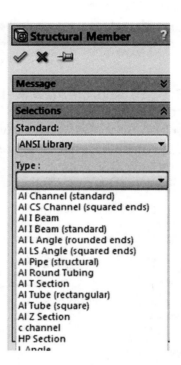

IMPORTANT: When adding a new location for weldment profiles, the names of the folders cannot be the same as the ones already in the default location, otherwise only one will be listed, that is why we used "ANSI Library" in the example above.

Exercise: Using the Weldment Profile library just made, build the next structure. Note the dimensions provided are in millimeters and are all external dimensions; this means that the finished structure cannot be bigger than these dimensions in any direction. After completing the structure, make a detail drawing including a cut list. This structure is not welded as it is usually assembled using special hardware.

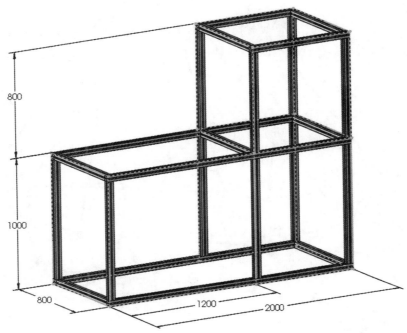

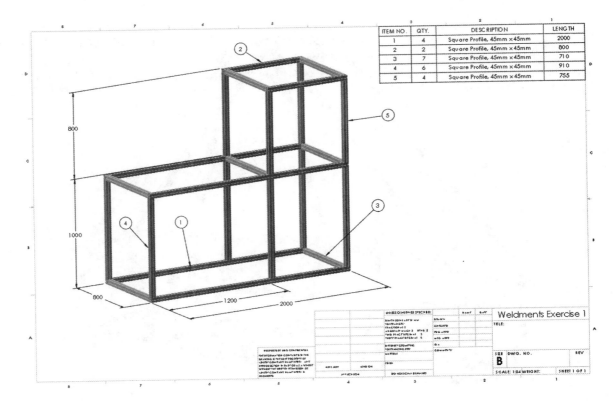

ITEM NO.	QTY.	DESCRIPTION	LENGTH
1	4	Square Profile, 45mm x 45mm	2000
2	2	Square Profile, 45mm x 45mm	800
3	7	Square Profile, 45mm x 45mm	710
4	6	Square Profile, 45mm x 45mm	910
5	4	Square Profile, 45mm x 45mm	755

Weldments Exercise 1

Exercise: Build the following dune buggy frame using a 1″ sch-40 pipe profile; trim all ends and add 0.125 weld beads to all joints. Open the *'Dune Buggy'* frame sketch from the accompanying files.

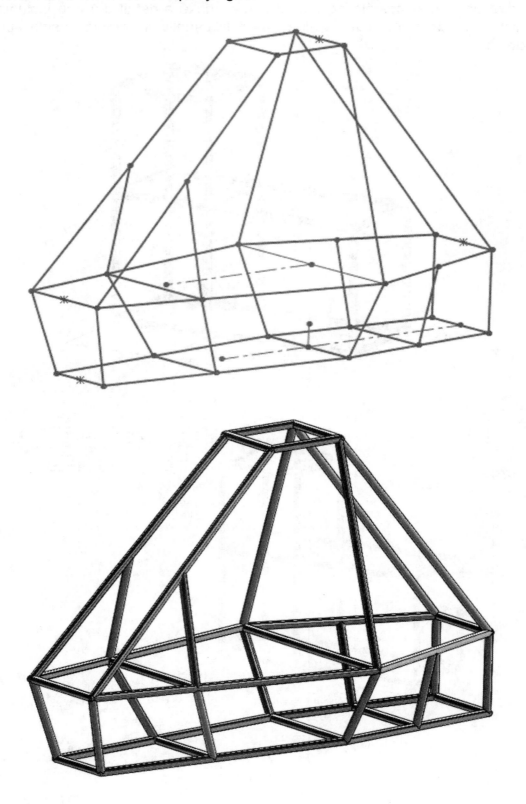

TIP: To correctly trim the unions where three or more elements meet, trim two at a time. You may have to turn off the "Allow extension" in the "Trim/Extend" command. If this option is turned off, the element is only trimmed and will not extend to the trimming boundary.

Original Structure (Continuous element transparent for visibility)	Trim both opaque elements using an "End Miter" to trim/extend each other
Trim both with the continuous element	Add the weld beads to all edges

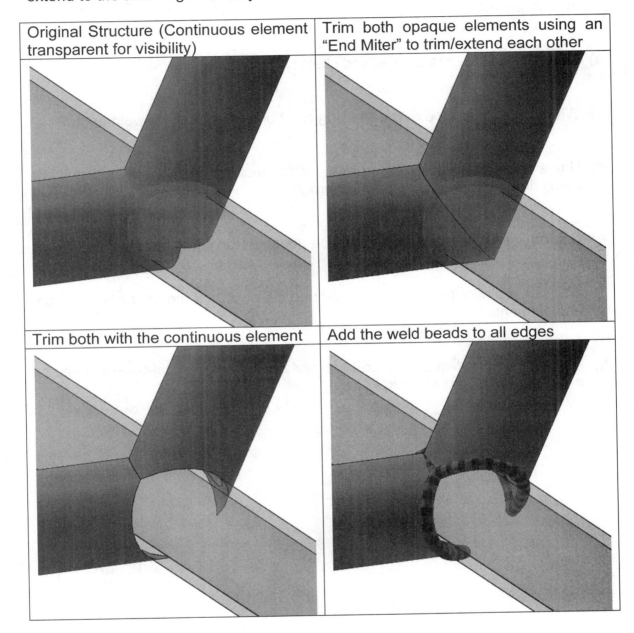

Review and Questions:

a) Name three geometric relations available in a 3D sketch that are not available in a 2D sketch.

b) Describe how to create a Derived sketch.

c) What is the difference between using a 2D or a 3D sketch for weldments?

d) Name the conditions to be able to select multiple sketch segments at the same time for a structural member group.

e) Can multiple gussets and end caps be added at the same time?

f) How can we generate a cut list to include the structural member's type, length and cut angles?

g) If welded part bodies are saved to an assembly, are identical bodies saved as a unique file or different files?

Answers:
a - Along X, Y, Z, On-Plane, Normal, ParallelYZ, ParallelZX.
b - Pre-select the sketch to derive and the plane/face to place the derived sketch and click in the menu "Insert, Derived sketch."
c - None. They both work the same.
d - They must be parallel or continuous connected by an endpoint.
e - Gusset's have to be added one at a time, end caps can be added multiple at a time.
f - Right mouse click in the "Cut list" folder and select "Update." Then import the Cut List table to a drawing.
g - Each body, even identical ones, are each saved to a different file always.

Surfacing and Mold Tools

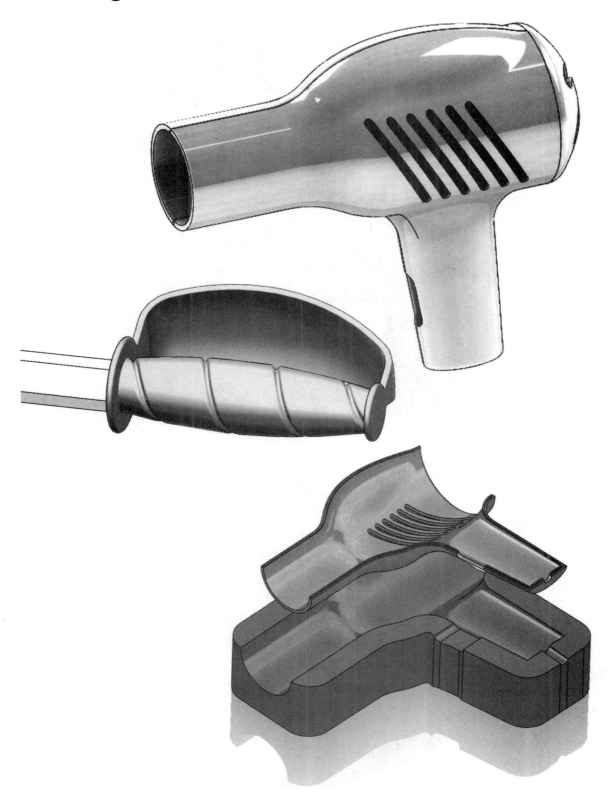

Surface Modeling

So far we have been working with single and multi body solid parts, and now we'll start working also with surfaces. By including surfaces in our models, we are able to create increasingly complex designs that would otherwise be very difficult to model just with solid bodies. Working with both solids and surfaces is called 'hybrid modeling', as we incorporate both solid and surface bodies in the same model. When we work in a hybrid model, the surfaces are usually used as support geometry, as stopping or trimming boundaries for solids and/or other surfaces, etc.

Background and a little history: The first computer models used for design (well over 40 years ago) were wireframe models capable of defining the edges of a part and had severe limitations, especially when it came to defining a three dimensional surface, as it could only be approximated using a mesh of lines with little detail. Think of a wireframe model as a 'stick figure.' A wireframe model is basically what we are able to do now with a 3D sketch – just lines.

Some years later, surface models were developed that allowed more definition of a component, including curved faces and were a huge improvement over the wireframe models, but still had limitations when it came to calculating a part's volume, weight, etc. When we talk about a surface model, think of a parade's floating balloon. It has the surfaces (fabric) and the wireframe (stitches), and is essentially hollow. Similarly, it's safe to say that a surface model *may* have imperfections, like faces not matching correctly creating gaps (think of the balloon with an opening at a corner) or overlapping faces. This would create a problem if we wanted to calculate volume, weight, etc., or use it for manufacturing.

Later, in the 1980's (give or take a few years…), solid models became available, but (just like surface modelers) were difficult to use, required special training and were available in high end (read 'expensive') workstations. A design station's cost, complete with hardware and software, would be in the tens of thousands of dollars, affordable only by big corporations or government agencies. Design tools were greatly improved, producing better results faster, making it easier to integrate with computer numerical controlled (CNC) manufacturing.

Some years later (circa 1990's), solid modelers became available to the Windows operating system, making them considerably more affordable to mainstream designers, and being native Windows applications made them easier to learn and use. These solid modelers incorporate many of the advantages surface and solid models offer, including better integration with manufacturing, analysis, animation and many other downstream applications.

At present, with the advances in technology making computers faster than ever before (and continuing to get faster every day), we are able to take advantage of it and create bigger, ever more complex designs in record time, helping us bring better and safer products to market faster than ever before.

A word about hybrid modeling: It is not uncommon to see that dedicated surface modelers usually available in high end CAD applications can generate superior and better looking surfaces for automotive, aerospace and consumer product design than many solid modelers (like SolidWorks), and they can also make changes easier and faster maintaining the intended shape, making it very attractive to create highly complex surfaces easier with those tools. With that said, it is common in those industries to generate complex surfaces in one software package and then import them into solid modelers like SolidWorks to complete the rest of the design, including tooling for manufacturing.

When we talk about surface modeling, we are referring to creating models using mainly surfaces, or the 'skins' of parts, and afterwards converting them into a solid or using them to build solids. The tools available to build surfaces include almost all the same features available to solid models like extruded, revolved, swept and lofted surface plus a few more tools specific to surface modeling, most of which will be covered in this lesson.

To start modeling with surfaces, we'll make a hair dryer using a few simple surfaces, and then we'll convert them into a solid model. After finishing, we'll split the model to make a file for each of the different parts to practice and learn more about multi body parts.

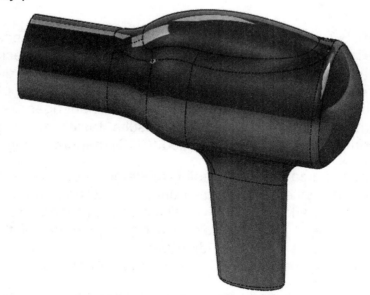

 If the Surfaces toolbar is not visible, activate it by right mouse clicking in a CommandManager tab, and selecting "**Surfaces**," and/or the menu "**View, Toolbars, Surfaces**."

444. – To save time and focus on the surfacing functionality, open the file *'Hair Drier.sldprt'* from the included files and open it. It has a number of pre-made sketches and planes for us to start working.

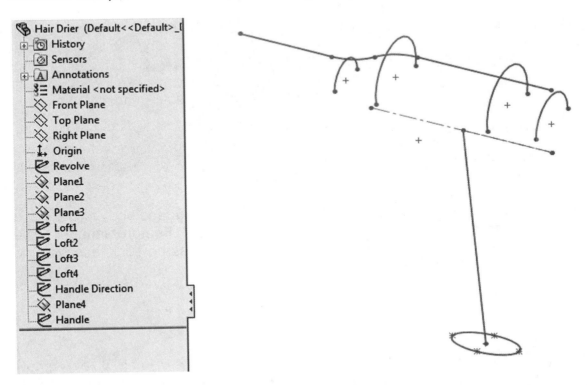

445. – The first feature will be a revolved surface. Select the sketch named *"Revolve"* in the FeatureManager and click in the "**Revolved Surface**" command, or the menu "**Insert, Surface, Revolve.**" The sketch's centerline is automatically selected. Revolved surfaces follow the same rules as a revolved feature.

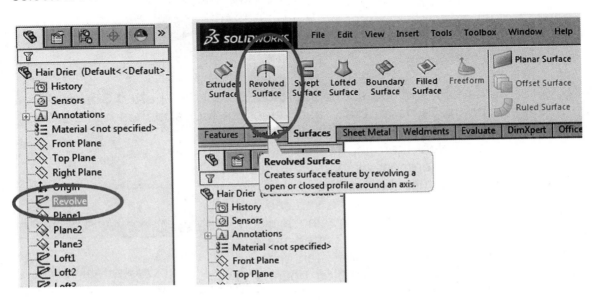

The revolved surface made will be 360 degrees. Click OK to complete it.

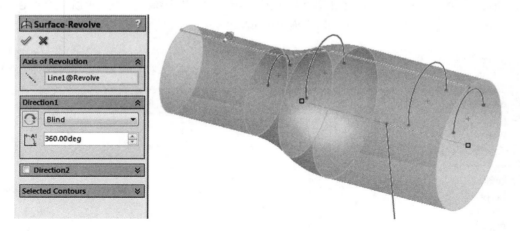

As opposed to when we work with solids, when the first surface body is created a **"Surface Bodies"** folder is immediately added to the FeatureManager. The *"Surface-Revolve1"* we just made is only a surface. It has no thickness, volume or mass; it's only a surface.

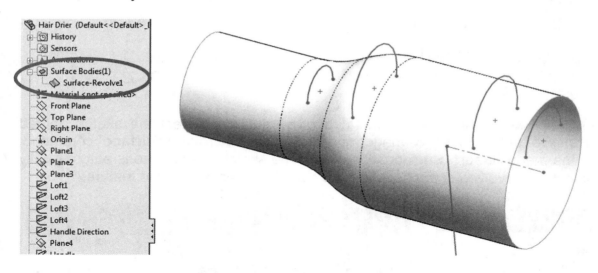

446. – The next feature will be a lofted surface. Select the **"Lofted Surface"** icon or the menu **"Insert, Surface, Loft,"** and select the four sketches named *"Loft1"* through *"Loft4"* on the screen or from the fly-out FeatureManager. Select them left to right or right to left, it makes no difference, as long as they are in consecutive order. Just like solid body lofts, we have to select the sketches near the vertices that will be connected by the loft, preventing twisting, and avoiding undesired results. A lofted surface can include guide curves, centerline, start and end constraints. Leave the rest of the loft options to their default value for this example and click OK to complete it.

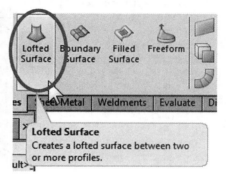

The main difference between a lofted surface and a solid body loft is that in a lofted surface we can use open *or* closed sketches, curves, edges, etc., to create irregular, complex surfaces.

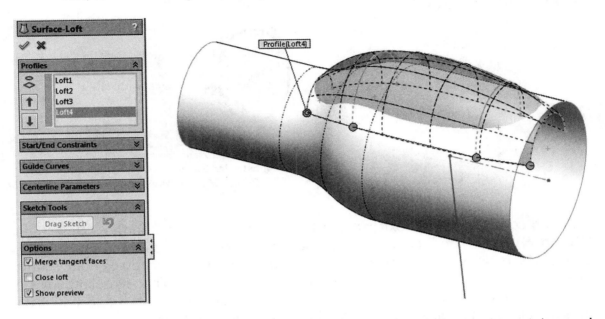

After the loft we have two surface bodies. Note that surfaces do not 'merge' automatically as solids do; they simply intersect each other and remain separate bodies.

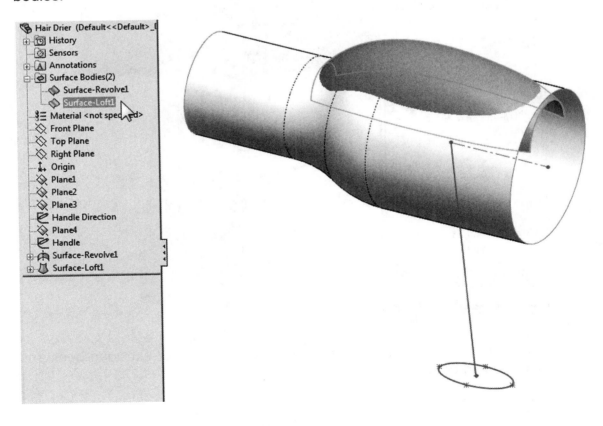

447. – For the next feature we'll make the hair drier's handle using an extruded surface with a draft along a defined direction. Select the sketch named *"Handle"* in the FeatureManager and click in the **"Extruded Surface"** icon, or the menu **"Insert, Surface, Extrude."**

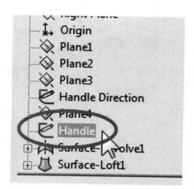
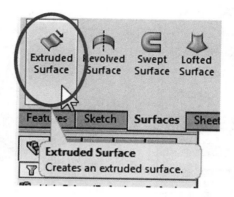

The end condition for the extrusion will be up to the revolved surface. Select the **"Up to Surface"** end condition and select the revolved surface. By default, extrusions (solid or surface) are made normal to the sketch plane, but in this case we'll make the extrusion along a defined direction. That direction will be defined by the sketch called *"Handle Direction."* Right under the end condition selection list is the "Direction of Extrusion" selection box. Click inside to activate it and then select the *"Handle Direction"* sketch as shown.

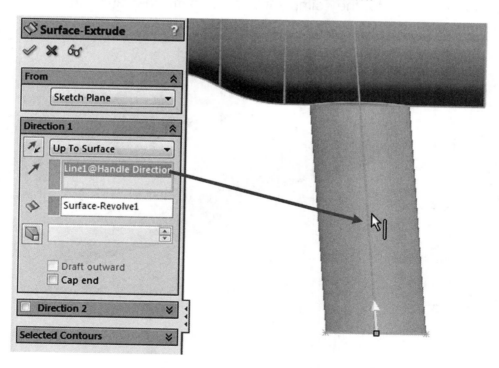

The "Direction of Extrusion" option is also available for extruded boss and cuts in solid bodies.

448. – Another option we'll use in the extrusion is the option to add a draft as we extrude. Adding a draft makes the body (surface or solid) grow out or shrink in with a defined angle. To turn the draft option on, click in the "**Draft On/Off**" icon; the angle box will be activated as well as the "**Draft outward**" checkbox. For our example, enter a 3 degrees angle and check the "Draft outward" checkbox. Click OK to finish the surface.

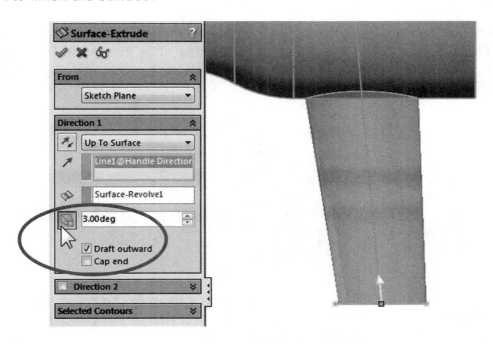

Now we have three surface bodies. Hide the *"Handle Direction"* sketch; we will not need it anymore.

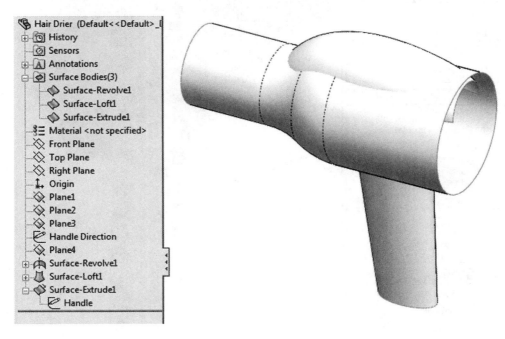

449. – The next step is to trim the surfaces to make them match exactly. By trimming the surfaces, we are going to make all surfaces touch each other at the edges only, with the ultimate goal to form a single, closed, continuous surface that will be converted into a solid body. There are several ways to trim surfaces. We can use other surfaces, planes, or a sketch as a trimming tool. The first step will be to trim the revolved surface with the handle.

Select the "**Trim Surface**" icon from the CommandManager, or the menu "**Insert, Surface, Trim**." Under "**Trim Type**" select "**Standard**"; this means that we'll use a trim tool (Sketch, Plane or Surface) to cut *other* surfaces. Add the handle surface under the "**Trim tool**" selection box.

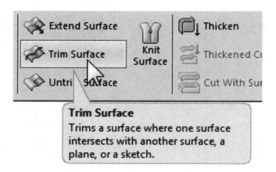

Use the "Keep selections" option and in the "pieces to keep" selection box, pick the revolved surface anywhere but the area where it meets the handle surface, because this is the part we want to keep after trimming.

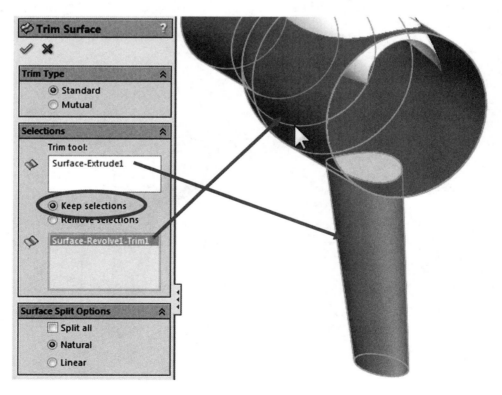

Optionally, if we use the "Remove selections" option, we have to select the elliptical surface inside to remove it. Use either approach, and click OK to complete. Trimming surfaces using this approach only cuts one surface. In the next step we'll see how to mutually trim two surfaces.

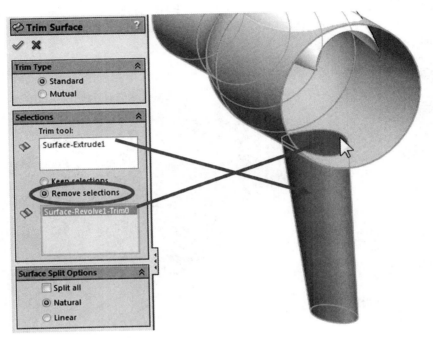

This is the result of trimming the first surface; we still have three surface bodies.

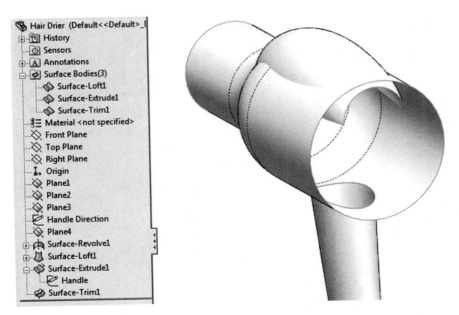

 When we use a sketch as a trim tool, the surface is trimmed by projecting the sketch normal to its plane onto the surface. To understand it better, think of the sketch being extruded as a surface with the "Through All" end condition in both directions first, and then used as a trimming surface.

450. – For the next feature, we'll trim two surfaces at the same time. This means that we'll select two (or more) surfaces, and they will be cut at their intersection(s); then we can select which bodies we want to keep or delete. Select the "**Trim Surface**" command, and use the option "Mutual" under "Trim Type." In the "Surfaces" selection box, pick the revolved and lofted surfaces. Use the "Keep selections" option and select the outside faces for both. They will change color when selected. Click OK to finish.

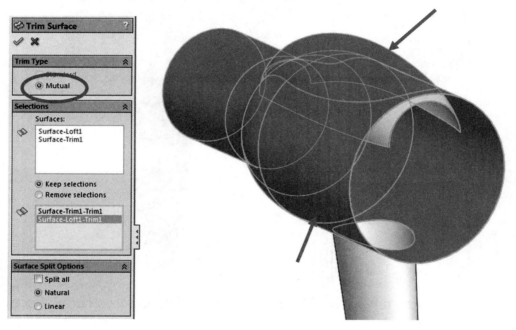

 When we trim two surfaces using the "Mutual Trim" option, the resulting surfaces are merged into a single surface.

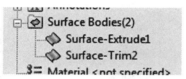

Our resulting model now looks like this:

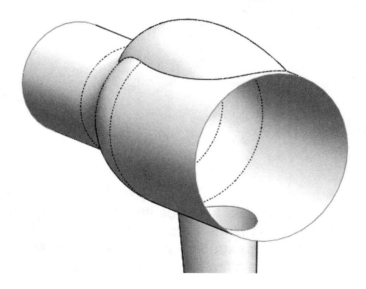

If we had selected the "Remove Selections" option, we would have selected the inside faces instead.

451. – Since we need to make a fully closed volume before we can convert it into a solid, we'll need to close the openings with surfaces. We have a couple of options to close them. For the opening at the bottom of the handle, we'll make a flat surface. Select the "**Planar Surface**"

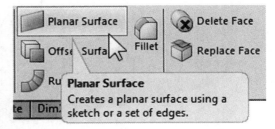

command from the Command Manager or the menu "**Insert, Surface, Planar**."

To make a planar surface, we can select a set of closed edges or a closed sketch. Turn the model over, select the edge at the bottom of the handle and click OK to add the new surface.

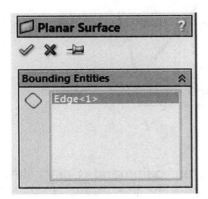

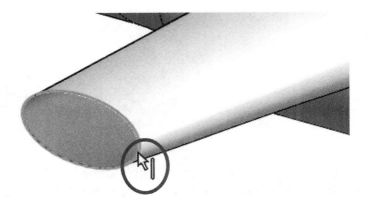

We added a new surface, and now have three surface bodies again.

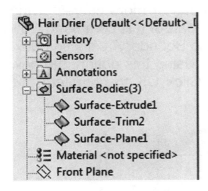

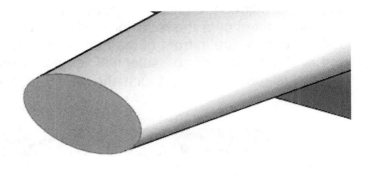

 Edges that are connected to other surfaces are colored black (default setting) and open surface edges that are not connected to other surface bodies are a different color. Surface bodies may have touching edges, but may not be necessarily connected, think of it as two pieces of fabric aligned, but not stitched together.

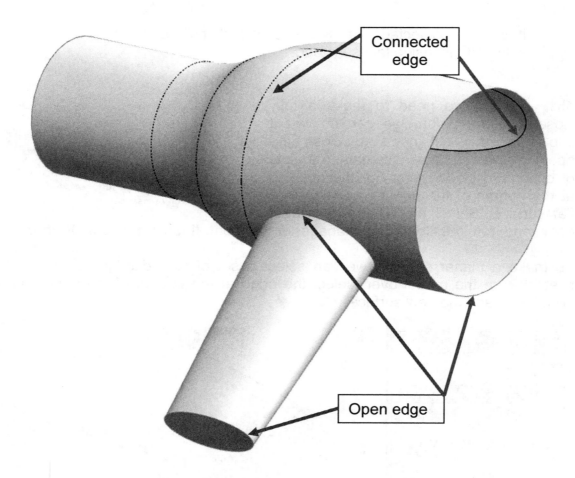

452. – For the front of the hair drier we'll use a different command. Select the "**Filled Surface**" command from the Command-Manager or the menu "**Insert, Surface, Fill**." This command is more flexible than the planar surface; it allows us to use planar or non-planar sketches, edges or curves, make the new surface tangent to adjacent faces, or constrain the surface to one or more curves giving us more control over the resulting surface. For the first part of this command we'll add the open edge in front of the hair drier to the "Patch Boundary" selection box. In the "Curvature Control" selection list pick "Contact." Using this option, the new surface is made to only touch the selected edges. Click OK to complete this surface.

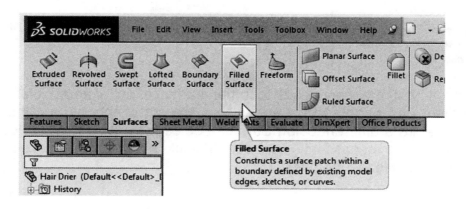

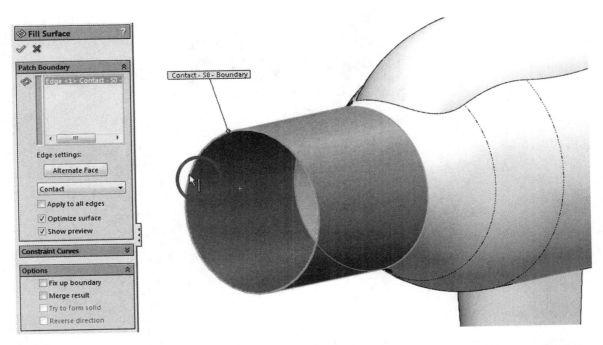

 The Fill Surface command has an option to merge the resulting surface to other surfaces; we'll use the default setting, unchecked.

This is our fourth surface in the model.

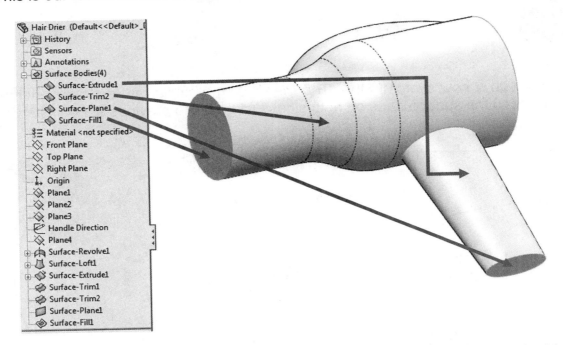

453. – The only place left to close in our model is the back of the hair drier. Before we close it, we need to add two sketches that will be used to control a "**Filled Surface**." Select the *"Front Plane"* and add the following sketch. Use a **3 Point Arc** or a **Centerpoint Arc**. Be sure to add "Pierce" relations between the endpoints of the arc and the surface's edge. Exit the sketch when done.

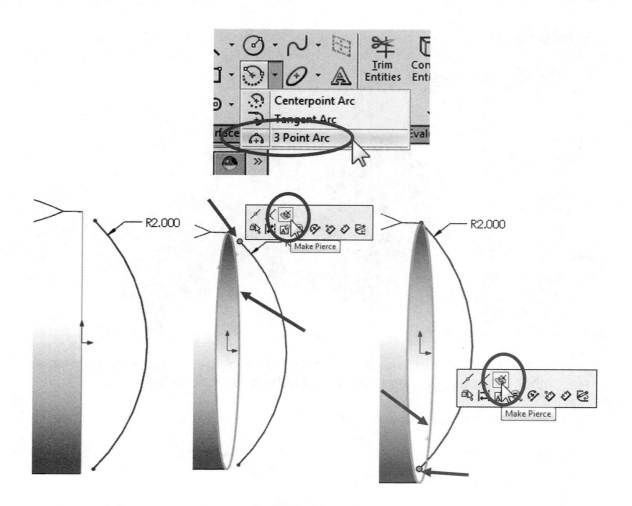

454. – Add a new sketch in the *"Top Plane"* with a similar arc and add "Pierce" relations to the surface edge on both sides just as in the previous step.

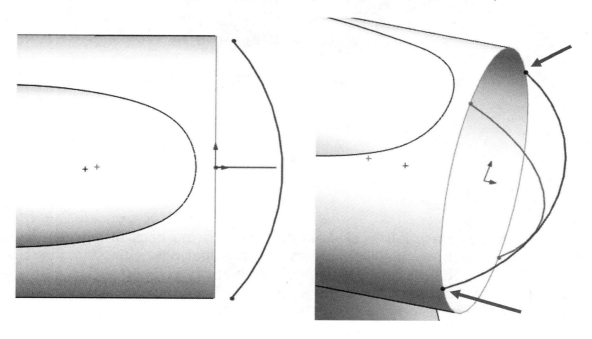

To make the arc coincident to the previous sketch, add a "Sketch Point" to its midpoint, and then a "Pierce" relation between the sketch point and the previous sketch. Exit the sketch when done.

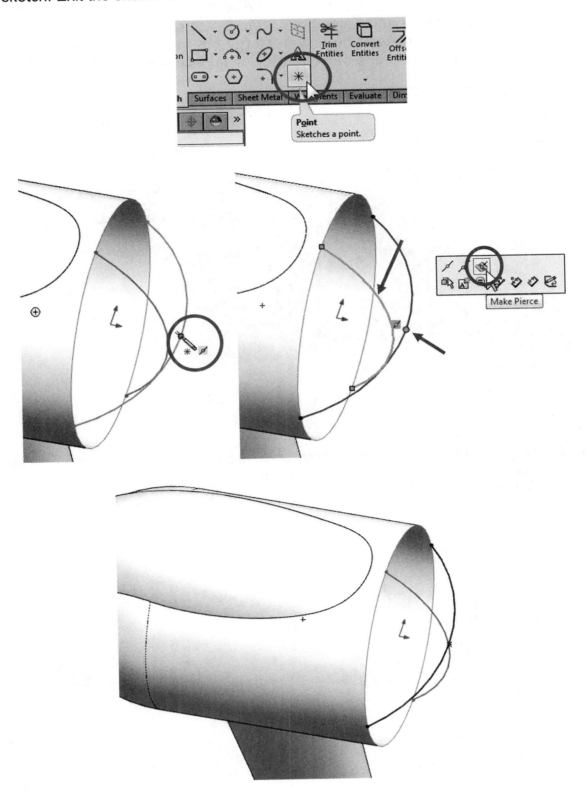

455. – Select the "**Filled Surface**" command from the Surfaces toolbar. Add the open edge in the back of the hair drier in the "Patch Boundary" selection box, and both of the previous sketches in the "Constraint Curves" selection box. The surface will conform to the curves and match exactly with the edge. Since we are adding constraint curves, changing the curvature control will make no difference in this case. Be sure to check the option "**Merge result**" at the bottom; this way the new surface will be merged with the revolved surface when we are finished.

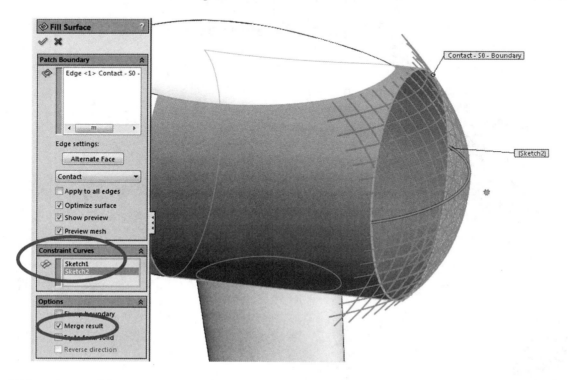

456. – It would now appear that we have a closed surface to convert into a solid, but we need to do one more step before, and that is to merge all four surface bodies into a single surface using the "**Knit Surface**" command. Note the open edges' color.

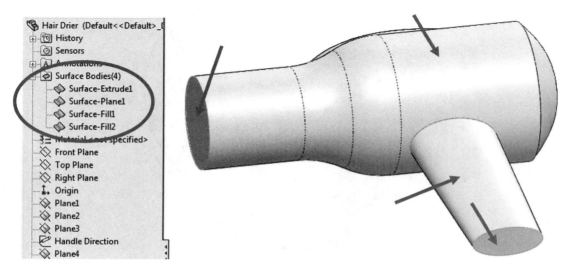

There are basically two ways to convert the surfaces to solid models:

a) We can make an open or closed surface solid by making the surface a specified thickness.

b) We can convert a closed surface into a solid if the surface is a single body defining a closed volume.

 When we refer to a *closed surface*, we are talking about a *water tight* surface. In other words, it has to be a completely enclosed volume, like a balloon.

457. – We'll show both of these methods to form a solid. The first option will be to make a surface solid by making it thicker. The first thing we need to do is to merge the surfaces into a single surface body; select the "**Knit Surface**" command in the Surfaces toolbar, or the menu "**Insert, Surfaces, Knit**."

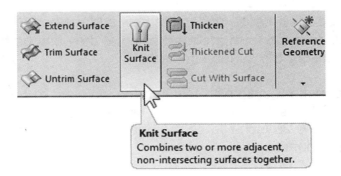

For our hair drier we will ultimately leave the front open, so we'll knit the other three surfaces into one. In the graphics area, select the handle, the flat surface at the bottom of the handle, and the main body surface. Using the "Merge entities" option will not make a difference in this case. Click OK to finish.

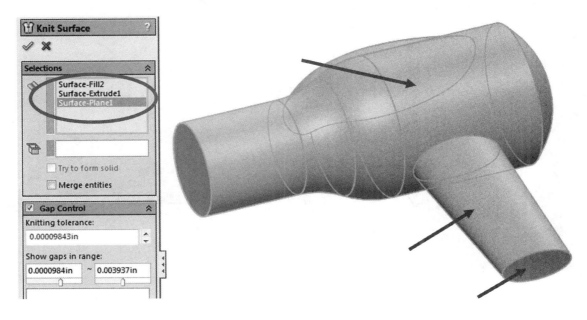

After knitting the surfaces our model has two surface bodies.

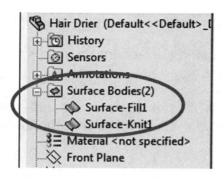

Things to keep in mind when knitting surfaces:

- Surfaces must be touching at the edges.
- Surfaces must not overlap.
- Surfaces knitted will be absorbed into a new surface body.
- If the surfaces selected form a closed volume, they can be turned into a solid body that will absorb the surface bodies.

458. – Before we make a solid with this surface we want to add fillets to our hair drier. We can add fillets to a surface just like a solid body. Select the "**Fillet**" command and add a 0.25 inch fillet to the back of the drier and a 0.125 inch fillet at the bottom of the handle.

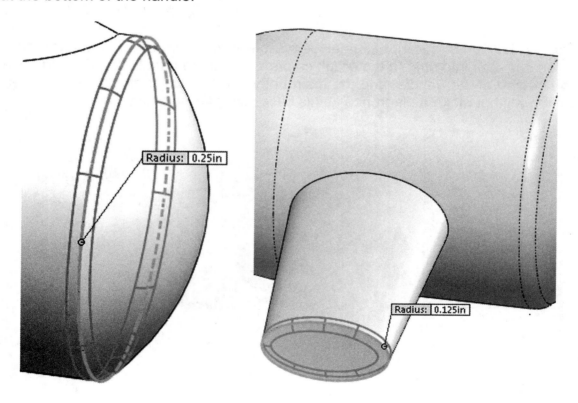

459. – To make the surface more aesthetic, add a 1″ fillet at the top of the hair drier. Make sure the "Tangent propagation" option is turned on.

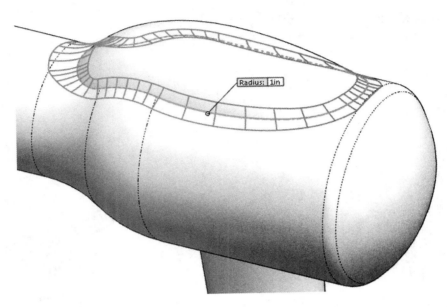

460. – For the last fillet, if we use a regular fillet with constant radius, we get a fillet that covers a large surface. Select the "**Fillet**" command again, but in this case select "**Face Fillet**" in the "Fillet Type" option and pick the two faces indicated, one in each selection box. We may need to click on the "Reverse Face Normal" option to make the arrows in the faces point to each other in order for the Fillet command to work. Make the fillet 1″ radius.

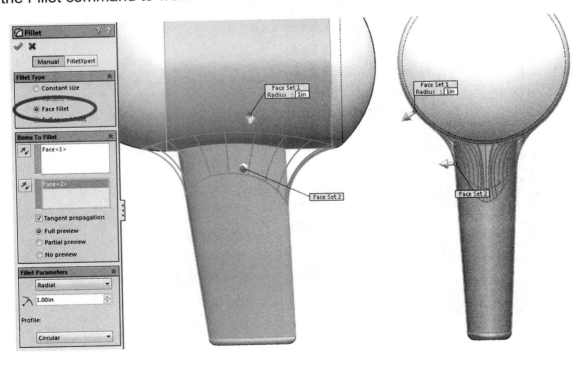

In the "Fillet Parameters" select the "**Chord Width**" option; notice the difference in the fillet's preview. The difference in the fillet is that instead of defining a fillet's radius, we are defining the length of the chord created by the fillet, making it a constant width all around the edge. Under the "Fillet Options" make sure the "Trim and attach" is selected; this option will ensure the surfaces are trimmed and merged with the fillet's surface. Click OK to finish.

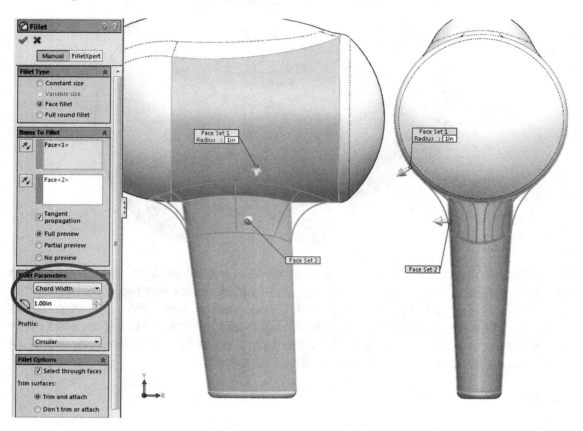

461. – Now we are ready to convert the knitted surface body into a solid. Hide the *"Surface-Fill1"* body to see the effect of thickening the surface. Select it surface in the graphics area or the FeatureManager and select the "Hide" icon.

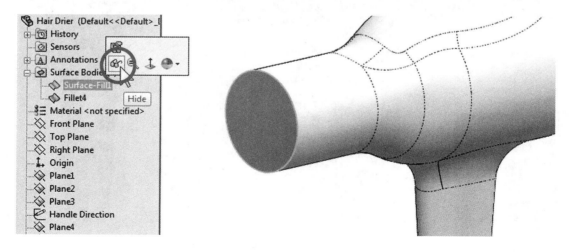

462. – The first method to make a solid is to thicken a surface. Select the "**Thicken**" command in the Surfaces toolbar, or the menu "**Insert, Boss/Base, Thicken**." In "Surfaces to Thicken," select the *"Fillet4"* surface. When we thicken a surface we can add material to one side, the other, or 'mid plane' to both sides of the surface. For our example, we'll make the surface 0.1″ thick inside to keep our model dimensions as outside dimensions. Notice the preview when we change the side to thicken. Be aware that using the "Thicken Both Sides" option adds the specified thickness to each side, making the end result twice as thick as the specified thickness. Click OK to finish.

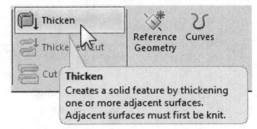

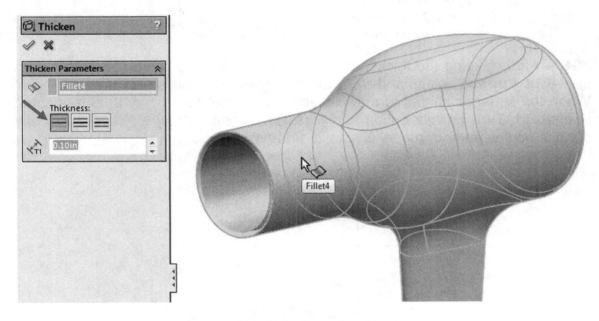

Now we have a solid body and a (hidden) surface body.

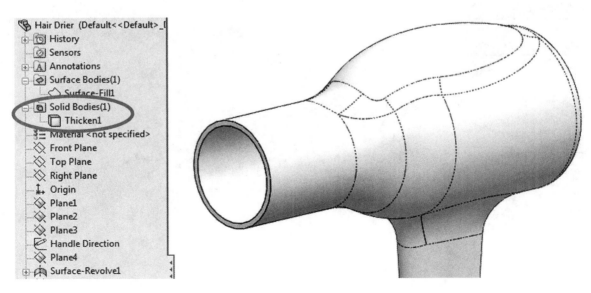

463. – To show the second method to convert surfaces to solids, delete the *"Thicken1"* feature, and show the surface that we had hidden before.

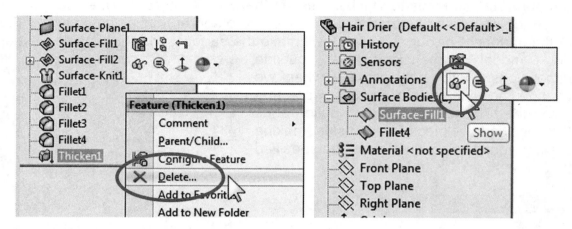

464. – After deleting the thickened body and showing the hidden surface, select the **"Knit Surface"** command from the Surfaces tab, and select both surfaces. When the surfaces knitted form a closed volume, the option "Try to form a solid" is enabled. Turn on this option and click OK to finish. Now we don't have any surface bodies and our model is a solid, non-hollow, body.

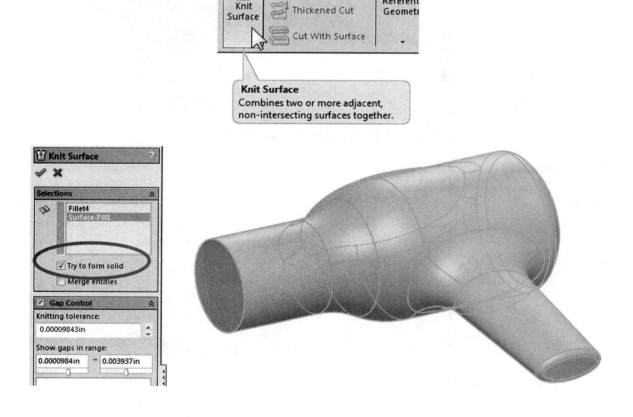

From this point on we can treat our model as a regular solid body; we don't have any surface bodies left and only one solid body. Make a 0.1″ thick shell feature removing the front face of the hair drier to complete the model.

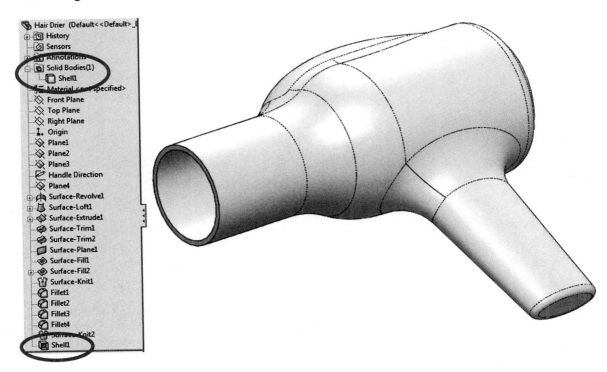

Choosing an option to convert surfaces to solid depends on many different factors; sometimes one method fails and we have to use a different approach. A common reason for errors when working with surfaces is when a face overlaps with another after thickening, or a face offsets into another after shelling the part. For example, if we change the fillet at the bottom of the handle to a 0.5″ radius, the "**Thicken**" command fails. In this case we would have to add the fillets *after* thickening the surfaces. There are no 'hard rules' to select one way or the other; just as both methods may work, sometimes we have to evaluate alternate ways and change the approach if the one we tried fails.

Notes:

Master Model

When making a design that is made of multiple parts that need to maintain an overall shape when assembled, like our hair drier, we use a technique called *Master Model*, where we start with a part that has the general shape and size of the finished assembly, and is later split (cut) into the individual components. The master model has all (or most) of the features that are common to all the components (usually the general outside shape), then it's split and the bodies are saved to individual parts as we did in the weldments section. From this point on, we add the features specific and unique to each (split) part.

The first body we are going to split from the hair drier's master model will become the back cover. To split a solid into multiple bodies we can use a plane, a surface (not necessarily flat) that crosses the entire model or a sketch (open or closed.)

465. – Add a sketch in the *"Front Plane"* and draw a line 0.125″ to the left of the origin as shown. When using an open sketch to split a model, the sketch *must* cross the model, and the split will be made by projecting the sketch normal to the sketch plane, much as if we made an extruded surface and cut our model with it.

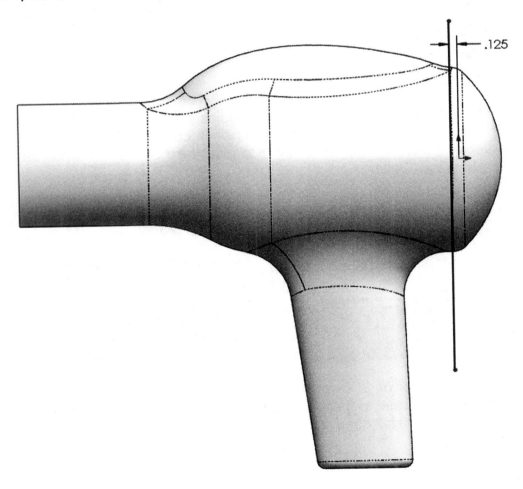

466. – While still editing the sketch select the "**Split**" command from the menu "**Insert, Features, Split**." The current sketch will be added to the "**Trim Tools**" selection box. Click in the "**Cut Part**" button to split the solid body (in our case) into two parts.

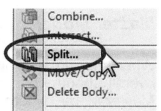

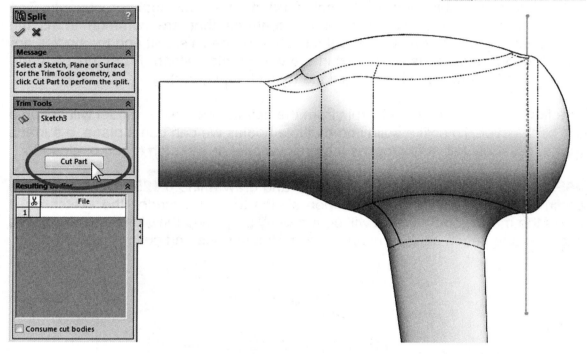

467. – After cutting the part we have two bodies listed in the "Resulting Bodies" list. With this split operation we'll make the back cover. Select the body for the back cover and double click in *<None>* to give the new file a name. Save this body as *'Hair Drier Cover'*.

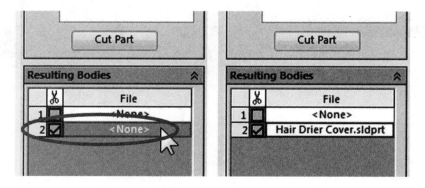

 If a system error is displayed after naming the split body saying "The above file name is invalid" you need to enable write access to the SolidWorks' install directory.

Before finishing the **"Split"** command, turn on the "Consume cut bodies" option. By doing this the body we are saving to an external file will be deleted from the master file. Click OK to split the part and finish.

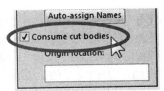

A *"Split"* feature is added to the FeatureManager of the master model, and the cover is now missing from it. The body we cut away is added to a new part with a single feature called *"**Stock_Parent_File**"*; it has an external reference to the *'Hair Drier'* part and its state is "In Context." Split parts follow the same external reference rules as top down design parts covered earlier in the book.

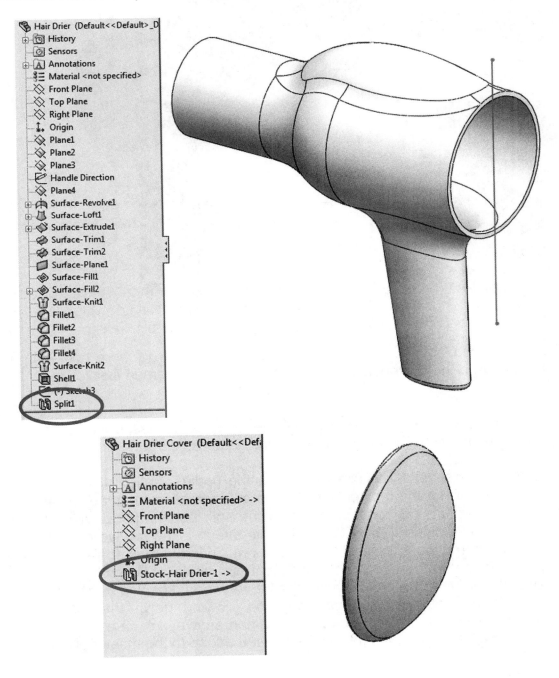

468. – The next split feature will cut the remaining body in two parts using the *"Front Plane"* of the part. Select the **"Split"** command again; add the *"Front Plane"* in the "Trim Tools" selection box and cut the part. Save both bodies and name them *'Hair Drier Left'* and *'Hair Drier Right'*. Turn on the "Consume cut bodies" option and click OK to finish.

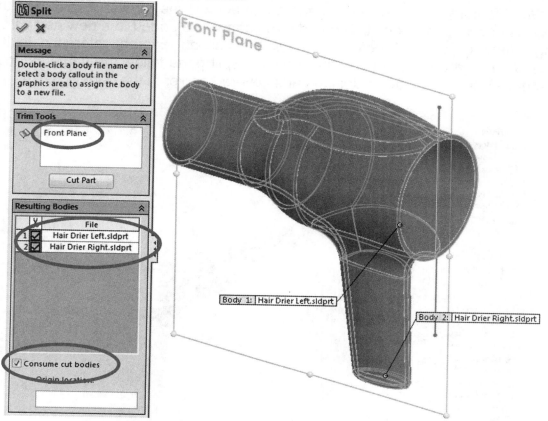

After completing the second **"Split"** command, there are no more bodies in our master part and the two bodies have been added to the named files.

 The "Split" command can use multiple surfaces, planes, or sketches at the same time to cut a part in multiple directions.

It is not necessary to use the "Consume bodies" option; the split bodies can remain in the master part, but be aware that features added *after* the "Split" feature will not propagate to the split bodies' parts. We used the option in this exercise to show the reader its effect. Likewise, we don't have to save to an external part; the split bodies can remain in the master file and we can keep adding features to each body as local operations. The problem with this approach is that the FeatureManager can be very long, keeping track of features will be difficult, rebuild times can be long slowing down the system, and eventually we'll have to save each body to a part anyway; might as well do it at a point where all the common features have been added to the master file.

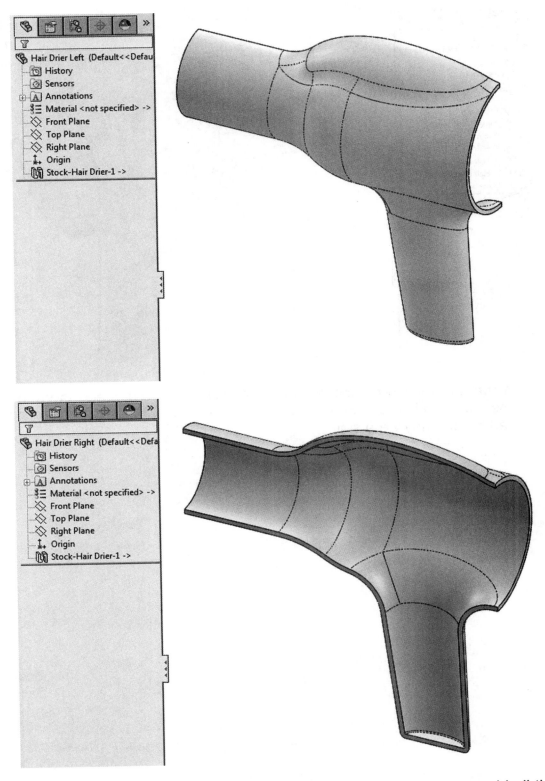

From this point on we can add more features to each part and then add all three parts to an assembly to finish our design.

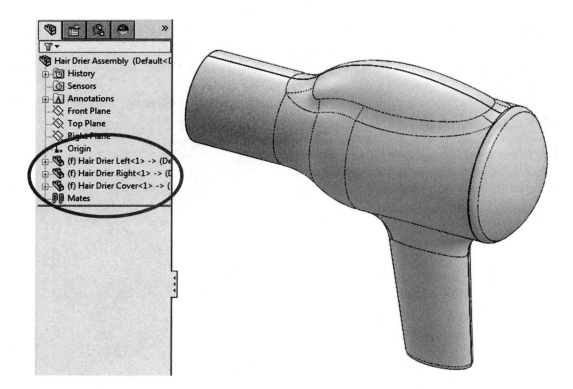

 When assembling components from a split part, we can add them to the assembly's origin and they will be fixed in the same place they were in the part before they were split.

CHALLENGE EXERCISE:

Using the hair drier files made in this lesson add the following features, select the best place to add them (in the master file before splitting the part *or* in the split body after saving them to a part). Remember, to add features before the split feature we have to roll back the FeatureManager.

Hint: Features present in more than one part can be added in the master file; features unique to each part can be added in the master file or the split part.

- Add vents to the back cover and sides (this is your chance to be creative ☺).

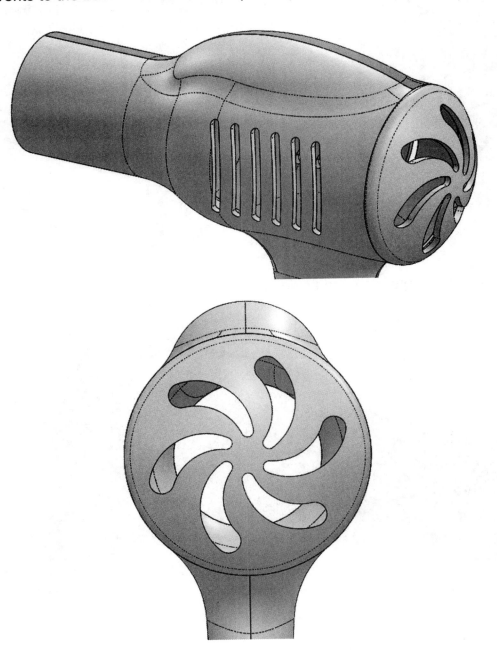

- Add a lip and cut extrusion to assemble the left and right sides, and between both halves and the back cover.

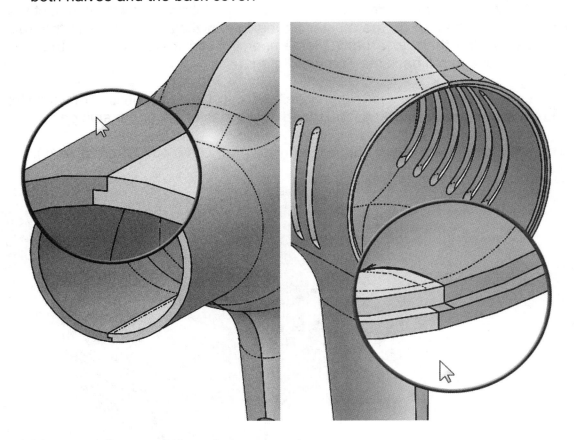

- Add cutouts for a switch and power cord.

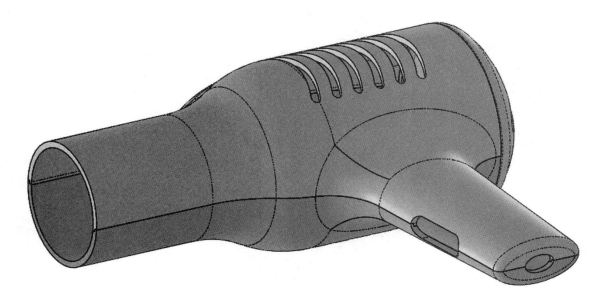

For our next example we'll design a bucket using surfaces and, at the end, convert them into a solid model. The process we'll follow to build it is not necessarily the most efficient way, but it will show the reader how to use several surfacing tools using a simple model.

469. – Make a new part and add the following sketch in the "*Front Plane.*" Don't forget the vertical centerline at the origin.

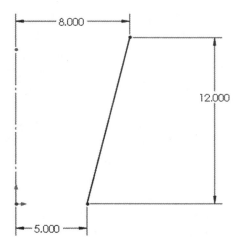

470. – Select the "**Revolved Surface**" command from the Surfaces toolbar, or from the menu "**Insert, Surface, Revolved**." Make the revolved surface 360 degrees. The surface will be revolved about the centerline.

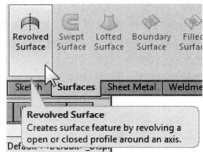

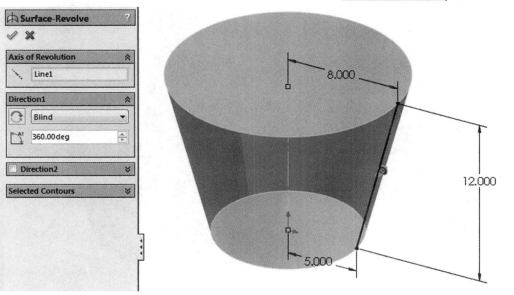

471. – Add a new sketch in the *"Front Plane"* and make a second revolved surface, this will form the bucket's spout. The top endpoint is horizontal to the top edge of the first surface.

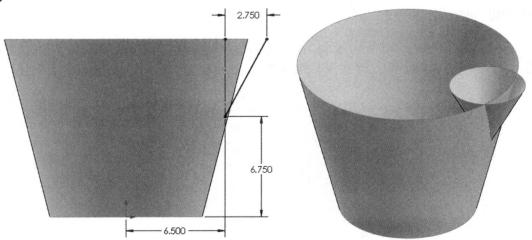

472. – After creating both surfaces we have to trim them. Select the "**Trim Surface**" command from the Surfaces toolbar, or the menu "**Insert, Surface, Trim**." In this step we'll use the "Mutual" trim option; this means that both surfaces will work as a trimming boundary. In the "Surfaces" selection box pick both surfaces. Activate the "Keep selections" option and select the main body of the bucket and the outside face of the second surface. Notice how the surfaces change color to preview the area of the surface that will be kept. Click OK to finish.

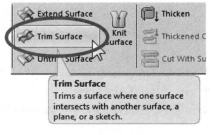

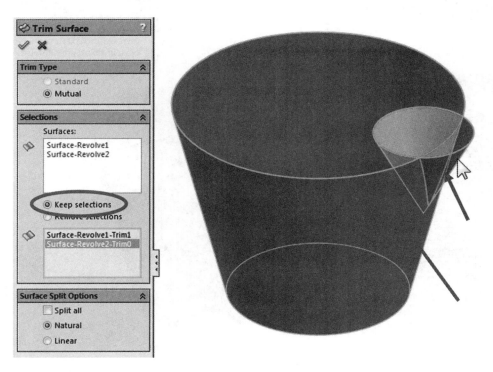

When surfaces are mutually trimmed they are merged into a single surface body. By default open surfaces have a light blue color on the open edges. Note in our surface the upper and lower open edges of the surface are blue and the edge where we the trimmed surfaces meet is black.

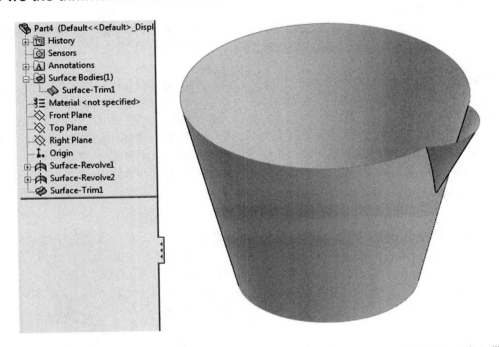

473. – Now add a fillet in the edge where both surfaces meet. Select the "**Fillet**" command, and add a 1″ radius fillet.

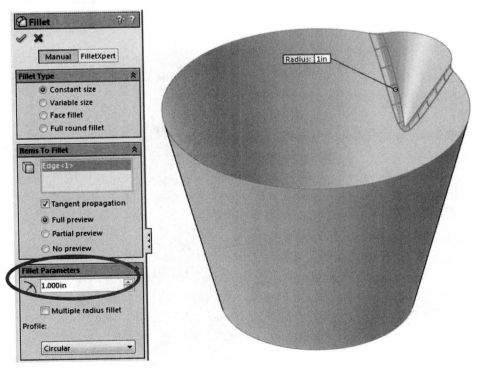

474. – We need to add a lip to the top of the bucket, but instead of creating a plane at the top of the bucket and making a sketch for the planar surface, we'll make a planar surface with the open edge, extend it, and finally trim it. Select the "**Planar Surface**" icon from the Surfaces toolbar or the menu "**Insert, Surface, Planar**."

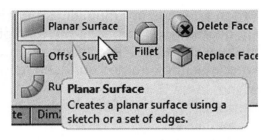

To make a planar surface, we need to select either a sketch that defines a closed area, or a closed loop of planar edges. To select the top edges (which are all tangent to each other), right mouse click in one edge and select the option "Select Tangency." This option automatically selects all tangent edges, which in our case are all the edges needed. The preview of the flat surface will be automatically shown covering the bucket. Click OK to complete it.

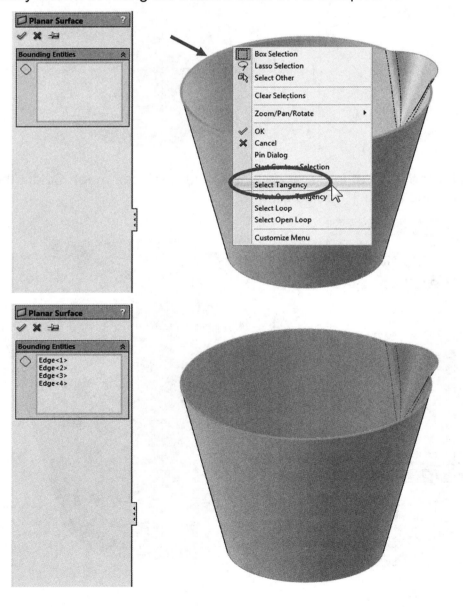

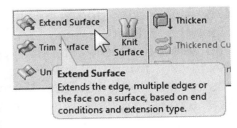

475. – This planar surface will be an intermediate step. What we'll do now is to make this surface bigger to build a lip at the top of the bucket, and then we'll trim and fillet the surfaces at the same time. Select the command "**Extend Surface**" from the Surfaces toolbar, or the menu "**Insert, Surface, Extend.**" The "**Extend Surface**" command helps us to make a surface bigger by extending its edges by a given distance, up to a point, or another surface. In case of a curved surface, we can extend its edges continuing the same surface, or linearly. Select the planar surface we just made and enter a distance of 0.875″. Since this is a flat surface, using the "Same surface" or "Linear" option makes no difference. Make sure the preview is going out and click OK to complete the surface.

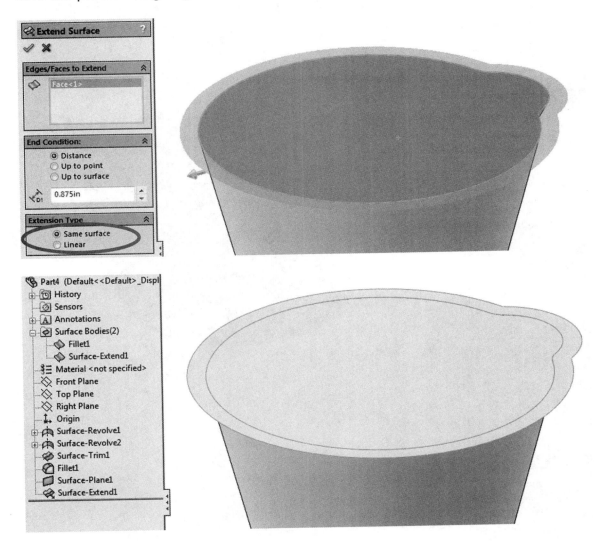

476. – The next step is to trim and round this surface to form the bucket's lip. Open the "**Fillet**" command and select the "Face fillet" type. Enter a 0.375″ radius and select the (now extended) planar face in the "Face set 1" selection box and the bucket's body in the "Face set 2" selection box. If a preview is not visible, click in the "Reverse Face Normal" buttons next to the selection boxes to make both arrows point towards each other.

Under "Fillet Options" make sure the "**Trim and attach**" option is selected; this option will trim the unused surfaces and knit the fillet with the other surfaces automatically merging them into a single body. Click OK to complete.

After completing the fillet, the planar face is trimmed, the fillet is blended along the perimeter of the bucket, and all surfaces are merged into a single body.

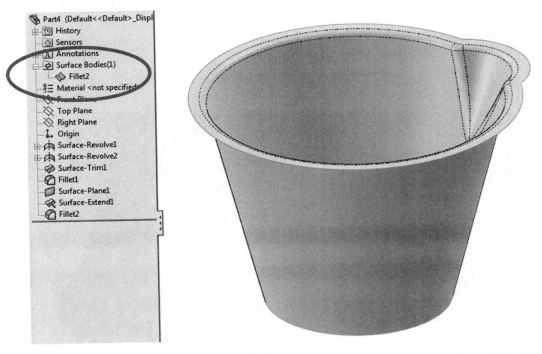

477. – Add a new planar surface at the bottom of the bucket using the "**Planar Surface**" command and selecting the round edge indicated. We'll have two surface bodies after this command. Click OK to continue.

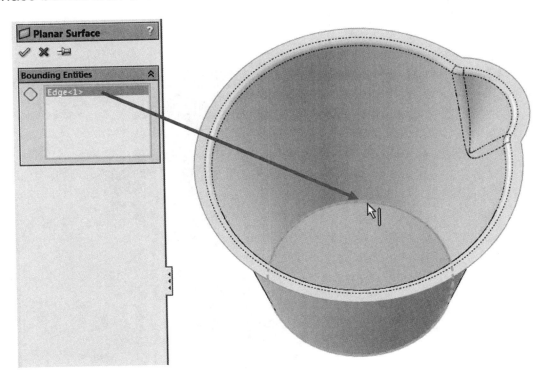

478. – Add a 0.625″ face fillet between the bucket and the surface just made using the same settings we used for the fillet at the top to trim and merge the surfaces. Make sure the normal direction arrows are pointing towards each other.

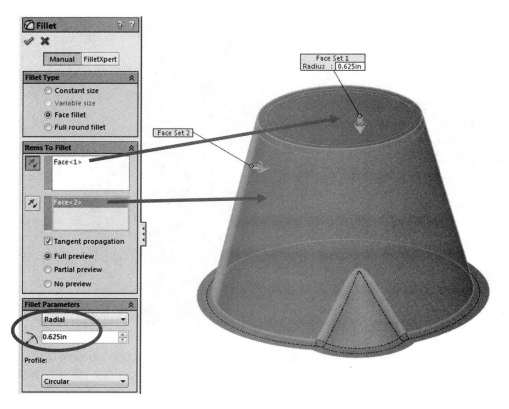

Now the bucket's surface is complete.

479. – To convert the surface to a solid, select the "**Thicken**" command from the Surfaces toolbar, or the menu "**Insert, Boss/Base, Thicken**." Make the surface 0.125″ thick, going *inside* the bucket and click OK to finish.

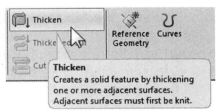

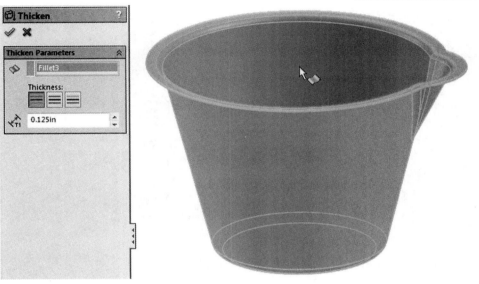

 Now our part is a single solid body (no "Solid Bodies" or "Surface Bodies" folder). The surface body was absorbed by the "**Thicken**" command.

480. – To complete the bucket's design, we'll add a couple of attachment points for a handle. To make the first extrusion we'll use the "**Extrude From**" option. Switch to a front view and make the following sketch in the *"Front Plane."* Notice the sketch is in the middle of the part.

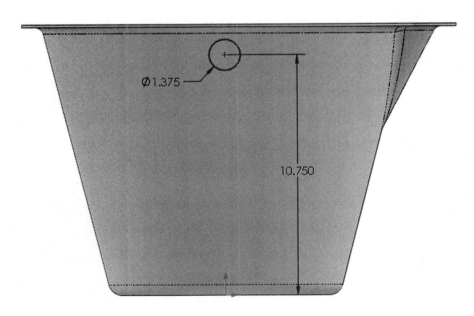

481. – Select the "**Extruded Boss**" command. In the "From" selection box, pick "Offset" and type 8.25″. In the "Direction 1" option box, reverse the direction if needed to point towards the bucket and use the "Up To Next" end condition. For this extrusion clear the "Merge result" checkbox, we'll model this feature as a multi body part. Click OK to finish.

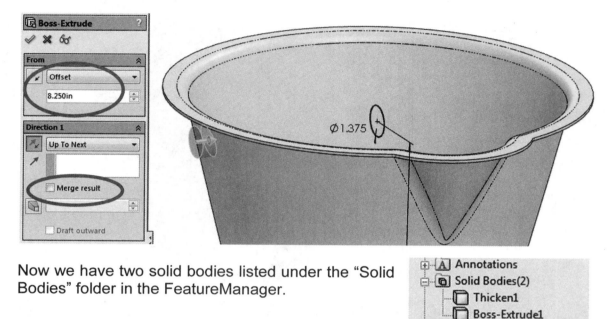

Now we have two solid bodies listed under the "Solid Bodies" folder in the FeatureManager.

482. – Add a new sketch in the front face of the boss we just made with a concentric circle 0.625″ in diameter. Make an extruded cut with a "Through All" option. Before finishing the cut extrude, under the "Feature Scope" options, clear the "Auto-select" checkbox and select the body made in the previous step, as we only want to affect this solid body with the cut and not the entire bucket. We could have added the hole in the extrusion, but chose to show how to do it this way instead to expand on multi body operations. Click OK to finish.

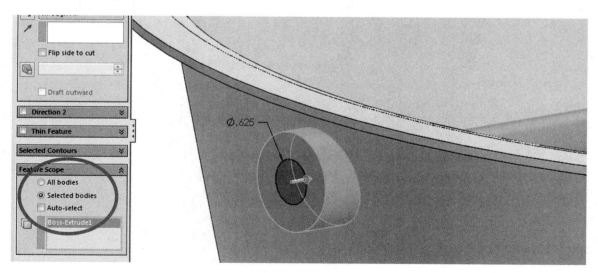

483. – In the next step we'll make a mirror of this body to have two attachment points for the bucket's handle. Select the "**Mirror**" command, and using the *"Front Plane"* as the mirror plane, select the *"Cut-Extrude1"* <u>body</u> to be mirrored in the "Bodies to Mirror" selection box. Do not merge the bodies at this time; we'll do that in the next step. Click OK to finish. Notice that selecting the "Bodies to Mirror" selection box automatically hides the features and faces to mirror boxes.

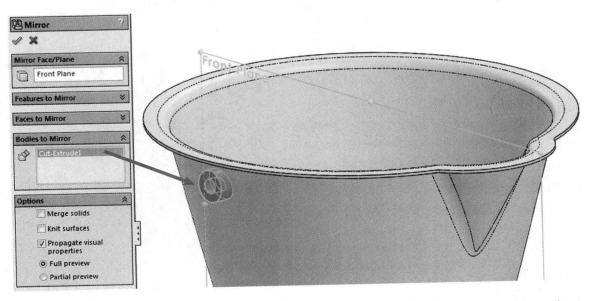

484. – At this point we have 3 solid bodies, and now we need to combine them into a single one. Select the menu "**Insert, Features, Combine**" or select all three bodies in the *"Solid Bodies"* folder and right mouse click to select the "**Combine**" option.

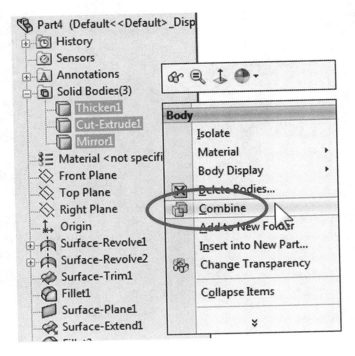

485. – Select "Add" under the "Operation Type" options and click OK to complete. This will merge all three bodies into one.

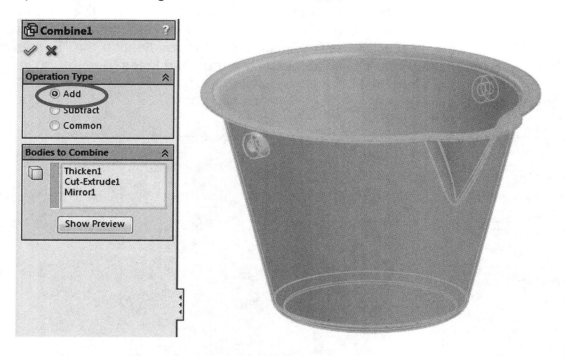

486. – The last step is to add a 0.25″ fillet to the bosses. We can select the face of the bucket or the two edges.

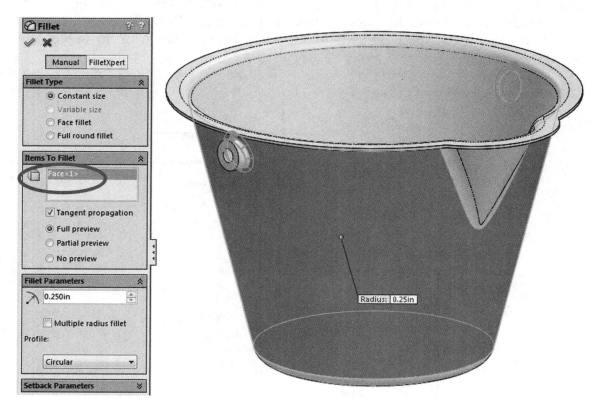

487. – Save the part as *'Bucket'* and close it; we'll use it later to make a mold. Our finished part now looks like this:

Notes:

More about Surfaces

When modeling more complex shapes, we may have to resort to surfaces in order to create auxiliary geometry as support for our models in the form of trim surfaces, start or end conditions for extrusions and cuts, to generate guide curves, etc. In the next model, we'll use a couple of surfaces to generate a curve that will be used for a sweep. Look at the following image of a sword's grip.

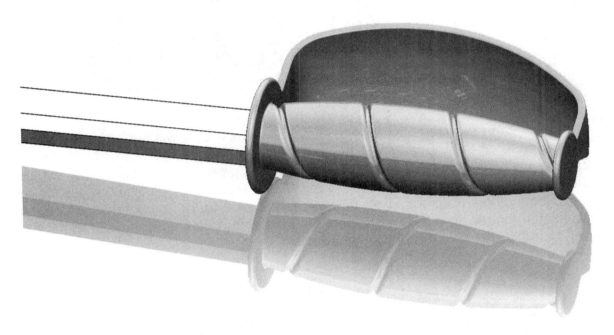

At first glance it seems to be a simple enough model, but if we look closer, we notice the groove in the grip is going in a spiral around an uneven surface, which makes for an interesting challenge which we are going to show how to create in the next few steps.

488. – The first thing we are going to do is to define the profile of the grip. Start a new part and make the following sketch in the *"Front Plane."* The horizontal line is a centerline (construction geometry) and the top is an arc.

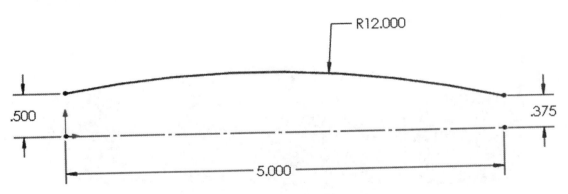

489. – Select the "**Revolved Surface**" command and make a 360 deg. revolved surface.

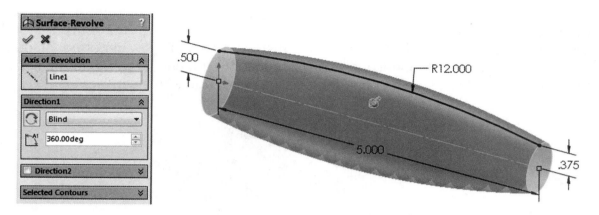

490. – For the next step, we'll make a swept surface that will twist along a path. The purpose of this surface will be to create an intersection with the revolved surface; we'll trim the surface and the resulting edge will be the path for a sweep cut. Make the following new sketch in the *"Front Plane"* and exit the sketch when done. It's a single horizontal line; this will be the path for our swept surface. Make the sketch longer than the part to make sure the surfaces intersect along the full length of the grip. Exit the sketch when done.

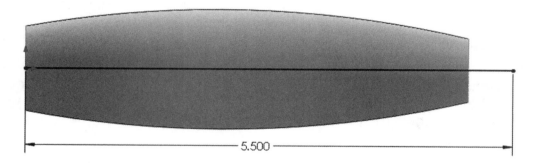

491. – Make a second sketch also in the *"Front Plane"*; this time it will be a vertical line as shown. Be sure to make this line higher than the revolved surface. Exit the sketch when done.

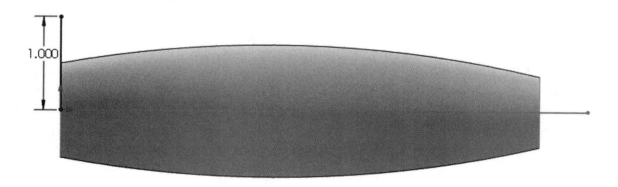

492. – Select the "**Swept Surface**" command from the Surfaces toolbar or the menu "**Insert, Surface, Sweep.**" Select the vertical line sketch as the "Profile" and the horizontal line sketch as the "Path." In "Options," under "Orientation/Twist" select "Twist Along Path," and under the "Define by:" list select "Turns" and enter a value of 4. The "Profile" sketch will twist along the length of the "Path" sketch 4 turns. Click OK to finish.

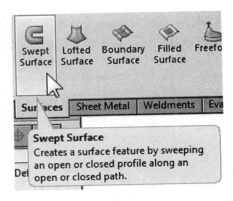

Swept Surface
Creates a surface feature by sweeping an open or closed profile along an open or closed path.

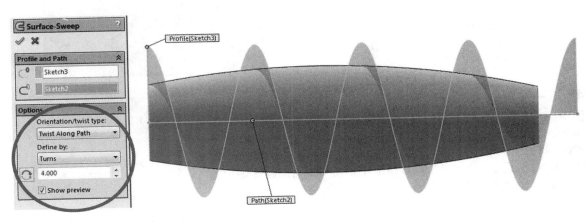

493. – What we are interested in is the curve along the intersection of both surfaces, so we'll trim the swept surface with the revolved surface, and use the edge as a path for a swept cut. Select the "**Trim Surface**" command; using the "Standard" trim type, select the revolved surface in the "**Trim tool**" selection box, activate the "Keep selections" option, and select the outside of the swept surface to keep it. Click OK to finish.

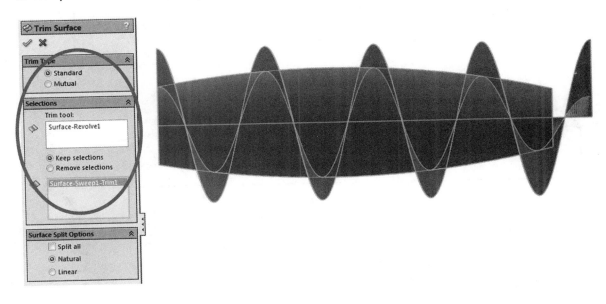

494. – Now select the revolved surface in the graphics area (or the Feature-Manager's *"Surface Bodies"* folder) and hide it.

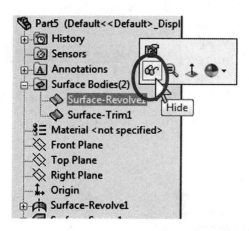

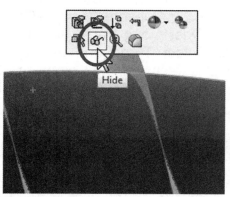

495. – Now we need to make a solid body for the grip. Show the sketch in the *"Surface-Revolve1"* and start a new sketch in the *"Front Plane."* Select the top line in the revolved surface sketch, use **"Convert Entities"** to project it into our new sketch and complete the sketch to form a closed contour.

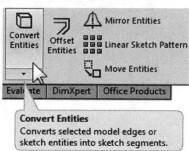

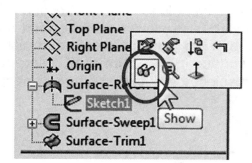

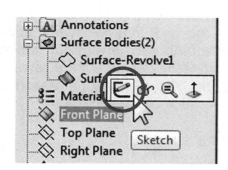

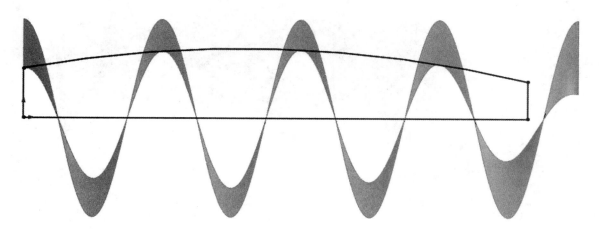

496. – Select the "**Revolved Boss/Base**" command from the Features toolbar (this will be a solid feature), and select the horizontal line to use as the Axis of Revolution to make the revolved feature about it. Click OK to complete it. Now we have a solid body and two surface bodies in our part. We can hide the revolved surface sketch after we are done.

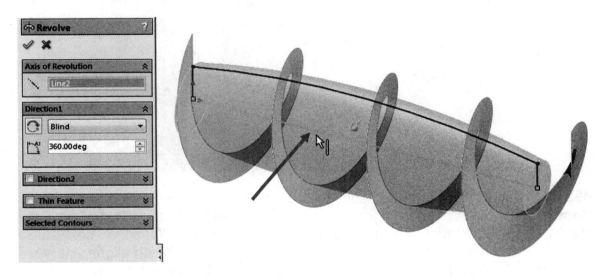

497. – After completing the solid we need to make the profile for a Sweep Cut. Add a new sketch in the *"Front Plane,"* close to the left of the swept surface, and draw it as shown. Draw an equilateral triangle (hint: make all three lines equal) and add the two construction lines as indicated, then add a "Pierce" relation between the endpoint of the horizontal construction line and the swept surface edge. Exit the sketch when done.

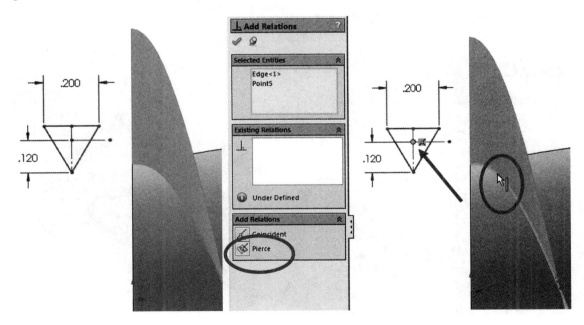

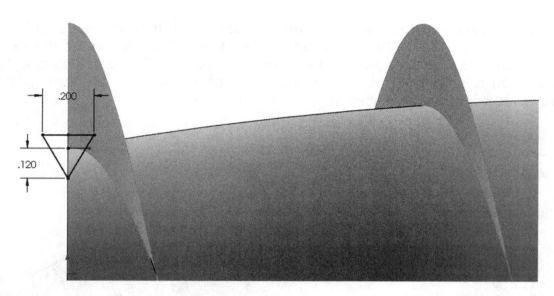

498. – Select the "**Swept Cut**" command from the Features toolbar or the menu "**Insert, Cut, Sweep**," and select the triangular sketch for the "Profile" and the inside edge of the trimmed surface for the "Path." In the "Options" box, select "Follow Path" in "Orientation/Twist type:" and "All Faces" in "Path alignment type." Turn off the "Align with end faces" option to prevent the swept cut from twisting and click OK to finish.

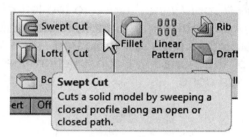

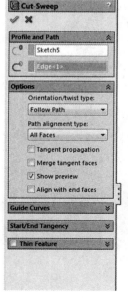

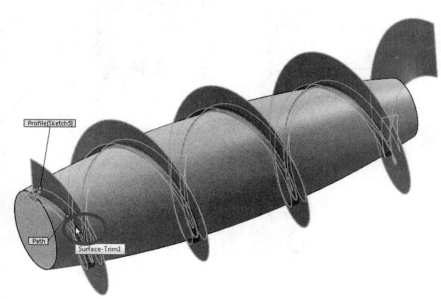

499. – Hide the trimmed surface to view the resulting cut in the solid body.

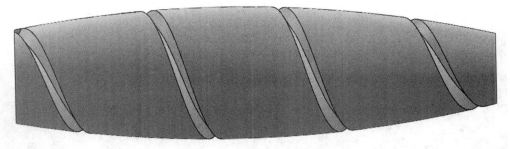

500. – Add a 0.035″ radius fillet at the bottom of the swept cut.

501. – Add a 0.075″ radius fillet to the two outer edges of the swept cut.

502. – As a final touch, add a guard and pummel to the grip using a revolved boss. Add a new sketch in the *"Front Plane"* and make a revolved boss. The Revolved Boss is overlapping the ends of the spiral cut to cover the ends.

503. – Save the file as *'Sword Grip'* and close it.

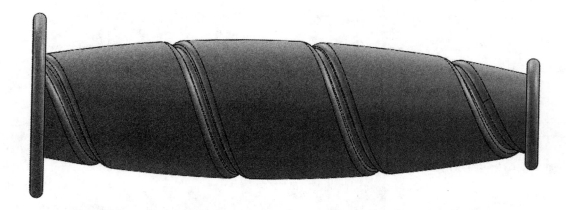

CHALLENGE EXERCISE:
Add a hand guard and a blade to the sword grip to finish your sword. Below is a suggestion. Use the "Display Pane" to change features, bodies, or part's appearance to your liking.

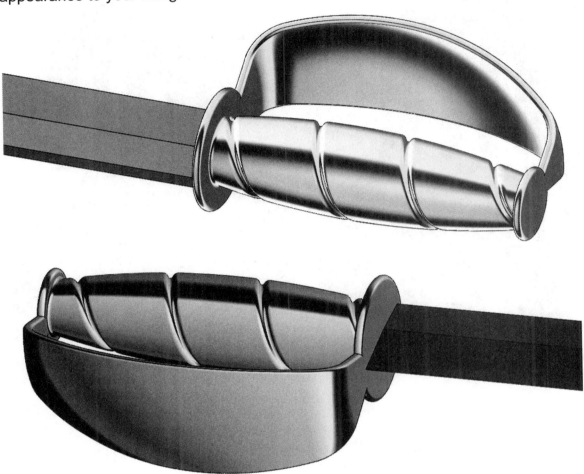

Exercise: Using the provided file *'Chair Surfaces.sldprt'* create the next model using the following instructions.

Sequence of operations:

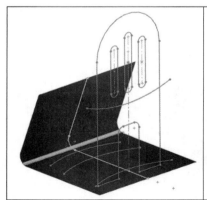

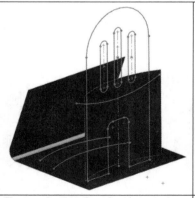

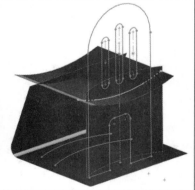

Extruded Surface using *"Legs Horizontal"* sketch, Mid plane, 28in	**Extruded Surface** using *"Legs Vertical"* sketch, up 20in	**Extruded Surface** using *"Seat"* sketch, going back 20in (direction 1), going front 4in (direction 2.)

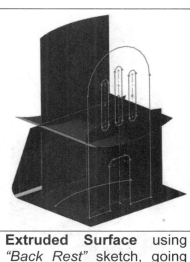	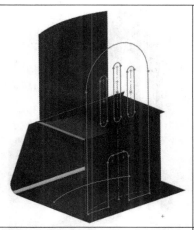	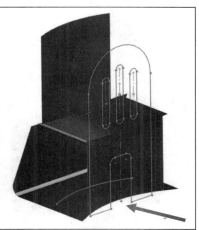
Extruded Surface using *"Back Rest"* sketch, going up 38in	**Trim Surface** Back Rest, Seat, Legs Vertical using Mutual trim	**Trim Surface** Legs Horizontal using previously trimmed surface
	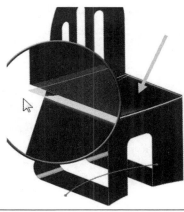	
Trim Surface using *"Chair Outline"* sketch	**Offset Surface**. Select the "Seat" surface and offset down 0.28in	**Trim Surface** Mutual trim legs and previous offset surface. Hide other surface

Trim Surface using *"Trim Under Seat"* sketch	Show hidden surface. **Knit Surface** both surfaces into a single surface	**Fillet** two edges 2in Radius

Fillet edge under seat in back, front legs at bottom (3 edges total) 1.5in radius

Thicken surface 0.25in thick, material going inside

Fillet corners under seat 1.5in radius

Fillet Select all tangent faces on both sides of the chair, fillet 0.1in radius

Save finished part and close.

Review and Questions:

a) What is the thickness of a surface?

b) When two surfaces intersect each other, they merge. True or False.

c) The option to make an extrusion along a specified direction is available **only** when extruding a surface. True or False.

d) Name two types of elements that can be used to trim a surface.

e) When we trim two surfaces using the "Mutual Trim" option, how many surface bodies do we end up with?

f) Name two types of geometry that can be used to create a planar surface.

g) Which of the following is a requirement to knit two or more surfaces:
   A - Surfaces must be touching at the edges.
   B - Surfaces must be open.
   C - Surfaces must be closed.

h) When thickening a surface, the option "Try to form a solid" can be used if:
   A - The surface is open.
   B - The surface is closed.
   C - The surface is a loft.

i) Name two elements that can be used to split a part into multiple bodies.

Answers:
a - Zero, it has no thickness.
b - False. When two surfaces intersect they need to be trimmed and knitted to be merged.
c - False. This option is available with any extrusion or cut feature, either solid or surface.
d - Another surface, a Plane or a sketch that will project onto the surface.
e - One, the mutually trimmed surfaces are merged automatically.
f - A closed 2D sketch or a group of closed edges on a plane.
g - A. Surfaces cannot be knitted if they are not touching at the edges.
h - B. When thickening a closed surface the option "Try to form a solid" will be enabled.
i - A Plane, a surface or sketch that extends past the solid body.

Notes:

Mold Tools

After designing plastic or cast metal parts, it is necessary to design a mold if the part is going to be made by plastic injection or forge. Mold making is a manufacturing specialty where experience plays as big a role as preparation and study, and it's not an easy trade to master.

When designing a mold, a large number of factors must be taken into consideration, including: the shape and size of the part, the material to be used (plastic, resin, metal, etc.), the process, the tooling necessary to make the mold, etc. In other words, it's a complicated process.

The purpose of this chapter is not to teach the mold making trade, as that by itself is enough to fill several books and there would still be much more to learn. Our intention is to show the user the tools available in SolidWorks for mold design, learn and understand how and why to use them, and briefly touch on design considerations that can affect manufacturing of a molded part, like draft and parting line selection.

Be aware that a complete mold design includes a large number of components starting with a mold base; from there we have to design cooling lines, runners, gates, add hardware components like springs, nuts, bolts, ejection pins, dowels, O-rings, slides, lifters, etc. Our focus in this chapter will be limited to showing how to create the core and cavity, side cores and inserts, as most of the time these are the most difficult tasks; adding the rest of the components and design elements to a complete, manufacturing ready mold design is the complex process we were just referring to.

In this chapter we'll learn how to make molds for a few of the parts we have previously made as well as a couple others where we'll learn different mold making techniques. This will give the reader a good idea of what needs to be done and, more importantly, how these techniques can be used in different situations. More complex parts will require additional steps, multi part molds, inserts, sliders and such, but that's precisely where the skill and knowledge of the designer comes into play. We are only showing how to use the most common tools available for that purpose; the rest of the design has to do with mold making knowledge, experience, and skill.

Notes:

504. – We'll start by making a simple two part mold for a business card holder. Locate the *'Card Holder.sldprt'* from the included files and open it.

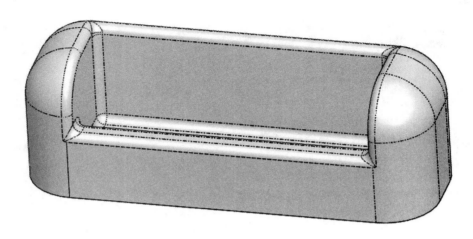

505. – One of the most critical and important details when making a mold is to know if we can get it out of the mold. The first thing we need to do is to make a **"Draft Analysis."** A draft analysis will tell us if we have the required draft in the 'vertical' faces to eject the part from the mold. By 'vertical' we are referring to the faces parallel to the direction the part will separate from the mold, or 'direction of pull.' Picture it this way: when we make a cake or a pie, the mold's vertical walls have an inclination, or **"Draft"** angle, to facilitate releasing the cake from the mold. If the walls were truly vertical, it would be very difficult to get the cake out of the mold in one piece. Even worse would be to have the walls inclined inside; this is a condition known as a negative draft, and the cake would never come out without breaking the part or the mold.

| Pull | Positive Draft (OK) | No Draft (Not OK) | Negative Draft (Not OK) |

The required draft to easily separate the part from the mold will vary depending on the part's size, the material used, whether the part has texture or not, the process used, etc. What we know for sure is that we <u>always</u> want to have a positive draft and as much as possible. If our part's design forces us to have a negative draft in one or more faces, we have to use a different approach, like a multi part mold, side cores or a combination of both. Later in this lesson we'll cover how to deal with this situation.

506. – To start, make sure the "**Mold Tools**" toolbar is open in the Command-Manager. (Right mouse click in a tab and select "**Mold Tools**" or turn on the **Mold Tools** toolbar.)

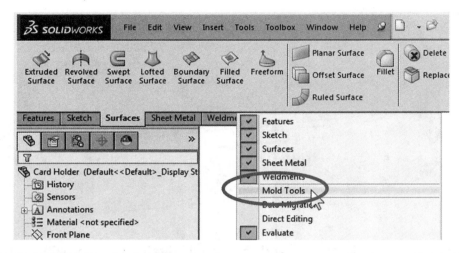

507. – In the Mold Tools toolbar we have some commands available in other toolbars, for example, from Features, Surfaces, Evaluate tools and a few mold specific tools.

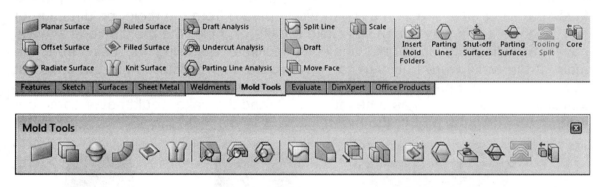

 When we design a mold, surfaces are used to create the Core and Cavity. A Core is defined as the 'male' half of the mold, and the Cavity as the 'female' half of the mold. Some molds only have either a core or a cavity.

508. – The first step is to analyze our part to make sure we have at least 3 degrees draft in the vertical walls. This is an arbitrary value that we'll use for our example, as values depend on material, geometry, etc. Select "**Draft Analysis**" from the Mold Tools toolbar or the menu "**View, Display, Draft Analysis**."

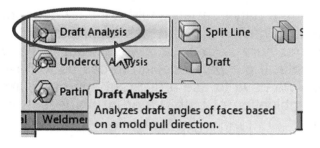

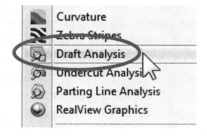

For the "**Draft Analysis**" tool we have to select a face or plane <u>perpendicular</u> to the desired direction of pull, or an edge along the direction of pull. In our case, we'll use the bottom of the *'Card Holder'* for our direction of pull (we could also use the *"Top Plane".*) Enter 3 degrees in the "Draft Angle" value box. Immediately after selecting the direction of pull, the part is color coded to let us know which faces have the required draft and which ones don't. Our part's faces are colored with:

Green	Positive draft, indicated by the arrow direction
Yellow	Faces with less than the required draft angle
Red	Faces with negative draft

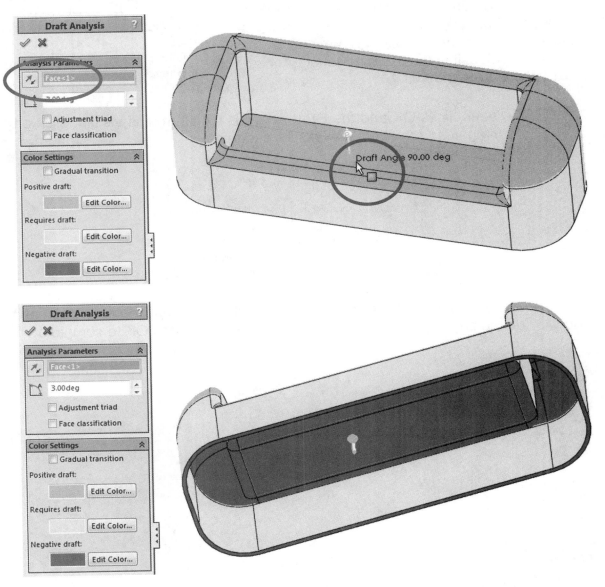

 Note that as we move the mouse around the part, the angle to the direction of pull is displayed next to the mouse pointer.

"**Positive**" and "**Negative**" are only telling us what side of the mold a face will be made with, and the direction can be reversed using the "Reverse Direction" icon next to the "Direction of Pull" selection box. When we have a Core and Cavity mold, "Positive" faces will be on one side, "Negative" faces will be on the other. In our case the green faces will be made in the Cavity, and the red faces in the Core. What we are interested in are the yellow faces that need to be given a draft to properly release the part from the mold, and these are the ones that have to be modified to be either "Positive" (green) or "Negative" (red). Once we have

identified the faces that need to be given a draft, Cancel the "**Draft Analysis**" tool. If we click OK, the draft analysis colors remain visible and we can see which faces need to be modified as we edit the part. The first feature to be modified will be the "*Boss-Extrude1*." Select it in the FeatureManager or the graphics area and edit it.

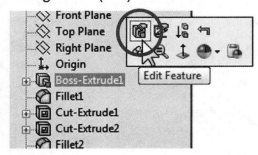

 If the user clicks OK and the Draft Analysis view remains visible, click on its icon again to turn it off; it works as a toggle.

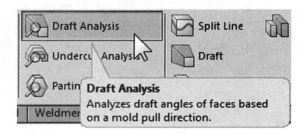

509. – In "Direction 1," activate the "Draft On/Off" option and enter 3 degrees. You will see the faces in the preview 'leaning' in the part. Click OK to complete.

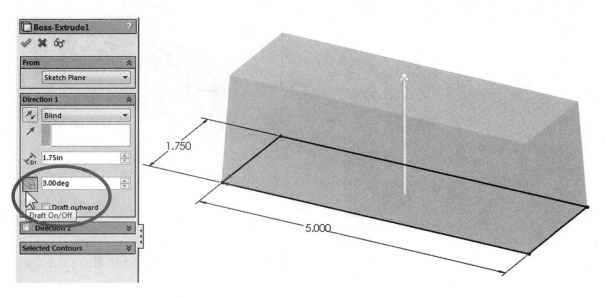

410

DRAFT ANALYSIS VIEW: After completing this change make a Draft Analysis again. We can see that only a few faces still require draft and are still yellow: some in the top (Cavity side) and some in the bottom (Core side).

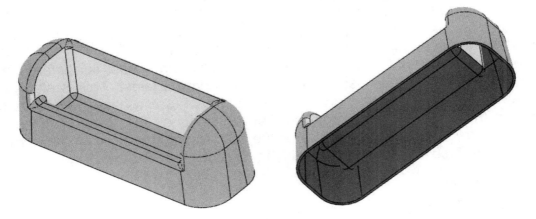

510. – Since not all faces can be given a draft at the time a feature is created, we

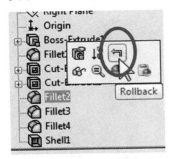

may have to add it as a secondary operation. To add a draft to the remaining faces that need it, we'll use the "**Draft**" command. Faces to be drafted must not be connected to a fillet or the draft will fail. Analyzing our model, we can see that we have a few operations, then fillets and a shell at the end to make the part hollow. What we need to do is to 'rollback' the model just before the *"Fillet2"* feature to be able to add the draft. Since the part is shelled at the end, fixing the faces in the top (Cavity side) will also take care of the faces in the bottom (Core side). Select the *"Fillet2"* feature, from the pop-up menu select "**Rollback.**" By doing this we go *'back in time'* in the FeatureManager, to the moment right before the fillets and the shell were created. By doing this, we are able to add a new feature at this position in the FeatureManager and when we are finished we can "**Roll Forward**" to finish building the part, essentially un-suppressing the remaining features. Notice the Rollback bar is now located just below the *"Cut-Extrude2"* feature and the remaining features are grayed out below it ("Suppressed").

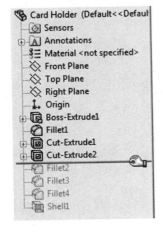

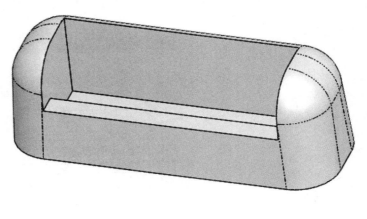

511. – Select the "**Draft**" command from the CommandManager or the menu "**Insert, Features, Draft**." The "**Draft**" command is available in the Features or Mold Tools toolbars. (To make selection easier, the "**Draft Analysis**" view is off.)

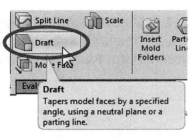

Draft
Tapers model faces by a specified angle, using a neutral plane or a parting line.

The faces we need to draft are the four inside faces, where business cards are placed. Select "Neutral Plane" in the "Type of Draft" selection box. The "Neutral Plane" is a reference flat face or plane that will be used to measure the draft against. This is also where the faces to be drafted will start to *'incline'*; think of it as a hinge where the face starts to move. Select the bottom face of the cavity as the "Neutral Plane" and enter 3 degrees in the "Draft Angle" value box. Notice the arrow in the corner of the selected face is pointing up; this is the "**Direction of Pull**" we want.

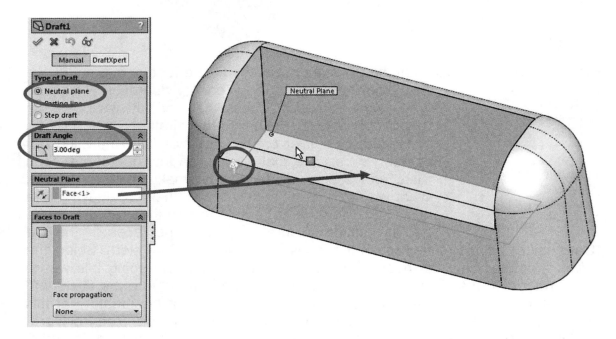

 The draft's "Direction of Pull" can be reversed using the "Reverse Direction" icon next to the "Neutral Plane" selection box. Reversing the direction of pull inverts the direction of the draft.

512. – Add the four vertical faces in the "Faces to Draft" selection box and click OK to finish. Notice the faces moving after completing the draft. Make sure the draft direction arrow is pointing up.

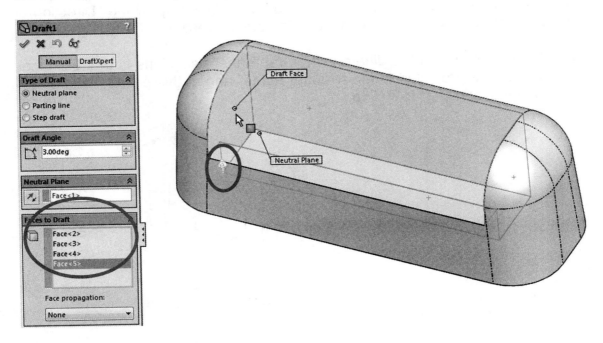

Top view before the draft feature...

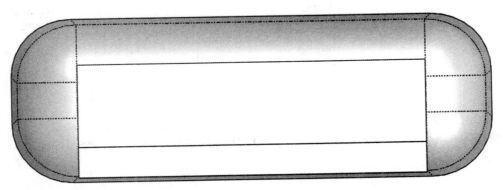

... and after the draft.

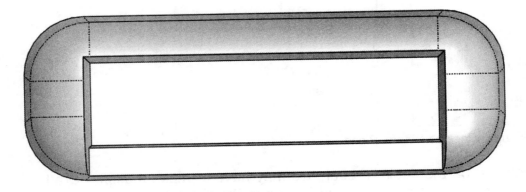

More information about the "**Neutral Plane**":

The plane we select as the "Neutral Plane" will define how the drafted faces will be inclined. Look at the following images and the neutral plane selections. In all cases the "Direction of Pull" is pointing up. If the "Direction of Pull" is reversed the draft will be added in the other direction.

The difference between the different "Neutral Plane" selections is that the faces to be drafted are projected to the "Neutral Plane" and at this point is where they will be 'hinged' and start to incline.

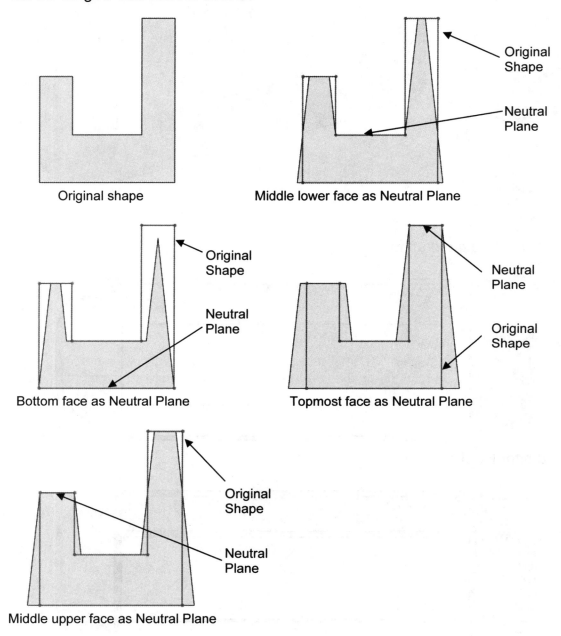

Original shape

Middle lower face as Neutral Plane

Bottom face as Neutral Plane

Topmost face as Neutral Plane

Middle upper face as Neutral Plane

Depending on the desired result, we may have to make multiple "Draft" features with different "Neutral Plane" and "Faces to Draft" selections.

513. – Now back to our part. Select the "Rollback bar" and drag it all the way to the bottom. If you get an error message, dismiss it for now to continue.

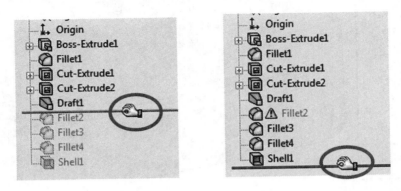

514. – By changing the faces of the model we get an error in the *"Fillet2"* feature. The error is caused because this fillet cannot find edges that were eliminated after we added the draft feature. To fix this error, edit the fillet and reselect the missing edges.

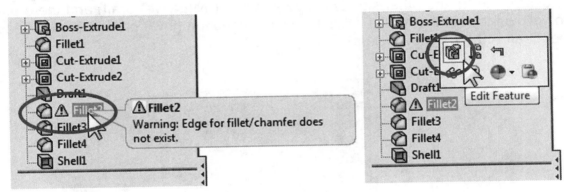

Notice the edges that are missing are highlighted with a faint red dotted line. Erase the missing edges in the selection box and click OK when done to finish.

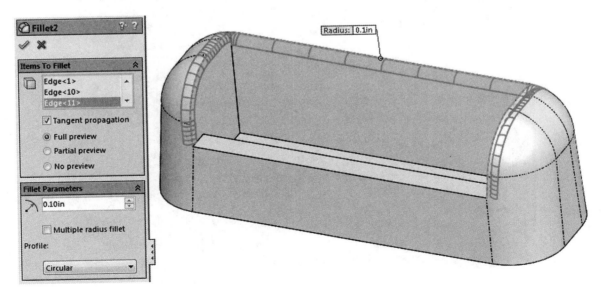

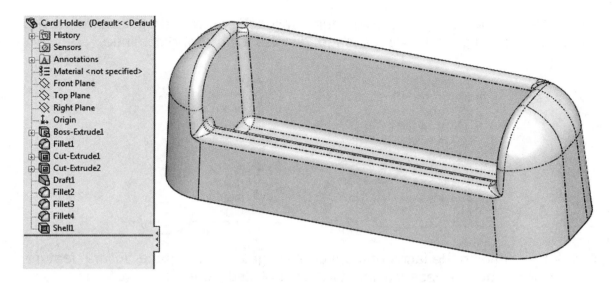

515. – Use the "**Draft Analysis**" view again and verify that all faces have the required 3 degrees of draft to continue. One side is now completely green ("Positive" draft), the other side is completely red ("Negative" draft) and there are no yellow faces in the model. Click Cancel to continue (if we press OK we'll keep the "Draft Analysis" view colors).

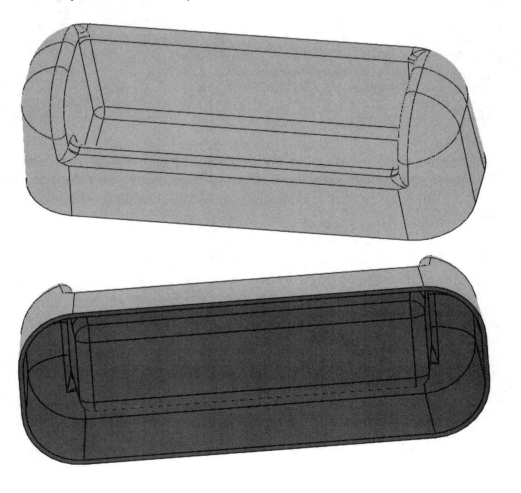

516. – When we make a mold for a plastic part, the designed part is usually scaled up to compensate for part shrinkage in the mold when the melted plastic solidifies. To scale the part (up or down), select the "**Scale**" command in the Mold Tools toolbar or the menu "**Insert, Features, Scale**."

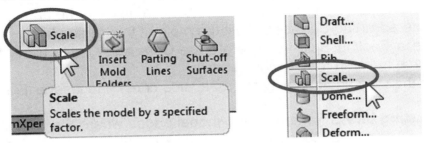

517. – Scale the part about the "Centroid" using a scaling factor of 1.03; this will make the part 3% bigger to compensate for mold shrinkage. Click OK to complete. Notice the part is slightly bigger when finished and a "Scale" feature is added to the FeatureManager.

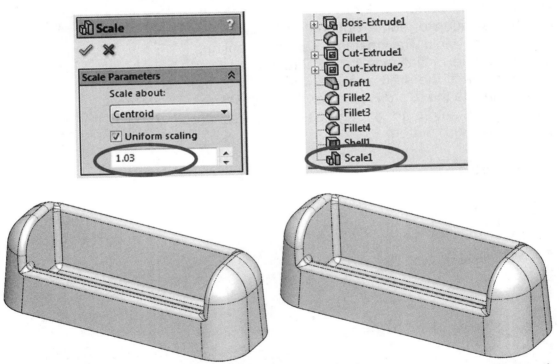

See the difference? Roll back and forward the *"Scale1"* feature to see the change; subtle, but visible.

The shrinkage values used to scale a part before creating a mold vary widely depending on the material used and the part's geometry. For example, if a part is elongated, it may shrink differently along one axis than the other, and in a case like this a non "Uniform scaling" option is used. As far as the material, the shrinkage value is a physical property of the material used in the molding process and is usually provided by the manufacturer.

518. – After scaling the part, we are ready to make the mold. In general terms, the process to make a mold in SolidWorks is as follows:

- Define a Parting Line where the mold halves meet.
- If needed, create Shut-off Surfaces to close holes in the part.
- Create a Parting Surface to split the mold at the parting line.
- Make a Tooling Split to generate the Core and Cavity using the Parting Surface made in the previous step.
- Split the core and/or cavity to create side cores and inserts if needed.

519. – A **parting line** is formed by the model edges where the two halves of the mold meet. An easy way to identify the parting line is using a "Draft Analysis." In simple terms, the parting line will be the edges where the faces with "Positive" draft (green) meet the faces with "Negative" draft (red). In a simple two piece mold (single core and cavity) like this, the parting line is usually automatically found and selected. Select the "**Parting Line**" command from the Mold Tools toolbar or the menu "**Insert, Molds, Parting Line**."

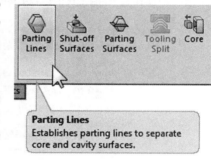

The command works similar to the "Draft Analysis," with the addition of some options and a selection box. In the "Mold Parameters" selection options, select a plane or flat face for a "Direction of Pull" like we did with the draft analysis. Select the same flat face inside the *'Card Holder'* or the *"Top Plane"*; set the draft angle to 3 degrees and press the "**Draft Analysis**" button.

Parting Lines
Establishes parting lines to separate core and cavity surfaces.

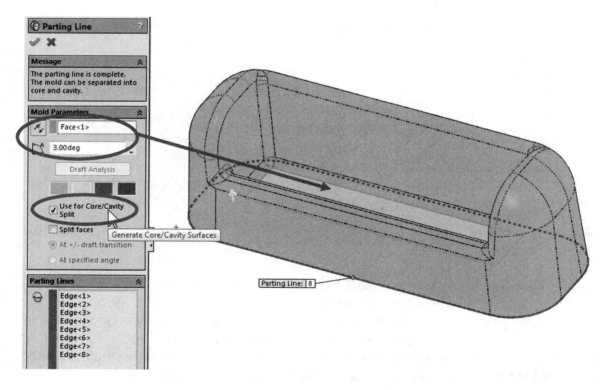

Make sure the "Use for Core/Cavity Split" option is checked. This option will automatically generate the surfaces needed for Core and Cavity creation (covered later). In the "Parting Lines" selection box the edges that make up the parting line are already selected, and the message "The parting line is complete" is highlighted in green at the top. Click OK to generate the parting line and finish.

520. – After completing the parting line, a new feature *"Parting Line1"* is added to the FeatureManager, and in the *"Surface Bodies"* folder two sub-folders are created, each with a surface for the Core and Cavity. These surfaces are automatically created with the **"Parting Line"** command. In this example, the *"Cavity Surface Bodies"* are the faces with a "Positive" draft (green) and the *"Core Surface Bodies"* are the faces with a "Negative" draft (red). The parting line remains visible on the screen and can be hidden like other features, if needed.

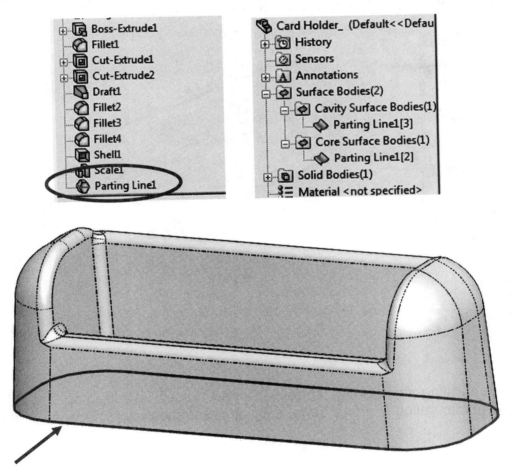

 The next step would be to close any holes in the model with **"Shut Off Surfaces,"** but since our part has no holes going from one side to the other, this step is not necessary.

521. – The next step is to generate the parting surface. The parting surface can be generated manually or automatically and serves as a boundary to separate the mold using the Core and Cavity surfaces. Think of it as the surface used to split the mold in two. The parting surface is connected to the parting line and (generally) radiates away from it perpendicular to the "Direction of Pull" used in the parting line. Select the "**Parting Surfaces**" command from the Mold Tools toolbar or the menu "**Insert, Molds, Parting Surfaces.**" In this part the parting line lies in a flat surface, and it's easy to generate. Select the option "Perpendicular to pull" (the parting line is automatically selected) and enter a value of 1″ for the "Parting Surface" distance value. The option "Knit all surfaces" automatically merges all the faces generated into a single surface. Click OK to finish when done.

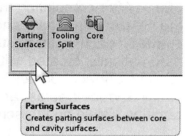

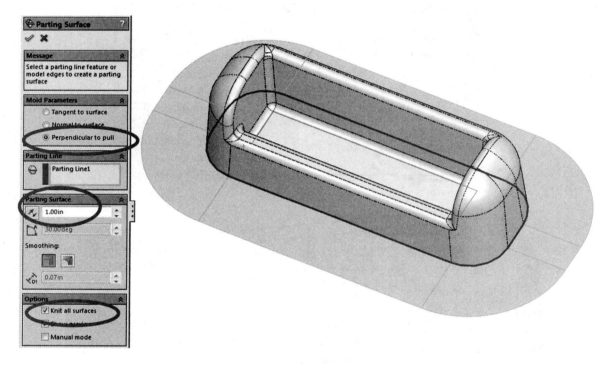

After generating the parting surface, a new folder is added to the *"Surface Bodies"* folder called *"Parting Surface Bodies"* and the new surface is listed here.

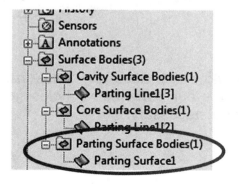

522. – The next step is to generate the tooling split. In this step we make the solid bodies for the core and cavity. To make these bodies, we need to make a sketch perpendicular to the "Direction of Pull" that fits inside the parting surface; the reason is that the parting surface will be used to split the bodies. The generated core and cavity bodies will be defined by this sketch. We can make the sketch first and then make the tooling split, or start the "**Tooling Split**" command and then make the sketch at that time. In this example, we'll make the sketch first. Select the parting surface and add a new sketch in it. Use "**Convert Entities**" to project the parting surface edges and exit the sketch.

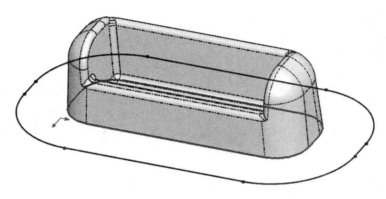

523. – Now, select the "**Tooling Split**" command from the Mold Tools toolbar or the menu "**Insert, Molds, Tooling Split.**"

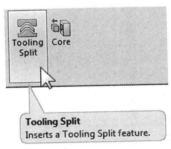

 The "**Tooling Split**" command is available after a parting line feature is added to the model.

524. – When we use the "**Tooling Split**" command, we are first asked for a plane or flat face to add a sketch, or an existing sketch (if a sketch is pre-selected this step is skipped). Select the sketch we just made in the graphics area or in the FeatureManager to continue.

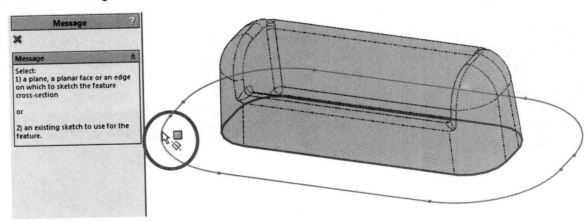

525. – After selecting the sketch, we have to enter the "Block Size" for the core and cavity measured from the sketch plane. In the "Core," "Cavity" and "Parting Surface" selection boxes the corresponding surface bodies are automatically selected. This is the reason why the surfaces are listed under the *"Surface Bodies"* sub-folders; alternatively we can manually select the surfaces if needed. Enter a block size big enough to completely enclose the *'Card Holder'* part, in our case 2.25″ up and 0.5″ down will be enough. Click OK to finish.

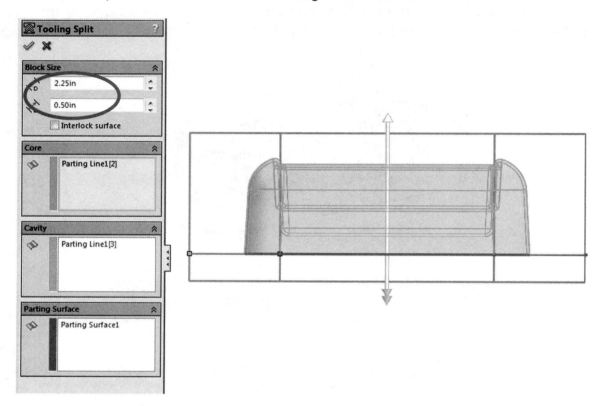

 Knowing the size of the mold base that will be used, or the sizes available, usually helps when deciding how big to make the blocks. This way they will be big enough to fit in the mold base and not bigger than needed.

526. – Now the core and cavity are finished. Hide the *"Parting Line1"* feature by selecting it and clicking in the "Hide" command. In the *"Solid Bodies"* folder we now have three bodies: one is the part, one is the core and the third is the cavity. At this point, the core and cavity bodies can be inserted into a new part file to continue designing the mold and adding the rest of the mold specific features, like cooling lines, injection ports, ejector pins, etc., using the same approach we used when working with a master model.

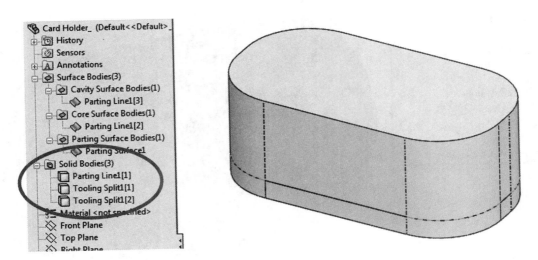

To finish our design, save the core and cavity bodies to a new part using the "Insert into New Part" command from the right-mouse-button menu. Name them *'Card Holder Core'* and *'Card Holder Cavity'* to continue. When finished save and close all the parts.

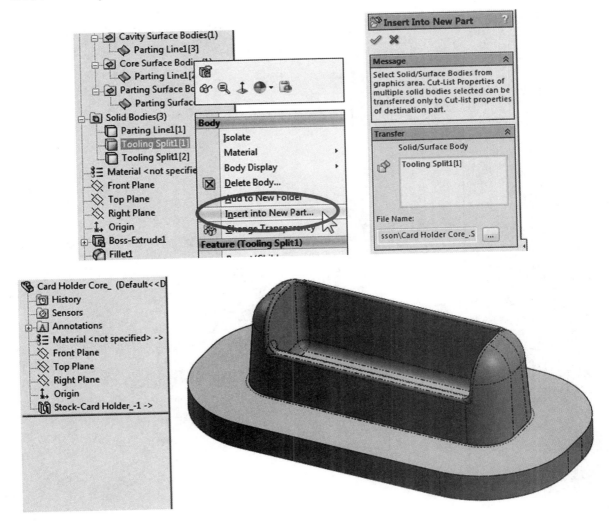

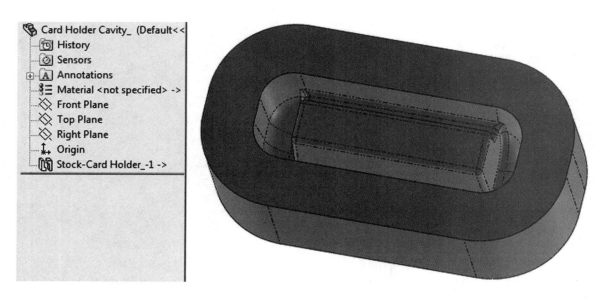

In the original part, the solid bodies and surface bodies can be hidden to see the result before saving the bodies to a new part file.

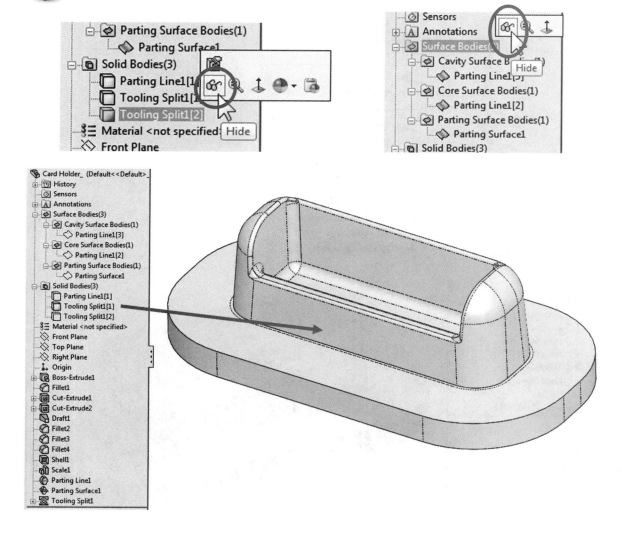

527. – For the second mold exercise, open the *'Hair Drier Cover-Mold'* part from the included files.

 This part's external references have been broken to avoid potential conflicts with user generated files.

528. – This part is similar to the previous one in the sense that it can be made with a simple two-part mold, but this one has holes going through the part (vents) that the previous part did not. This will help us learn how to work with a part that has holes in it. The first thing we need to do is to verify if our part's faces have at least three degrees of draft. (Remember this value is arbitrary for our exercises.) Select the "**Draft Analysis**" command from the Mold Tools toolbar and select the *"Right Plane"* for the "Direction of Pull."

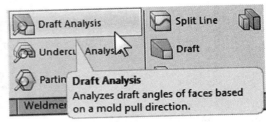

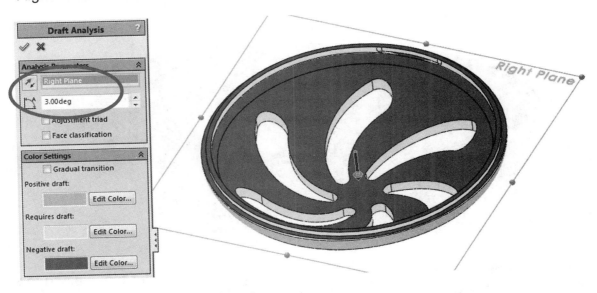

From the analysis we can see that the ventilation holes and the extrusion built to assemble the cover, *"Boss-Extrude1"* don't have the required three degrees of draft, and therefore, we need to fix them before we can make a mold. Click Cancel to continue. (If we click OK, the Draft Analysis colors will remain visible.)

529. – Select the *"Boss-Extrude1"* feature and edit it.

530. – Activate the "Draft On/Off" option under the "Direction 1" group and enter a value of 3 degrees. Click OK to finish.

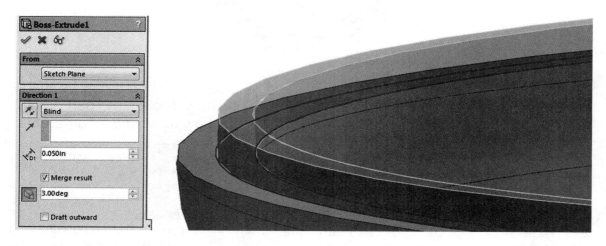

531. – Making a "**Draft Analysis**" again reveals a small cylindrical (yellow) face on the inside that needs to be drafted.

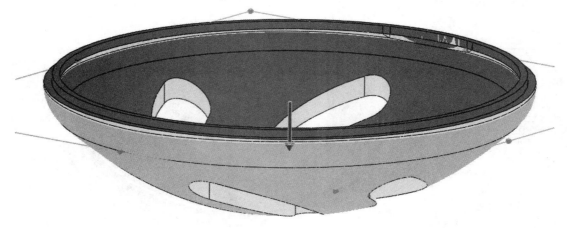

Upon closer examination we can see that there are a few smaller faces along the internal surface that also need to be drafted. Cancel the analysis to continue.

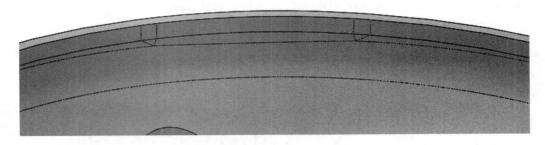

532. – To fix the draft and the smaller faces, we'll add a face fillet. Select the "**Fillet**" command using the "Face fillet" option. For the "Face Set 1," select the inside of *"Boss-Extrude1"* and for "Face Set 2," the rounded face inside. Using a 0.25″ radius fillet will blend both faces and eliminate all the faces between them.

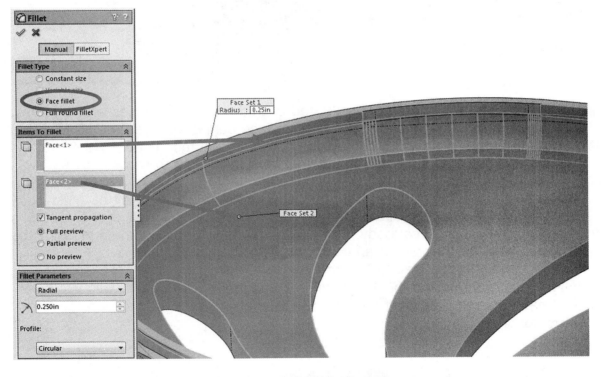

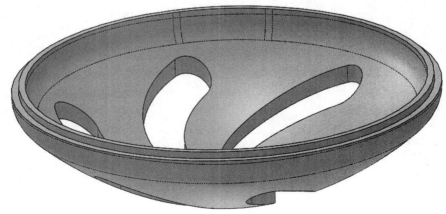

533. – Run the "Draft Analysis" tool again to reveal the remaining faces that need to be drafted using the same "Direction of Pull."

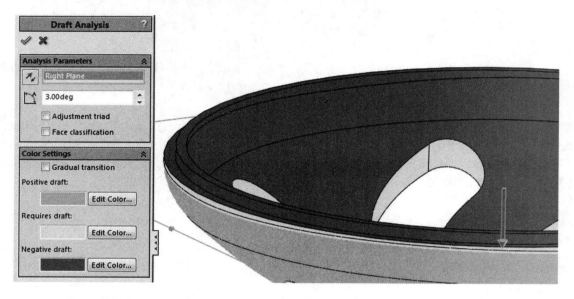

The only faces that remain to be drafted are the ventilation holes and a narrow cylindrical face on the outside. The outside face cannot be fixed as the previous one using a face fillet. In this case we'll use a different technique; we'll delete the faces, add a new one with the correct draft, and finally knit the faces into a solid.

534. – In SolidWorks we can delete surface body faces, as well as solid body faces. In the case of a solid, when we delete a face, the rest of the model is converted into a surface body. Select the "**Delete Face**" command from the Surfaces tab in the CommandManager, or the menu "**Insert, Face, Delete.**" Using the "Delete" option, locate and select the four outside faces that don't have the required draft and the fillet, and click OK to continue.

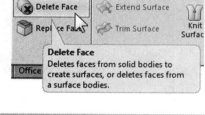

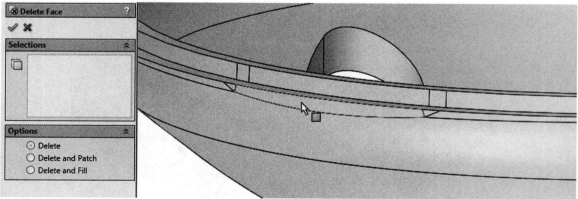

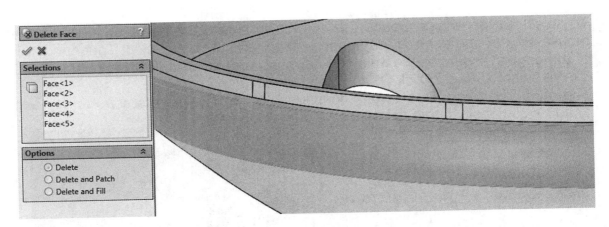

 The difference between "Delete and Patch" and "Delete and Fill" option is that "Patch" will add faces and trim them to fill the gap. "Fill" will add a single surface to fill the gap. In our example, neither option produces the desired result.

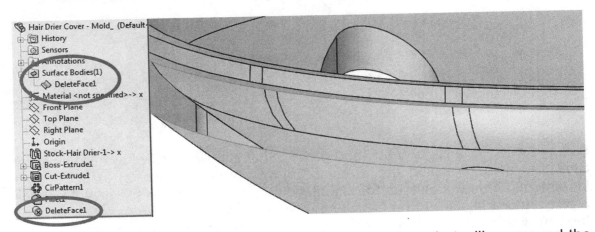

535. – The gap created will be filled with a swept surface that will go around the part connecting both sides of the opening. When the previous faces were deleted the resulting edge is a combination of edge segments, and a swept surface needs a continuous edge or curve for either a path or a guide curve. Select the "**Composite Curve**" from the "Curves" command or the menu "**Insert, Curve, Composite**."

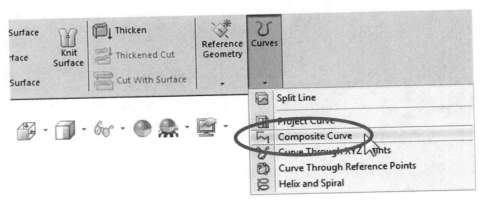

To select all the edges around the surface to create the curve we need, right-mouse-click in one edge of the upper surface, and click in "Select Tangency." All of the edges tangent to the selected edge will create a single, continuous curve for our swept surface. Click OK to finish the curve and continue.

The other open edge is a single, continuous edge, and therefore we don't need to add a composite curve to use it as a path or guide.

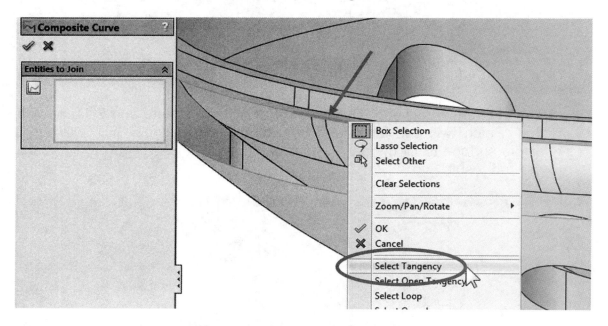

536. – Switch to a Right View, and press "Shift + down arrow key." Holding the Shift key while pressing the arrow keys will rotate the screen in 90 degree increments. This will give us the orientation we need to make the swept surface profile. Add a sketch in the "*Top Plane*" and draw the following sketch close to the left side of the surface. The straight line and the arc are tangent to each other, and the construction line is added for the angular distance to the line.

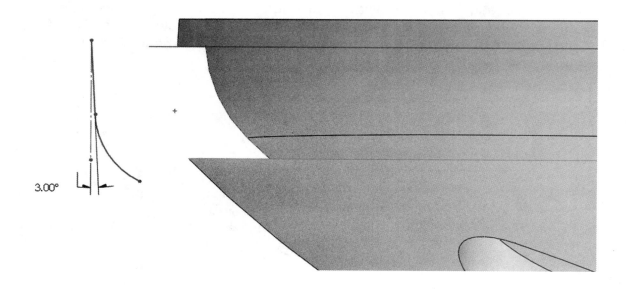

537. – In order to make the sketch arc tangent to the lower surface, we need to project the surface's curvature to the sketch. To do this we'll use the "**Intersection Curve**" command, located in the drop-down menu in the "Convert Entities" command. The intersection curve will add sketch entities at the intersection of the selected face and the current sketch plane giving us the curve we need. Select the lower face and click OK to add the curve and close the "Intersection Curves" command.

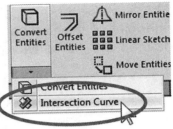

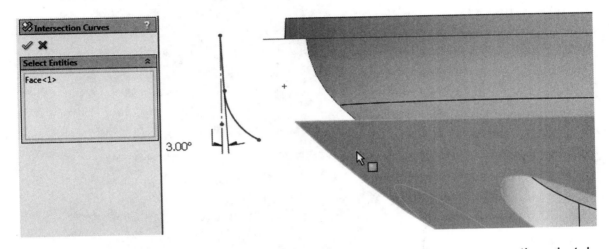

The resulting curves will be added everywhere the surface crosses the sketch plane. Delete the other sketch segments that will not be needed and convert the segment closer to our sketch to construction geometry.

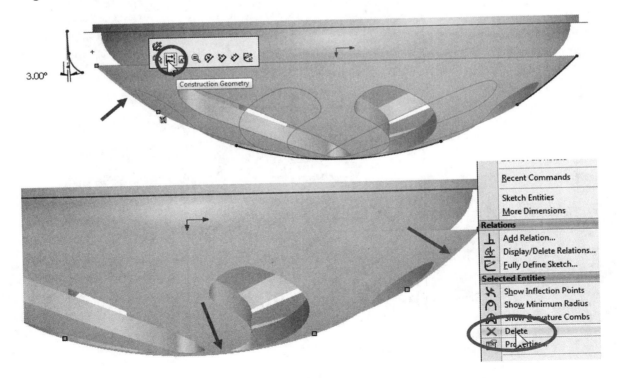

538. – Make the arc coincident and tangent to the construction line, and the top endpoint of the vertical line pierce the composite curve. The sketch will be fully defined with these two relations. Exit the sketch when finished.

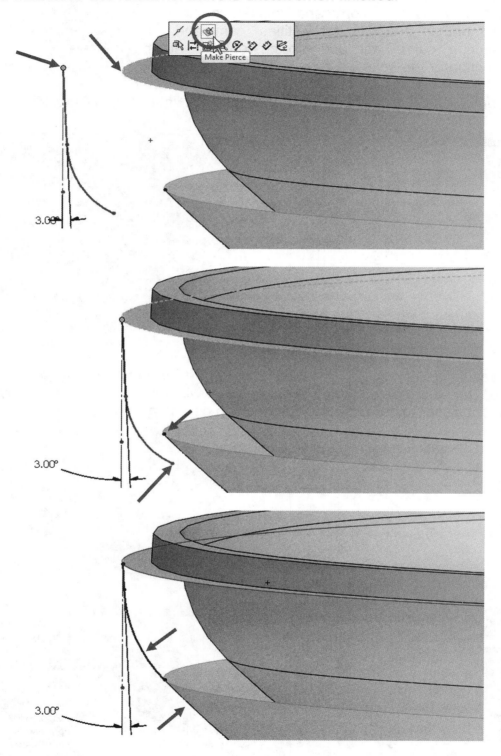

539. – Select the "**Swept Surface**" command from the Surfaces tab, or the menu "**Insert, Surface, Sweet...**" using the previous sketch as a profile and the Composite curve as a path.

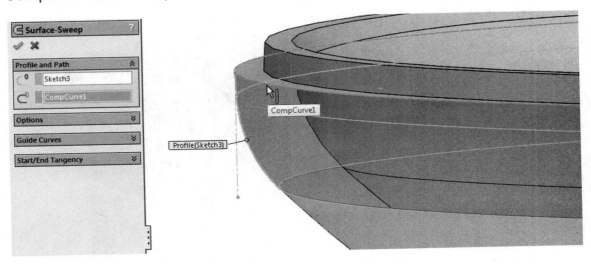

 We could have used the lower edge as a path instead of creating the composite curve, but we chose to show the reader the added functionality of the composite curve and take advantage of the learning opportunity.

In this example the path is circular and the resulting surface is what we expected. If the path and/or the lower open edge are an irregular shape, we could add the lower open edge as a guide curve to better control the resulting surface. In that case we would also need to change the "Orientation/twist type" to "Follow path and 1st guide curve" to obtain the correct result. Use either approach and click OK to finish the swept surface.

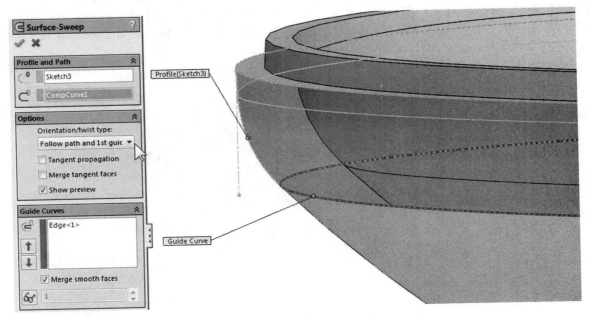

540. – After adding the swept surface we have two surface bodies. Knit both surfaces into a single body. When the knitted surfaces form a closed volume, the option "Try to form solid" is enabled. Check this option and click OK to continue.

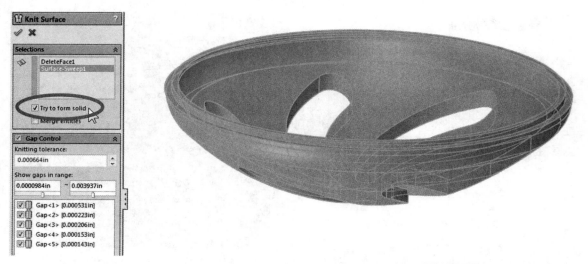

After knitting the surfaces we have a single solid body and no surface bodies.

541. – For the ventilation holes we need to decide if they will be made in the Core or the Cavity, in order to add the draft with the correct "Direction of Pull." To make our mold easier to build, we'll add the holes to the Core side (male part) and make the Cavity part completely round inside, which is also easier to make. If we turn on the "Draft" option in the *"Cut-Extrude1"* feature as we did with the first extrusion, the holes will become smaller than what we had originally intended, because the cut feature starts in the face where the part was originally split, and adding a 3 deg. draft will shrink the holes too much as it moves farther away from the sketch plane. Cancel this change as this is not the result we are looking for.

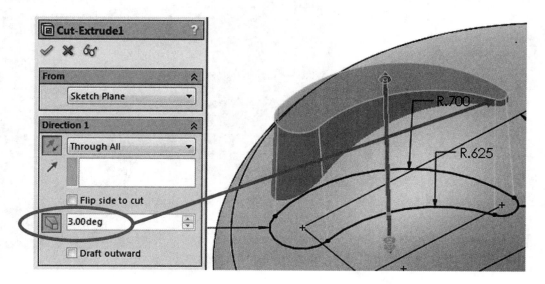

542. – To add a draft to the vent cuts, we'll rollback the FeatureManager before the circular pattern, add draft to the original hole, edit the pattern to add the draft feature to it and lastly roll forward. Rollback before the *"Fillet1"* and select the **"Draft"** command; in this case we'll add draft to the vent hole faces using the "Parting Line" option.

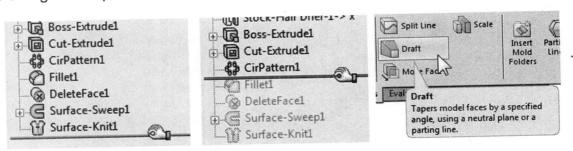

543. – In the "Type of Draft" select the "Parting Line" option. This way of adding the draft will allow us to draft the faces starting at a selected parting line, which in our case will be the top edge of the hole. For the "Direction of Pull" select the *"Right Plane"* making sure the arrow is pointing down, as we intend to make the holes with the core; this will be easier to visualize when the mold is finished. In the "Parting Lines" selection box select the top edge of all the holes using the "Select Tangency" option. Click OK to add the draft and continue.

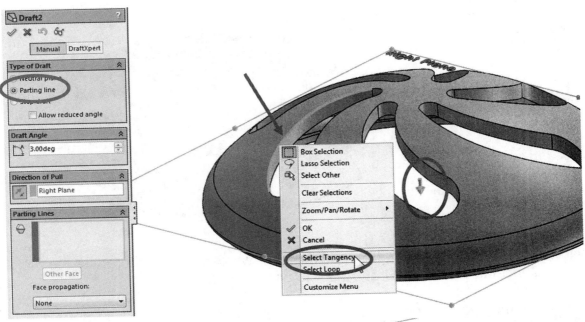

After selecting the edges, a faint arrow is pointing to the face that will be drafted. If an arrow is pointing to the wrong face, select the edge in the "Parting Lines" and click in the "Other Face" button.

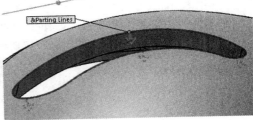

As of the writing of this book, adding the draft to the first hole and using the pattern to copy the drafted holes does not work as expected, that is the reason to draft all the holes at the same time.

544. – After adding the draft to the vent holes, move the Rollback bar to the bottom.

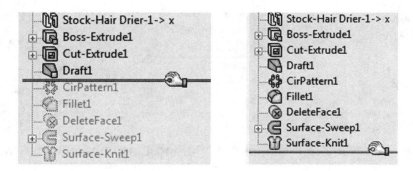

By adding the draft this way, the holes are wider inside and narrower outside, allowing us to make the holes in the Core side as we intended.

545. – Make a "**Draft Analysis**" to check all the faces have the required draft.

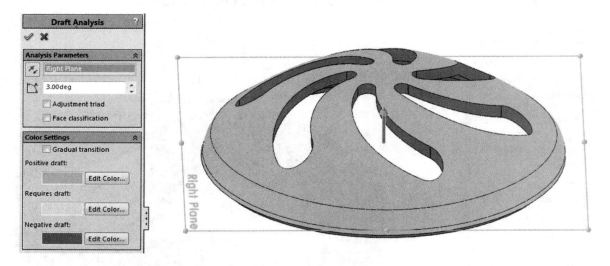

Now that all the faces have the required draft we can proceed to make the mold. Click Cancel to close the "Draft Analysis" view and continue.

Sometimes it's difficult to identify which faces need draft from looking at this image. In this case we can use the "Face Classification" option in the Draft Analysis tool. When we activate this option the faces can be hidden or shown by color code, and at the same time we get a count for each case. After activating it we'll see a new color indicating the "Straddle faces." Straddle faces are defined as faces that have both positive and negative draft, typically cylindrical or curved faces, as we'll see in a later example.

In the next image some faces did not have a draft to show how the "Face classification" works. To turn a group of faces On or Off, click in the Show/Hide icon next to each color. By turning off the positive and negative draft faces we can clearly see the faces that don't meet the required draft.

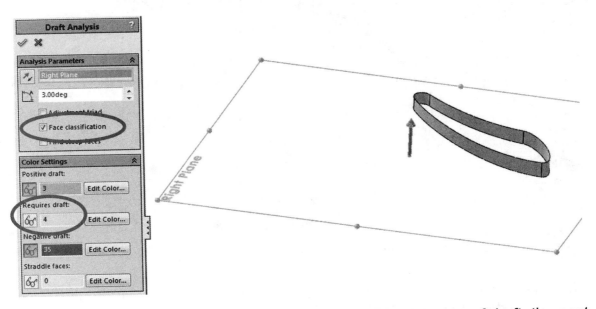

546. – Now that all the part's faces have the required 3 degrees of draft, the next step is to generate the parting line. Select the **"Parting Line"** command from the Mold Tools toolbar or the menu "**Insert, Molds, Parting Line.**" Select the *"Right Plane"* or a flat face parallel to it for the "Direction of Pull," enter 3 degrees as before, and click the "**Draft Analysis**" button.

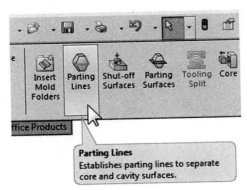

If the parting line edges are not automatically selected, we have to select them manually. When a part has faces that need draft along the parting line, the edges may not be automatically selected because there may not be a clear choice as to which side of the mold a face should go. In our case the parting line edges are correctly selected, but if they are not in your model, manually select the edges where the red (or green) faces meet the yellow faces. Notice the perimeter is made of multiple edges; be sure to select them all *or* right mouse click in one edge and use the "Select Tangency" option as we did with the draft in the holes.

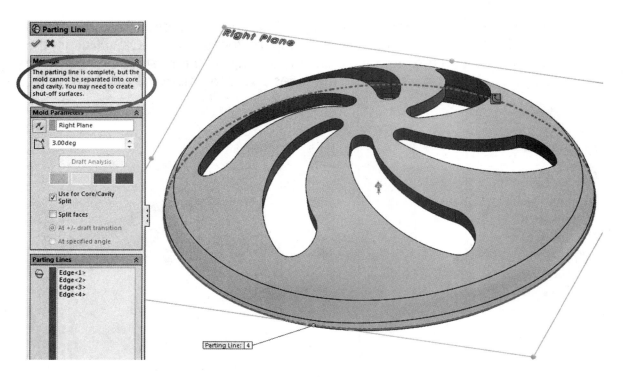

After the parting line edges have been selected, the message at the top of the command (still with yellow background) says that the parting line is complete, but the mold cannot be separated into Core and Cavity, and we may need to create shut-off surfaces to close the holes in the part. Click OK to finish the parting line.

547. – After completing the parting line we need to close the holes in the part to be able to split it into Core and Cavity. Select the "**Shut-Off Surfaces**" command from the Mold Tools toolbar or the menu "**Insert, Molds, Shut-Off Surfaces.**"

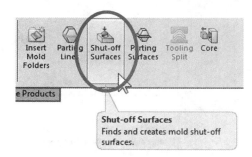

After selecting the "**Shut-off Surfaces**" command, the open loops are automatically selected based on the "Direction of Pull" in the parting line settings. Be sure to leave the "Knit" option checked to have the new faces automatically merged with the rest of the Core and Cavity faces when we finish this command. By turning on the "Show Preview" option we can see the resulting surfaces. The message at the top of the command now has a green background and says that the mold can be separated into Core and Cavity. If a hole had not been automatically selected, we'd have to manually select its edges. Click OK to complete the command.

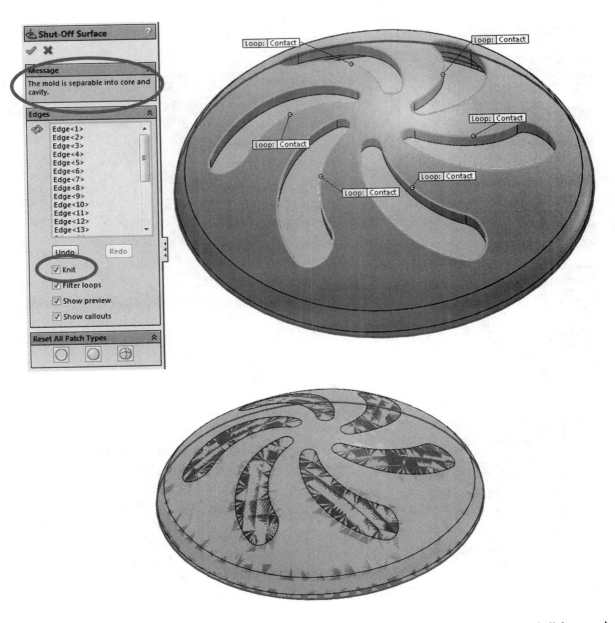

When finishing the shut off surfaces, all surface bodies are made visible and overlapping, that is the reason for the multiple colors of the part.

548. – The next step is to generate the parting surface to separate the mold halves. Select the "**Parting Surfaces**" command from the Mold Tools toolbar or the menu "**Insert, Molds, Parting Surfaces**." The option "Perpendicular to pull" option is automatically selected, as well as *"Parting Line1."* Enter a value of 1″ for the "Parting Surface" distance and be sure the "Knit all surfaces" option is checked. Click OK to finish.

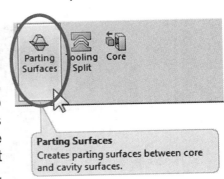

Parting Surfaces
Creates parting surfaces between core and cavity surfaces.

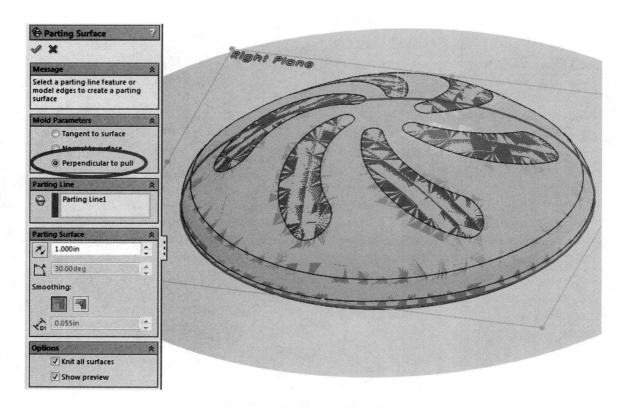

After completing the parting surface, we can see the *"Core," "Cavity"* and *"Parting Surface"* folders listed under the *"Surface Bodies"* folder, and a single solid body. If the surfaces are visible, they can be manually hidden. By making the surfaces visible, we can see them overlapping each other. In the holes we see the green (core) and red (cavity) surfaces overlap. Because the shut-off surfaces are knitted to both the Core and Cavity surfaces, the green and the part overlap in the Cavity side, and the red and the part overlap in the Core side.

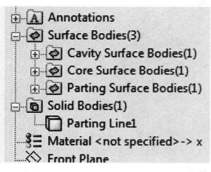

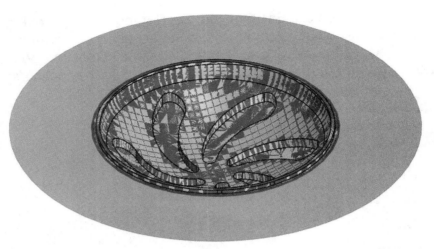

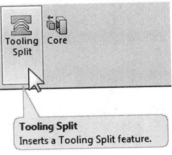

549. – Now we are ready to make the core and cavity. Select the "**Tooling Split**" command from the Mold Tools toolbar. In this example we'll make the tooling sketch at the same time we make the tooling split. Select the parting surface to add a new sketch in it, use "**Convert Entities**" to project the parting surface's edge and exit the sketch to continue.

Tooling Split
Inserts a Tooling Split feature.

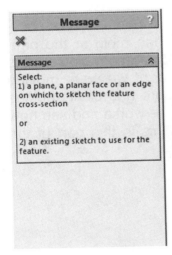

Message

Select:
1) a plane, a planar face or an edge on which to sketch the feature cross-section

or

2) an existing sketch to use for the feature.

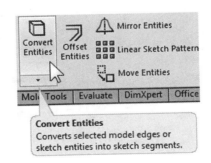

Convert Entities
Converts selected model edges or sketch entities into sketch segments.

550. – After using "**Convert Entities**" on the Parting Surface, exit the sketch to return to the "**Tooling Split**" command; the necessary surfaces are automatically selected in the "Core," "Cavity" and "Parting Surface" selection boxes. Make the "Block Size" 1″ up and 0.5″ down. Click OK to complete the tooling split.

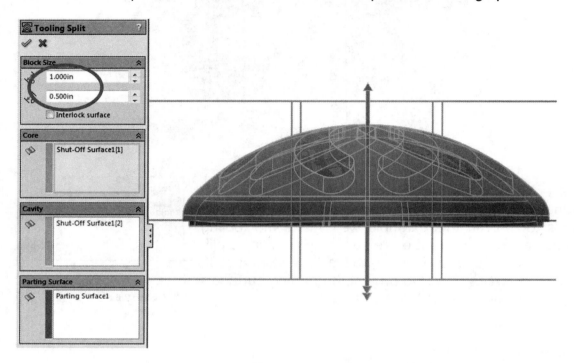

551. – To view the core, cavity and part bodies as an open mold in the part, select the menu "**Insert, Features, Move/Copy….**" This command allows us to copy and/or move surfaces and solid bodies in the part as if we were moving parts in an assembly. In our example we'll separate the core and cavity bodies away from the design part body only to show how this feature works and see how the bodies look. In reality there is no good reason to do this in this part of the process other than visually inspecting the resulting bodies.

Select the top body (Cavity) and either enter a value to move it along the "X" direction, or drag the direction arrows in the graphics area, similar to exploding an assembly. If we see the "Mate Settings" option to move the bodies like in an assembly, select the "Translate/Rotate" button at the bottom of the properties to continue.

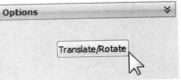

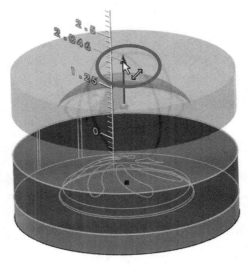

552. – Repeat the **Move/Copy** command to move the lower body (Core) down. Make sure the "Copy" option is not checked.

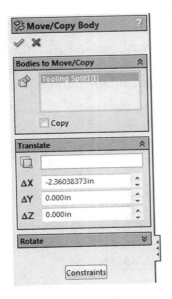

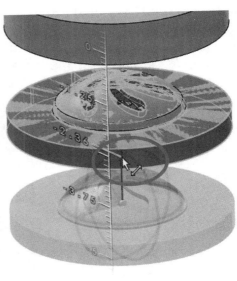

 If needed, hide all the surfaces at the same time by selecting the *"Surface Bodies"* folder and picking "Hide."

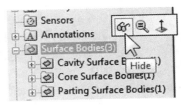

 When making multi body operations like combining bodies where one or more bodies are absorbed, it's often useful to copy bodies in the same locations as the originals to retain the original bodies.

553. – The next step is to save the core and cavity bodies to separate files as we did with the previous part. Select the core and cavity bodies and save them to new parts as *'Hair Drier Cover-Core'* and *'Hair Drier Cover-Cavity.'*

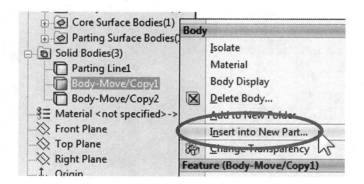

 If we save a core/cavity/body to a new part after moving them with the "Move/Copy" command, the body in the new part will be located in the same position relative to the origin it had before saving it to a new part. To retain the original location suppress the "Move/Copy" commands *before* saving them to a new part.

554. – Open the cavity part. If we notice, the holes of the cover are outlined in the cavity. A part like this is usually made with a Computer Numerical Control (CNC) machine, and for manufacturing purposes, it's better to have a single surface than a surface split in multiple areas like in this case. One way we can fix this is by deleting the faces from the solid body, and patching them with a new one. Select the "**Delete Face**" command from the Surfaces toolbar or the menu "**Insert, Face, Delete**."

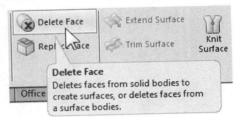

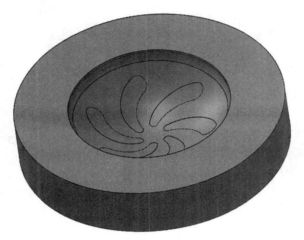

Select all the faces inside the Cavity that outline a hole. Use the "Delete and Patch" option and click OK to finish. This way we'll remove the outlines and patch the surface merging it with the cavity to produce a smooth surface.

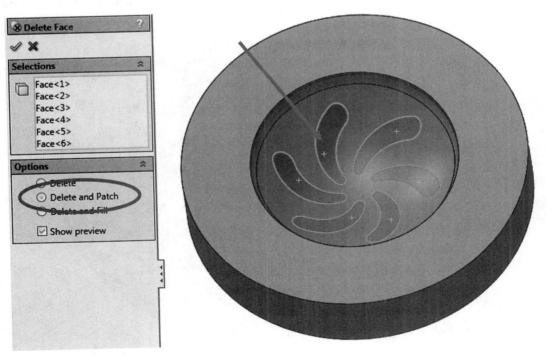

And now we have a single surface in the Cavity that can be easily machined with a CNC. Save and close the Cavity.

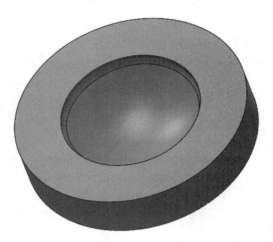

555. – Another common task is to add inserts to a mold. Opening the Core part, we see the protrusions that will make the holes in the cover extending past the top surface; these can be made adding inserts to the core, or building the entire core as a single piece. The problem with a single piece core is that it's harder to make, and if these protrusions wear out, it's more difficult to repair and expensive to replace, whereas an insert can be replaced easier, cheaper and faster. This is a decision that can be answered with help from the tool maker and depends on the volume of parts to be molded, cost of making it one way or the other, etc. We'll learn how to cut an insert if needed from the Core by using a split feature. Make a new sketch in the bottom of the core and use "**Convert Entities**" to project the outline of one insert onto it (use "**Select Tangency**" to make selection easier).

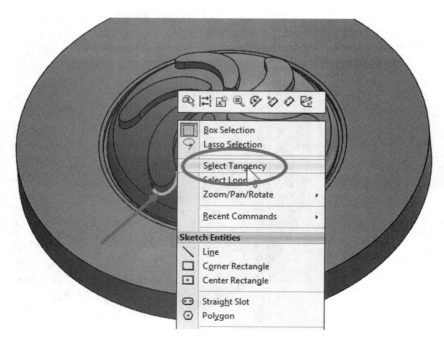

 We cannot select the top face of the insert with "**Convert Entities**" since it's smaller than the base because of the draft we added to release it from the mold, and using it to convert entities would give us an incorrect result.

556. – We can split the part directly using the sketch as we did when splitting the main body of the *'Hair Drier'*, but we'll show a different way to split a solid body using a surface. From the Surfaces toolbar select "**Extruded Surface**" or the menu "**Insert, Surface, Extrude.**" Extrude the surface past the top of the Core as shown and click OK to finish.

557. – Select the menu "**Insert, Features, Split,**" select the previous surface in "Trim Tools" and press the "Cut Part" button to generate the different bodies. Double click the insert to give it a name and save it as *'Hair Drier Cover-Core Insert.'* Use the "Consume cut bodies" option to remove it from the part. Click OK to finish and hide the surface used to split the part.

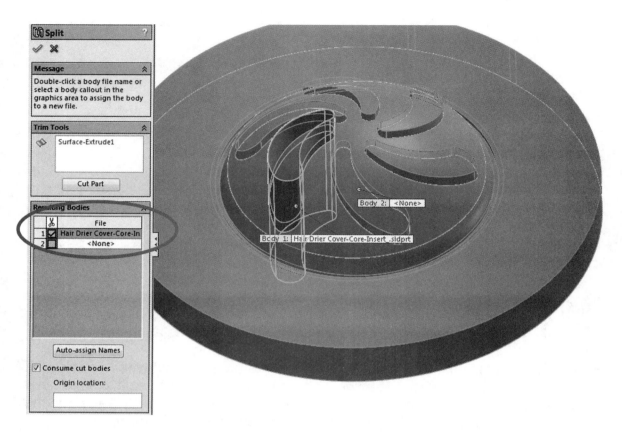

558. – To make the rest of the inserts, we'd follow the same procedure, but since all the inserts are equal in this part we can make one insert and use a cut extrude feature to cut out the rest of the holes.

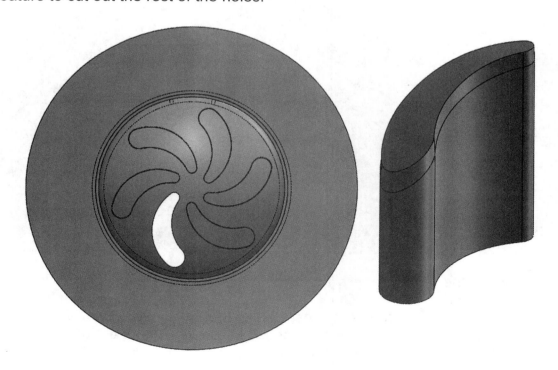

When designing molded parts keep in mind how they are going to be molded, assembled, and if at all possible, involve the toolmaker in the design process to get his/her input before it's too late. He'll be glad you did, and you will be too. ☺

Save and close all parts.

559. – For the next exercise, we are going to use a part that will not generate the correct parting line and we'll have to manually select it. Locate the *'Hair Drier Mold Exercise'* from the included files and open it.

*This part is finished including vents, power cord and switch cutouts, drafts have been added to most faces and external relations have been broken to prevent possible conflicts with user generated files.

560. – In this part we'll make further use of the "Face Classification" option in the draft analysis. Select the "**Draft Analysis**" command using the *"Front Plane"* for the "Direction of Pull" and enter 3 degrees draft. Turn on the "Face classification" checkbox to identify the model faces as "Positive Draft," "Negative Draft," "Requires draft" or "Straddle faces."

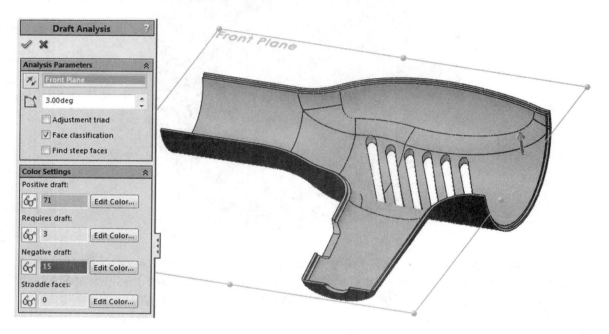

561. – Click in the Hide/Show icons next to the green and red boxes to turn off both "Positive" and "Negative" draft faces, and isolate the faces that require draft. Here we can see that only three faces do not have the required three degrees draft. At this time we'll proceed to make the core and cavity ignoring these faces. The reason we are not adding draft to these faces is because adding draft would generate the correct parting line automatically, which in general is good; however we are overlooking this detail to force an incorrect automatic parting line to cover additional functionality. Cancel the "**Draft Analysis**" to continue.

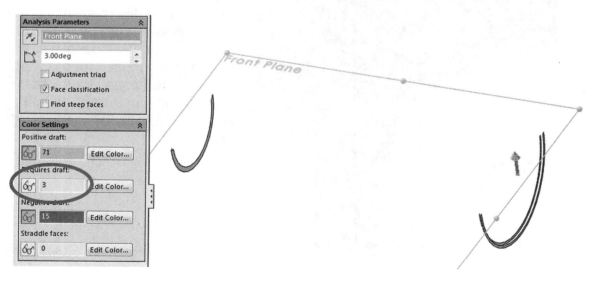

562. – To compensate for shrinkage in the mold when the plastic cools down, scale the part, and make it 2% bigger. Select the "**Scale**" command; use the option "Uniform Scaling" about the "Centroid" and enter 1.02 as the scaling factor. Click OK to finish.

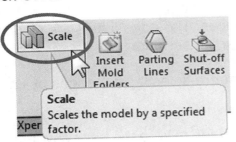 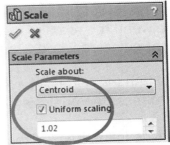

 Remember that material shrinkage depends on process, materials, geometry, etc. This is just an example.

563. – Select the "**Parting Lines**" command; use the "*Front Plane*" for the "Direction of Pull," enter 3 degrees for the draft angle, and press the "Draft Analysis" button to calculate it. A parting line is automatically selected, but since we left three faces with no draft, we need to review and make sure the parting line is where we want it; be aware that experience always helps in making these decisions.

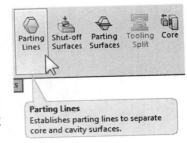

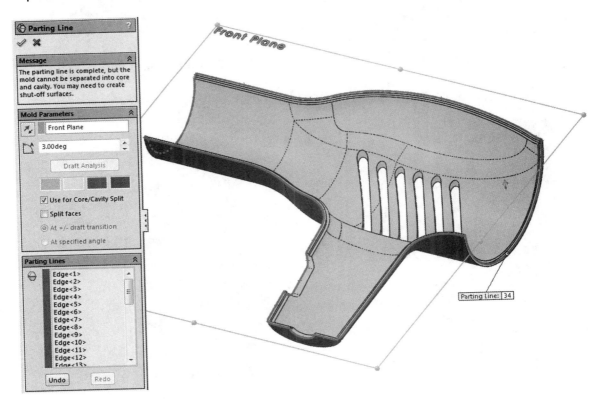

Upon closer examination, in the front and back of the *'Hair Drier'* we see the parting line going across the faces that need draft. Since it may be easier to mold the yellow faces that need draft made in the Core side (green) we have to manually select the edges we want the parting line to follow. Unselect the edges between the green and yellow faces in the front and back of the part, and select the edge between the red and yellow faces. The edge labels are displayed when the edges are selected in the "Parting Lines" selection box.

If needed, re-run the "Draft Analysis" after re-selecting the edges by changing the draft angle to reset the analysis button.

- One side in the back of the *'Hair Drier'* (where the back cover fits)

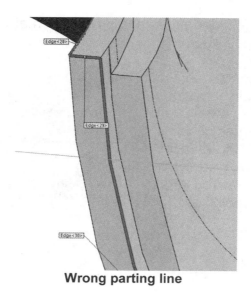

Wrong parting line

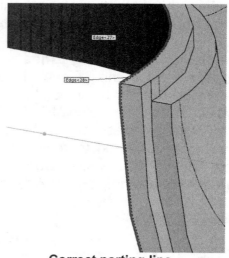

Correct parting line

- The other side in the back of the *'Hair Drier'*.

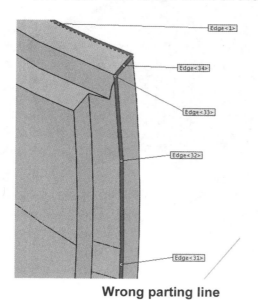

Wrong parting line

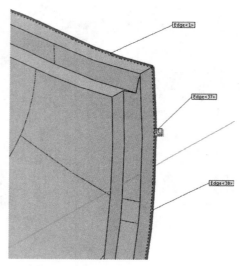

Correct parting line

- One side in the front of the *'Hair Drier'*…

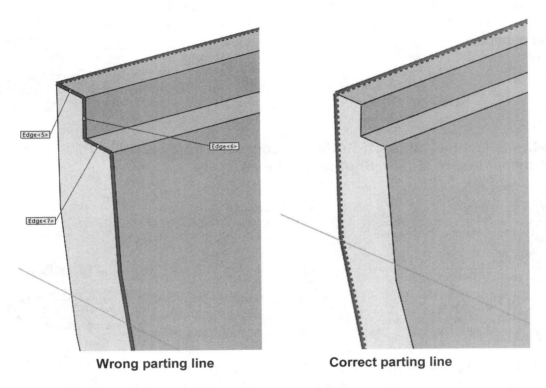

Wrong parting line   **Correct parting line**

- The other side in the front of the *'Hair Drier'*.

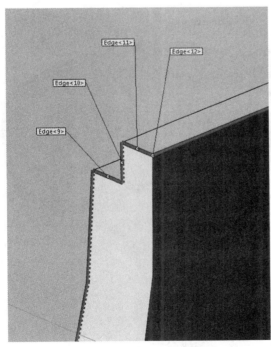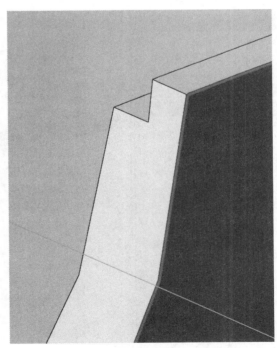

Wrong parting line   **Correct parting line**

The message in the "Parting Line" command is in yellow and says that the parting line is complete but we still need to create shut-off surfaces. Click OK to add the parting line and continue.

564. – Select the "**Shut-off Surfaces**" command from the Mold Tools toolbar or the menu "**Insert, Molds, Shut-off Surfaces**."

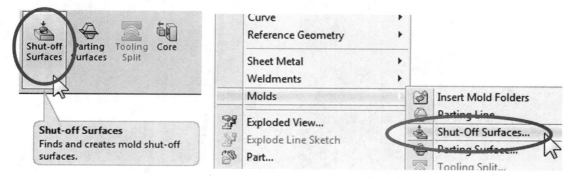

After selecting the command all the ventilation holes are automatically selected and closed; our mold can now be separated into core and cavity. Click OK to continue.

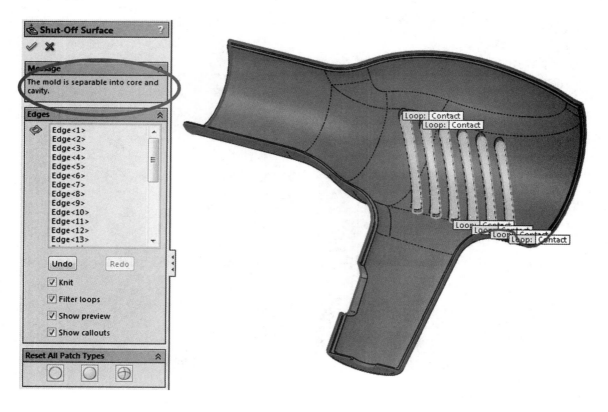

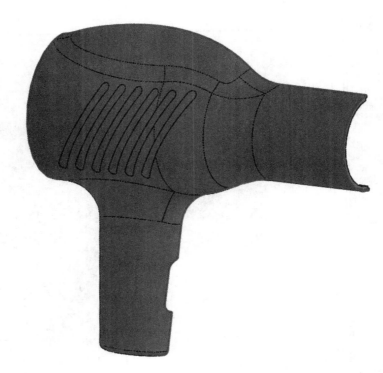

565. – The last step before creating the core and cavity is to create a parting surface. In the previous examples the parting surface was generated automatically saving us time. In this case an acceptable parting surface cannot be automatically generated with any combination of parameters and options as of the writing of this book (this situation *may* change depending on future software updates) and therefore we have to manually modify and complete it. Select "**Parting Surfaces**" from the Mold Tools toolbar or the menu "**Insert, Molds, Parting Surface**." Use the option "Perpendicular to pull," enter a surface distance of 0.75″ and turn on the "Knit all surfaces" and "Manual mode" options. Change to a front view for visibility.

Enabling the "Manual mode" allows us to manually manipulate the resulting parting surface.

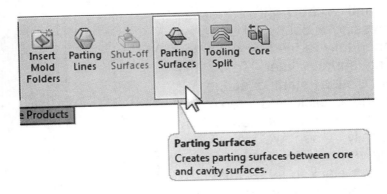

 Feel free to explore the different combinations of parameters to see the (unacceptable) resulting surfaces before continuing, and the reason to manually modify it.

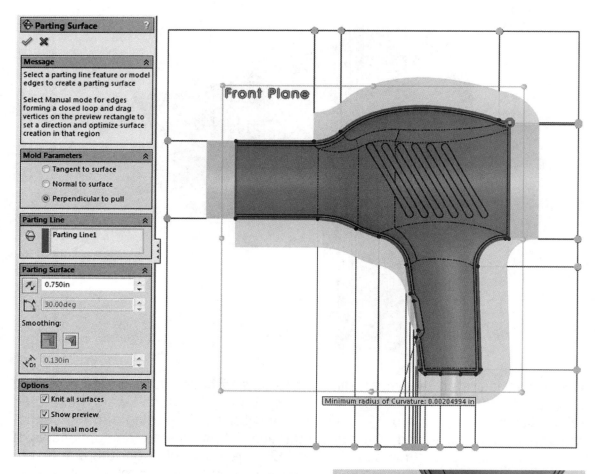

When the "Manual mode" is activated every vertex in the parting line edges are connected to a handle in an outside frame, where we can click and drag the handles to define the direction of the surface at each vertex. In the area around the switch opening we can see that all handles are pointing down and in the front of the hair drier we are missing a couple of surfaces that will be manually added and knitted with the parting surface later. To fix the surfaces at the switch, starting at the left most switch handle, click and drag each handle along the outside frame horizontally and then vertically, when we get to the projection of the vertex in the vertical frame the handle will turn black to let us know we are aligned with the vertex. A preview will let us know which vertex we are working with.

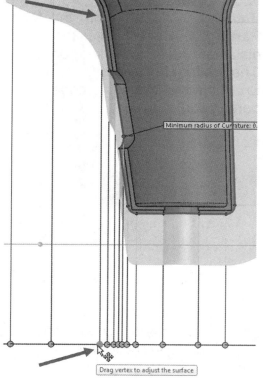

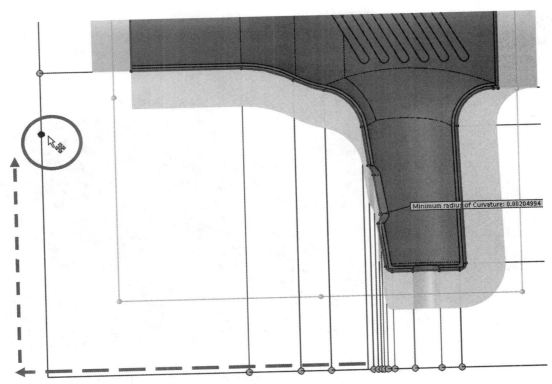

Move the rest of the handles to the vertical frame to fix the surface as shown.

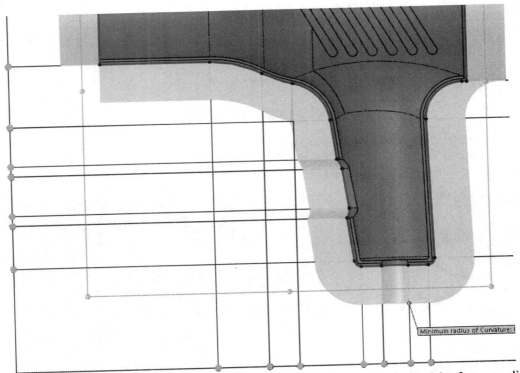

After correctly locating the switch cutout handles on the left side frame click OK to generate the parting surface. The next step will be to add the missing surfaces.

566. – To complete the lower part of the surface add a sketch in the indicated surface, and add a tangent arc as shown. This sketch will be used as a guide for a loft surface. Be sure to capture relations to the vertices and Exit the sketch.

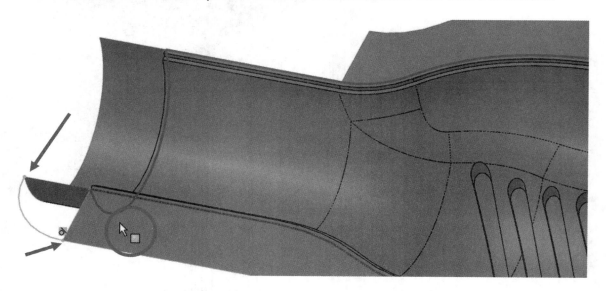

567. – Select the "**Lofted Surface**" command to complete the next surface. In the "Profiles" selection box select the two surface edges indicated near the outside of the surface, and the linear edge and the previous sketch in the "Guide Curves" selection box. Click OK to finish and generate the new lofted feature.

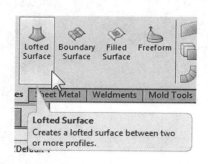

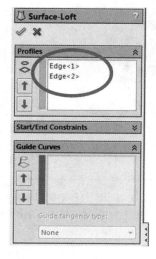

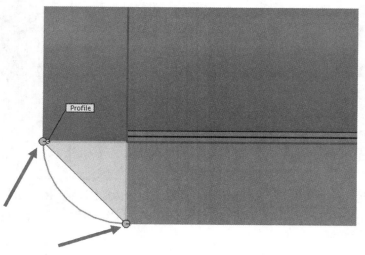

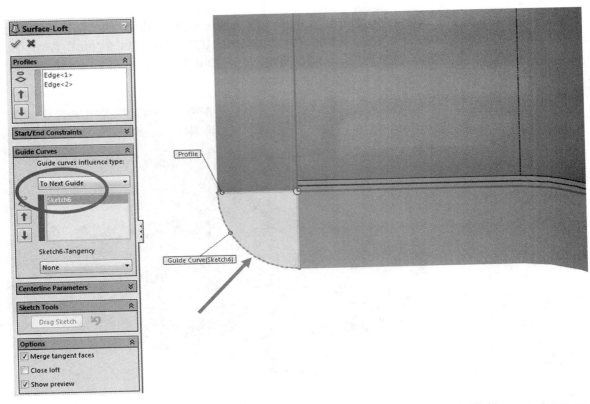

568. – To make the next surface add a new sketch in the parting surface just as in the previous step and add a line and arc as indicated. Make the arc's center point coincident to the edge at the front of the surface to fully define the sketch. This sketch will also be used as a guide curve. Exit the sketch when done.

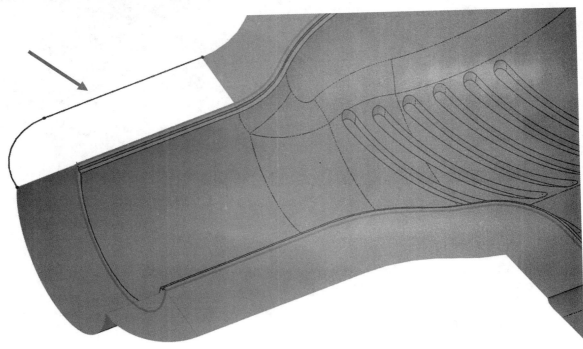

569. – Select the "**Lofted Surface**" command again and select the two existing surface edges in the "Profiles" selection box and the previous sketch and the opposite model's edge in the "Guide Curves" selection box. Click OK when finished.

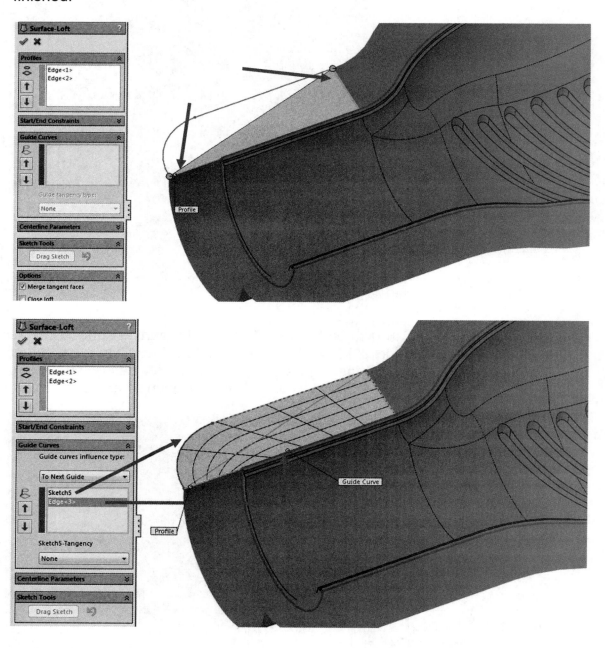

 If after selecting the surface edges in the "Profiles" selection box we cannot see a preview, it's probably because the edges were selected in opposite ends causing the loft to twist. To correct it drag the green dot in one of the edges to the other vertex.

After adding these two surfaces we have a total of six surface bodies.

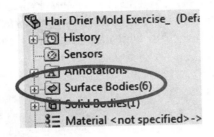

570. – To use these surfaces for a tooling split we have a couple of options:

- They can to be added to the *"Parting Surface Bodies"* folder.
- They can be selected at the time of making the tooling split.
- We can knit them to the existing *"Parting Surface1"* to have a single surface and then add it to the *"Parting Surface"* folder.

We are going to use the last option to show additional functionality and make it easier to create the tooling split sketch. Select the **"Knit Surface"** command and knit the two new lofted surfaces with the two *"Parting Surface1"* bodies.

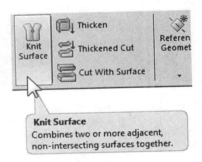

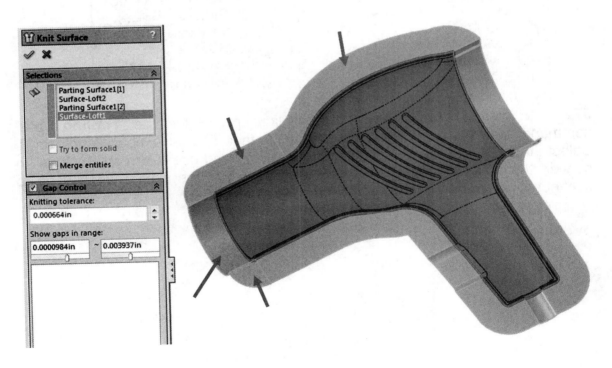

571. – After knitting the surfaces the new lofted surfaces become part of the parting surface and we need to add the new knitted surface to the *"Parting Surfaces"* folder. Expand the *"Parting Surface Bodies"* folder (the Core and Cavity surface bodies will become visible) and drag the *"Surface-Knit1"* surface into it.

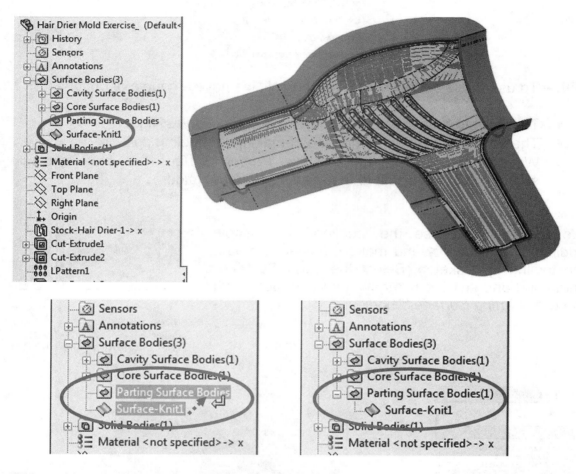

572. – Now we are ready to make a tooling split. Select the **"Tooling Split"** command; when asked select the parting surface to add a new sketch on it. Select the entire outline using "Select Open Loop" and use **"Convert Entities"** to convert the entire outline of the parting surface. Exit the sketch when done to continue the **"Tooling Split."**

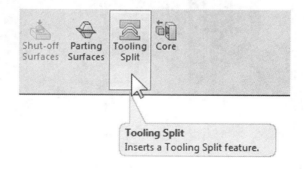

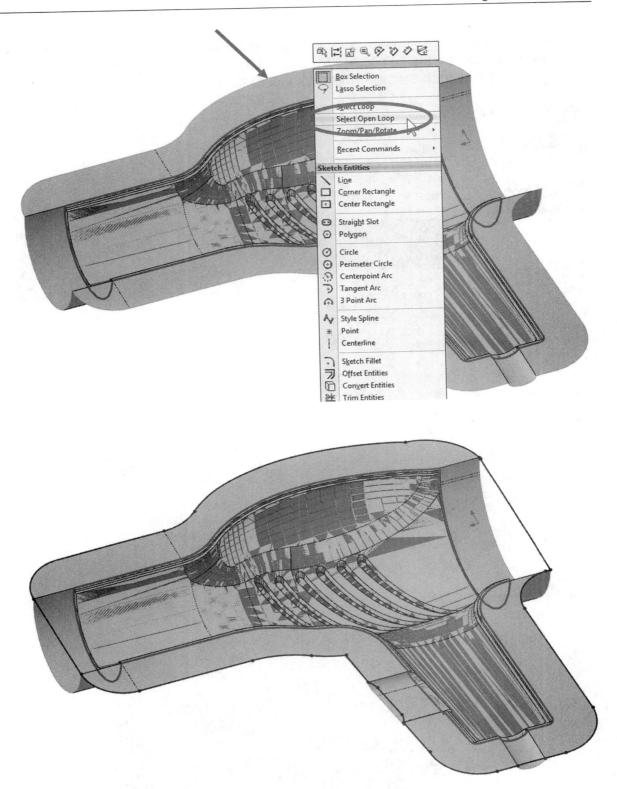

 The tooling sketch must be bigger than the parting line and as big as or smaller than the parting surface.

573. – After exiting the sketch **c**hange to a Top or Bottom view to see the size of the tooling block. In the **"Tooling Split"** command, make the "Block Size" big enough to fully enclose the part and click OK to continue.

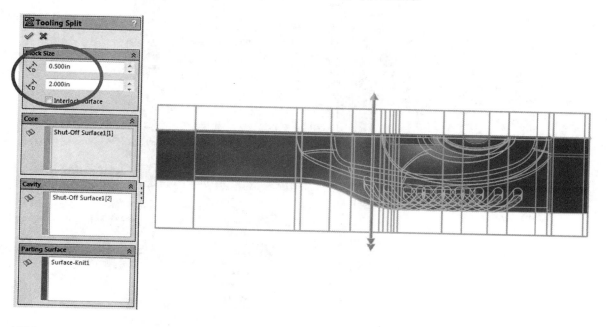

574. – Hide the *"Parting Line1"* feature and all *"Surface Bodies"*…

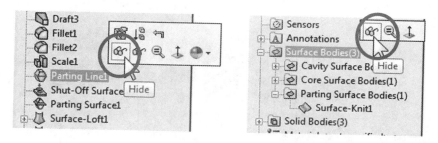

… and then hide the body at the top of the mold to see the cavity and the part.

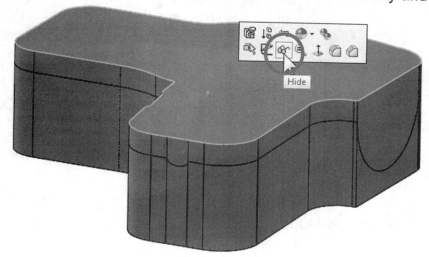

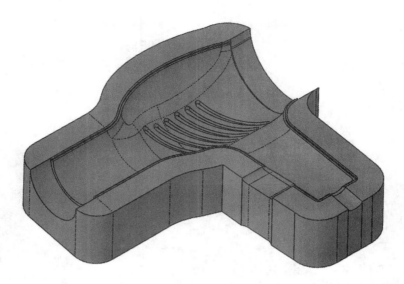

575. – Optionally we can change the color of the molded part's body (just the body's color) to make it easier to identify. Select the molded part, and from the pop-up menu select the "**Appearance**" icon. From the drop down we can choose to change a face, feature, body or the entire part's appearance. Select the "Body" option and change its color for visualization purposes.

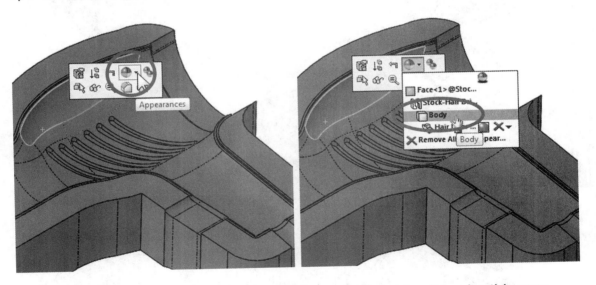

Or select the body in the *"Solid Bodies"* folder and change its color this way.

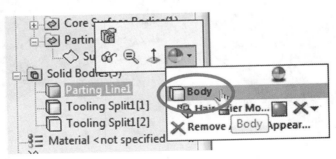

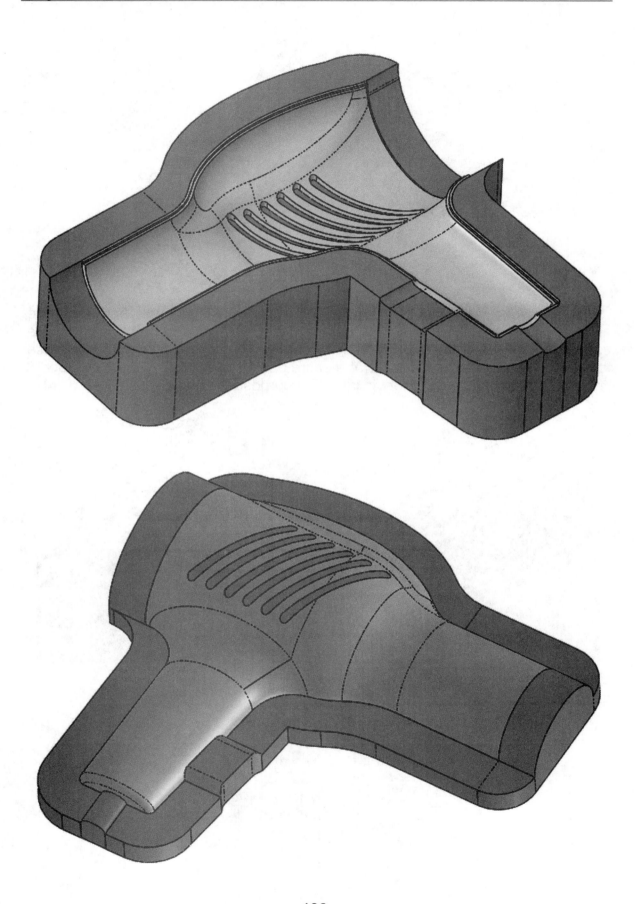

Exercise: Create new part files from the Core and Cavity bodies.

- In the Cavity part, eliminate the vent hole markings to make a smooth surface (like we did with the back cover).

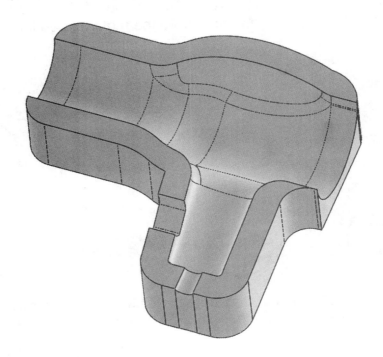

- In the Core part, cut an insert for one of the vent holes.

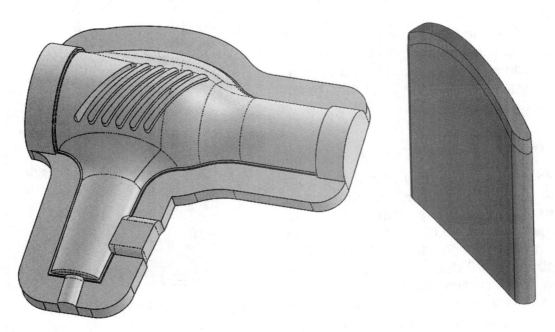

- Make a Core and Cavity for the other half of the *"Hair Drier."*

576. – When designing plastic parts its common practice to add reinforcement ribs to make parts sturdier. Reinforcement ribs can be added using regular modeling tools like extrusions, cuts, drafts, etc., however using this approach can be very time consuming. The SolidWorks' "**Rib**" feature automates and simplifies the creation of reinforcement ribs. To add a couple of ribs to our part, rollback before the *"Scale"* feature to add them before the mold related features.

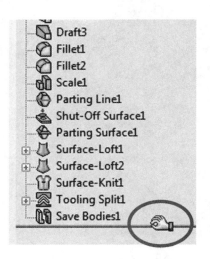
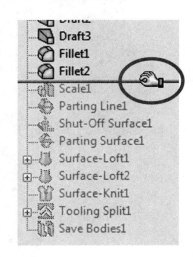

 Using this approach we can add the ribs and other features, then roll forward to the end and have the mold's core and cavity update.

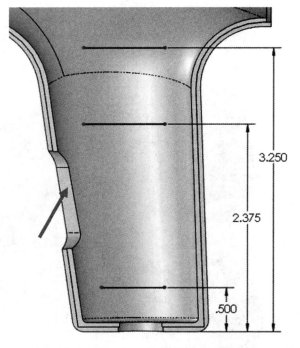

577. – Add a new sketch in the flat face of the switch cutout; draw the next three lines and dimension them. Notice that the lines do not go all the way to the edges of the part. The rib feature will extend these lines until they reach the part.

578. – Select the "**Rib**" command from the Features toolbar or the menu "**Insert, Features, Rib**." Set the "Thickness" to 0.075in using the mid plane option, activate the "Draft" option and make it 2 degrees. After activating the "Draft" option we'll see the option to select where to start the draft. We'll use the default setting "At sketch plane," meaning the sketch plane will also be the "Neutral Plane" for the draft. (The option "At wall interface" makes the neutral plane at the bottom of the rib.) The "Extrusion Direction:" will be set to "Normal to Sketch." If the direction arrow is not pointing into the part turn on the "Flip material side" checkbox, otherwise the rib feature will fail because the material could not be added. The "Draft Outward" option makes the rib feature grow in the direction of the arrow. Click OK to complete the first rib feature.

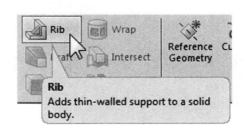

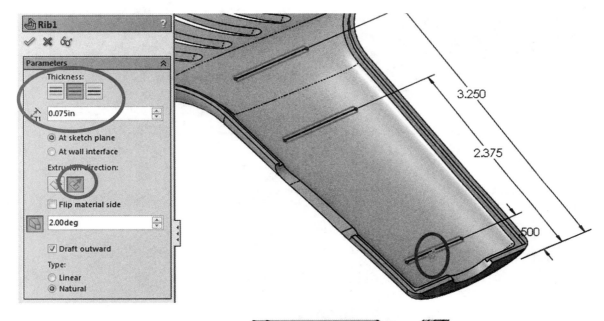

The finished rib will add material using an *'up to next'* end condition in the direction of the extrusion and extending the sketch lines until they reach either a part's wall or another rib, making it a powerful feature because it can match the geometry under the rib feature. An important thing to know about the rib feature is that if we don't have a stopping face for the rib in either direction, the feature will fail.

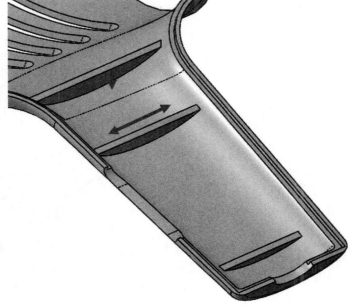

579. – For the second rib, add a plane parallel to the *"Right Plane"* 6 inches to the left and add the next sketch in it. Again, the sketch is not touching the edges.

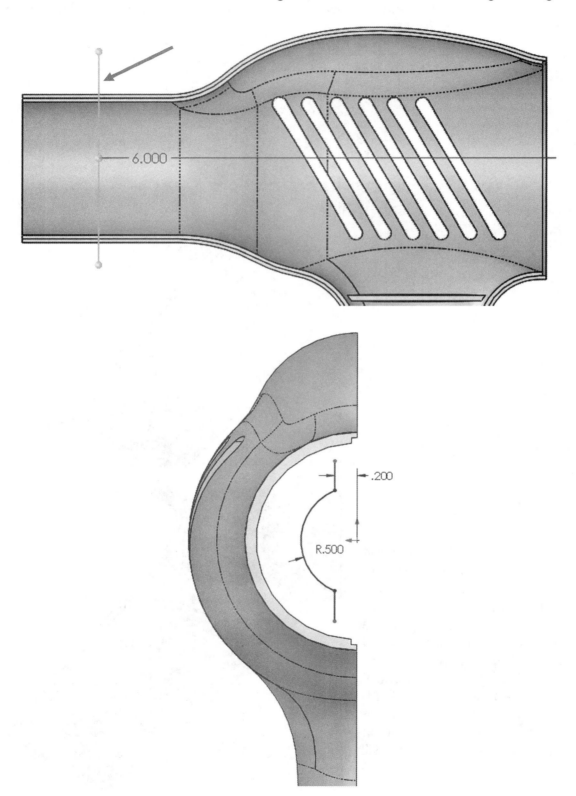

580. – Select the "**Rib**" command and use the same settings as the previous rib, but instead we'll use the option "Parallel to plane" which is automatically pre-selected. After activating the "Draft" button, a new button labeled "Next Reference" allow us to select the sketch segment that will be used to start measuring the draft. Click in it to change the reference to one of the straight lines. Make sure the material direction is pointing into the part. Click OK to continue.

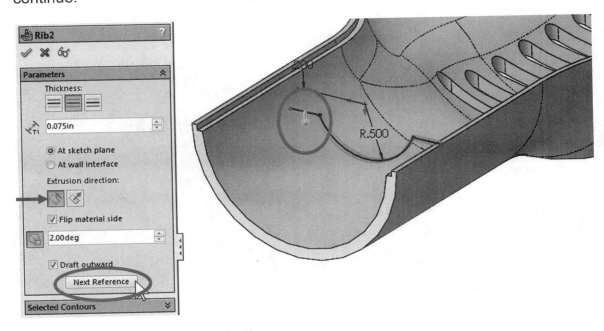

The second rib is complete.

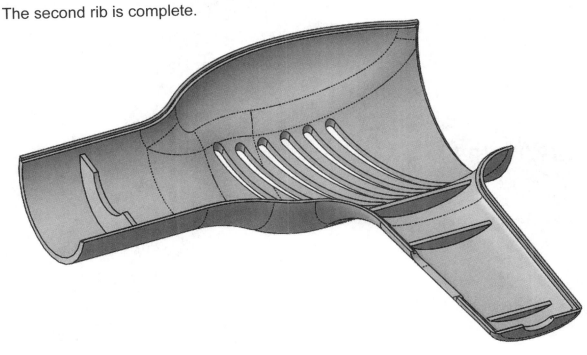

581. – For the next step we'll create a linear pattern of the second rib and use an option to skip instances of the pattern so that it's not a *'regular'* pattern. Make a linear pattern of the second rib with four instances and a spacing of 0.625 inches. Before finishing the pattern, expand the "Instances to Skip" selection box. Here we can *'turn off'* pattern instances by selecting the unwanted instance in the screen; notice the preview is gone. Click OK to finish the pattern and notice the missing instance and how the rib feature pattern conforms to the part's geometry.

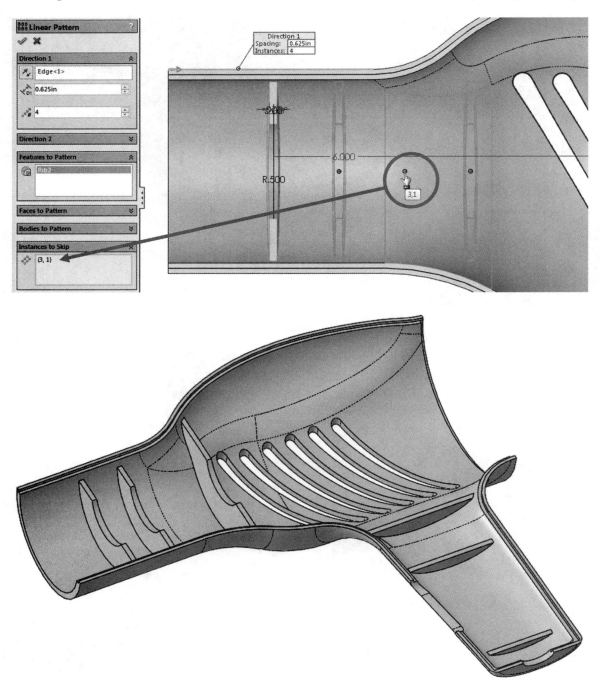

582. – As a final touch add a 0.020" fillet to both rib features and the pattern.

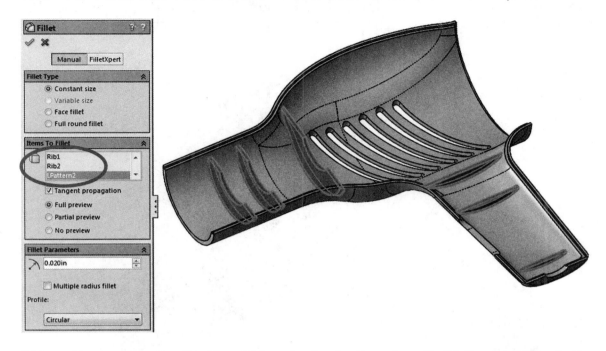

583. – Now move the "Rollback" bar to the bottom to rebuild the mold features. After rebuilding these features the mold features will (should) rebuild correctly because we didn't modify any edges along the parting line or parting surface.

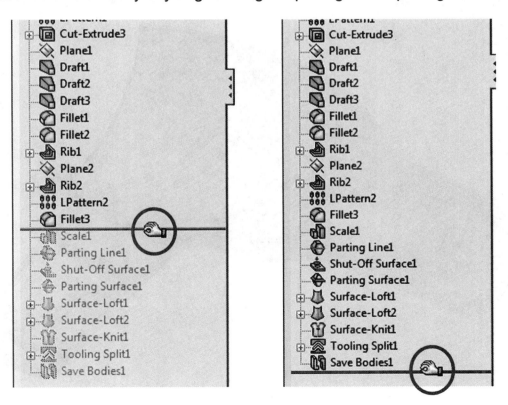

584. – While the finished part is open, load the core part made in the previous exercises, and see how the changes added in this section are propagated. The Cavity would be exactly the same as only the core side was changed.

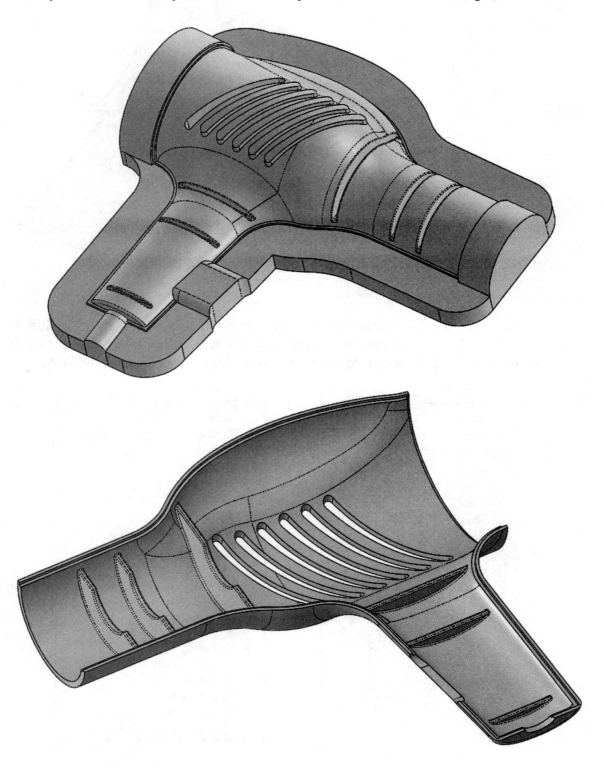

585. – In our last mold exercise we'll learn how to make a "**Side Core**." Think of a side core as a part split from a core or cavity to help us make a feature that would otherwise prevent the part from being released from the mold, like a hole perpendicular to the "Direction of Pull," an undercut, or a negative draft face in the positive draft side of the part.

Open the part *'Bucket Mold'* from the included files (or use the bucket we made in the Surfacing lesson) and run a "**Draft Analysis**" with 3 degrees using the *"Top Plane"* for "Direction of Pull." Use the "Face classification" option to show the "Straddle faces" in and around the holes for the handle. The blue faces will not allow the part to release from a two part mold, making this part a good example of why we need to make a side core. After a part is molded, the side core is the first part pulled away from the mold base assembly, the mold is then opened, and the molded part can be released. Cancel the *"Draft Analysis"* to continue.

586. – From the draft analysis we see the lip at the top of the bucket needs a draft. Add a 3 degree draft to the outside face of the lip at the top of the bucket in the negative draft direction (down) to mold the lip with the cavity. Use the "Along Tangent" face propagation option to select the rest of the faces that need a draft. Click OK to complete it.

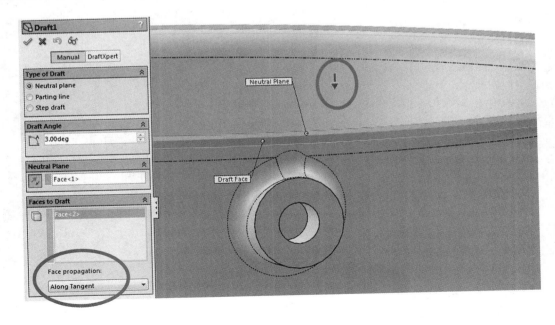

587. – The next step is to add a draft to the side cores, since they also need to have it in order to release them from the part, but in this case the "Direction of Pull" is along the boss' axis. Edit the *"Boss-Extrude1"* feature and activate the "Draft" option with 3 degrees and turn on the "Draft outward" option. Click OK to continue.

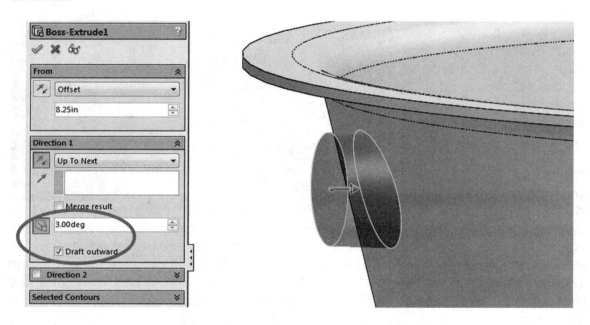

579. – Now edit the *"Cut-Extrude1"* feature and also activate the "Draft" option but in this case with a 2 degrees draft. Leave the "Draft Outward" option off.

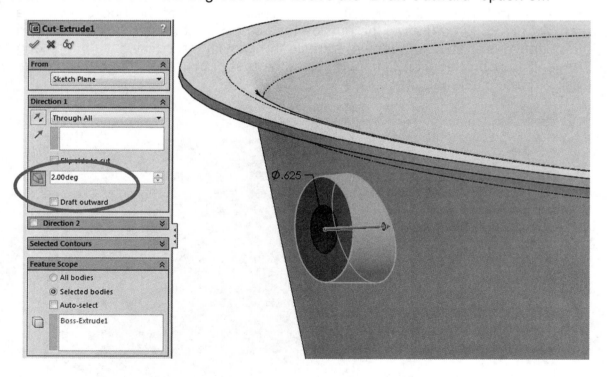

580. – Run a "**Draft Analysis**" using the flat face in the boss for "Direction of Pull" and 2 degrees. Now we can see the boss and hole have the draft angle required for the side core to release properly, and since the boss on the other side is mirrored it will also be correct. Cancel the "**Draft Analysis**" to continue.

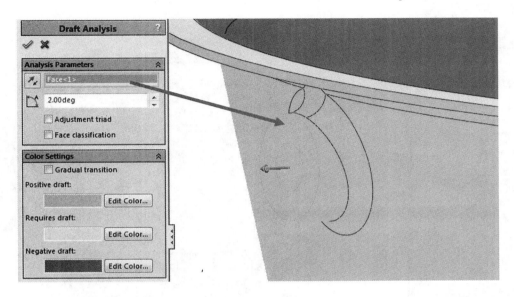

Using this face for the direction of pull will tell us that there are faces that need draft, but since we are only interested in the faces that belong to the boss, using this face for direction of pull the side core has the required draft and it will work as expected.

581. – Add a "**Parting Line**" using the *"Top Plane"* or the top most face for "Direction of Pull" and click OK to continue.

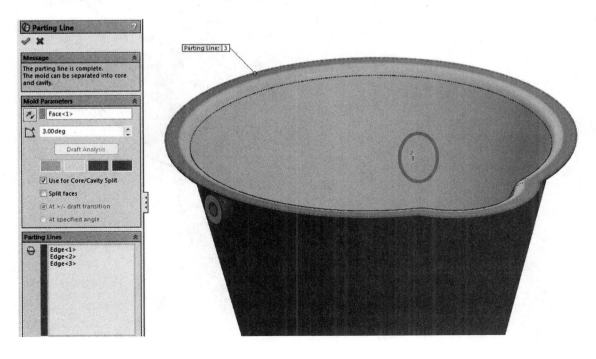

582. – Add a "**Parting Surface**." Make the surface 2 inches "Perpendicular to pull" using the "Smooth" option. This option smoothes out the transition between adjacent surfaces, giving us a better result in certain situations like this.

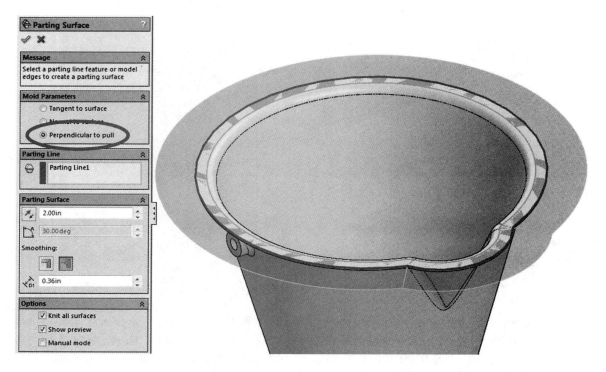

583. – Using the "**Tooling Split**" command, add the sketch in the parting surface; using "**Convert Entities**" project the parting surface into the sketch.

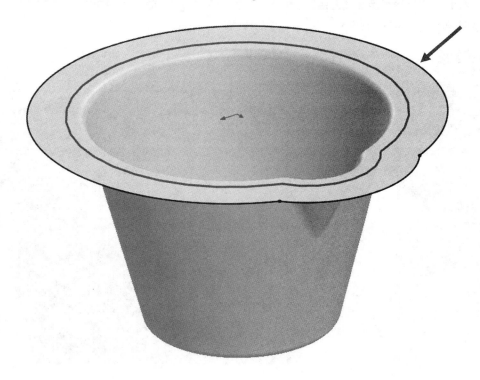

Exit the sketch to continue the tooling split and make the block size 1″ up and 13″ down as shown to fully enclose the bucket. Click OK to finish. If needed or wanted, hide the *"Parting Line1"* feature.

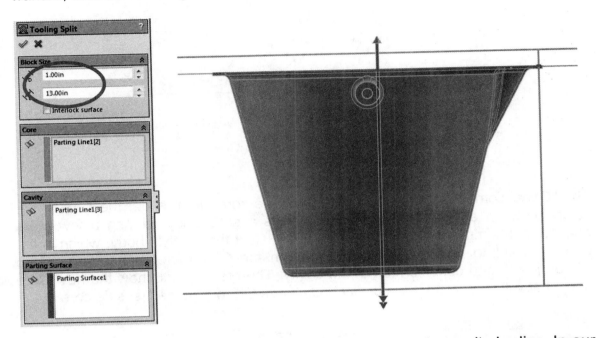

584. – After creating the tooling split we have a core and a cavity bodies. In our part the side cores will be made by splitting them from the cavity. Switch to a Front view and hidden lines visible mode to see the area we are interested in. Add a new sketch in the *"Front Plane"* as shown. The sketch is a closed profile starting at the parting line. The objective is to split a side core from the cavity body using this sketch as a cutting tool. Exit the sketch when done to continue.

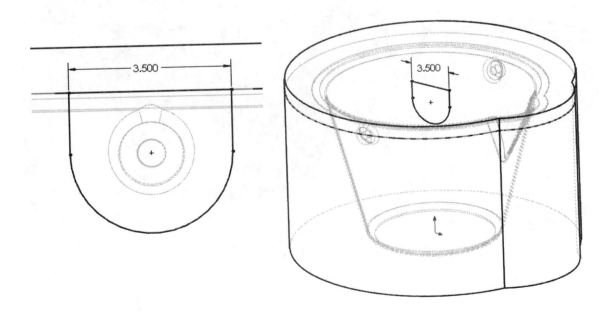

585. – Select the "**Core**" command from the Mold Tools toolbar or the menu "**Insert, Molds, Core**," and select the sketch we just made.

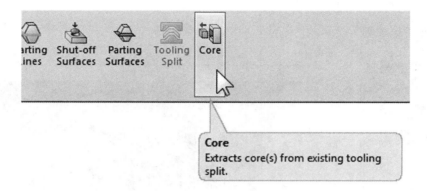

The "**Core**" command allows us to cut the side cores from either the core or the cavity body; in essence, this command splits a solid body offering a few more options than simply using a split command. Select the cavity's body, which is the one we want to make a core from, and in the "Parameters" options use the "Through All" end condition to cut the cavity. The preview will show the new Core body that will be split from the cavity body. Click OK to split the first side core.

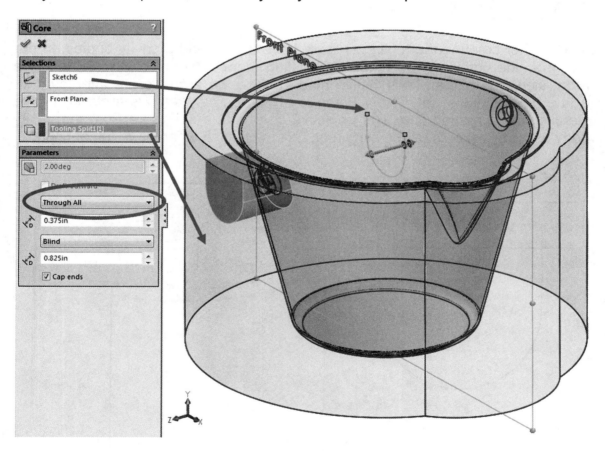

The core is split and added to a *"Core bodies"* folder in the *"Solid Bodies"* folder.

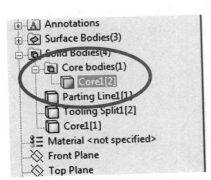
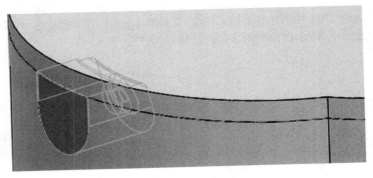

586. – Make the same sketch again in the *"Front Plane"* and make a second core, but in this case going in the other direction. When done, click OK to finish the second core. A sketch used in a Core feature cannot be reused for other features; that's why we need to make a new sketch.

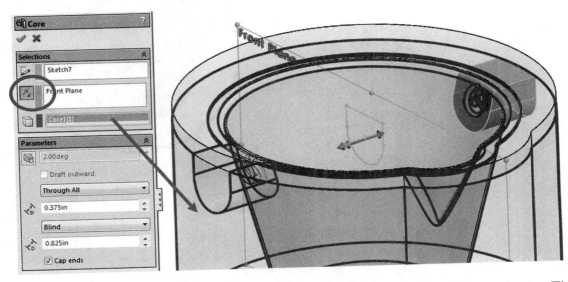

The two side cores are highlighted after splitting them from the cavity body. The bucket, all surface bodies and the main Core solid body have been hidden.

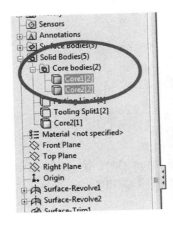
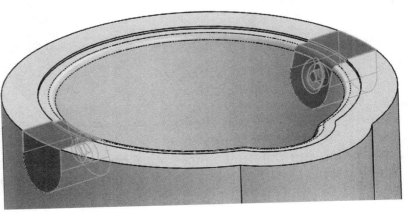

587. – Hide all the surface bodies, show the solid bodies and use the "**Move/Copy**" command (menu "**Insert, Features, Move/Copy**") to separate and see the different bodies of the mold including the side core. The bucket's body color was changed for visibility.

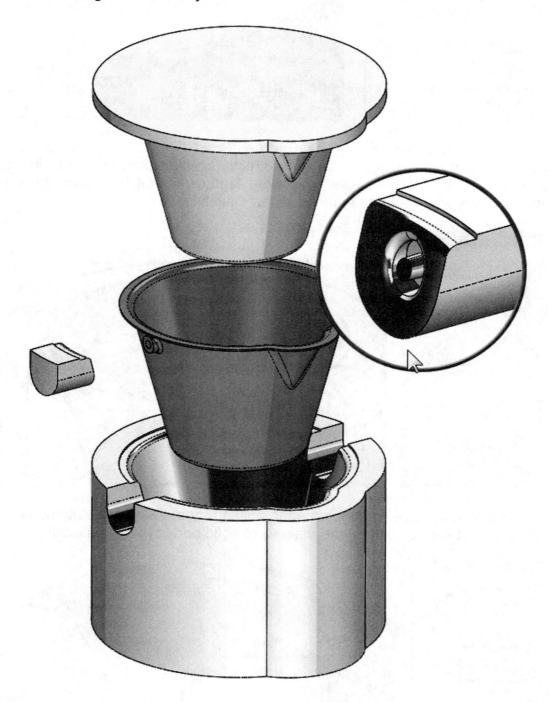

At this point, the different bodies can be inserted into a new part to add the remaining mold specific features to each one. Save the part and close.

CHALLENGE: Save the core, cavity and side cores to external files and make an assembly with all the components.

Exercise: Open the provided part *'Remote Control Master.sldprt'*. Save the top body and the bottom body each to a new part, make a tooling split for each half of the remote control, and create side cores for the trapped areas of the mold.

'Remote Control Master' (Provided File, 2 solid bodies)

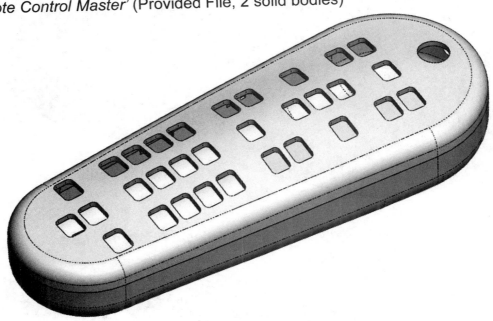

Completed parts and molds

'Remote Control Bottom'

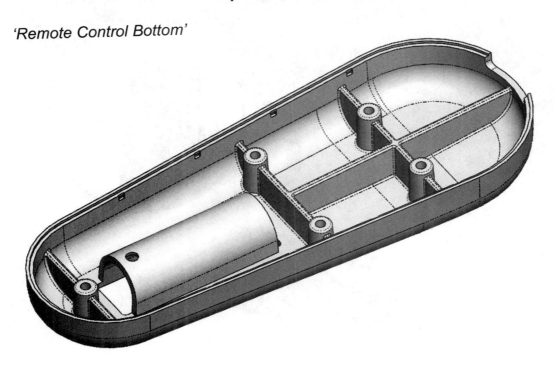

'Remote Control Bottom Core'

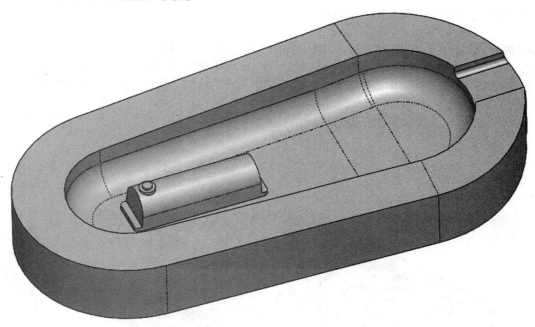

'Remote Control Bottom Cavity' and *'Remote Control Bottom – Insert'*

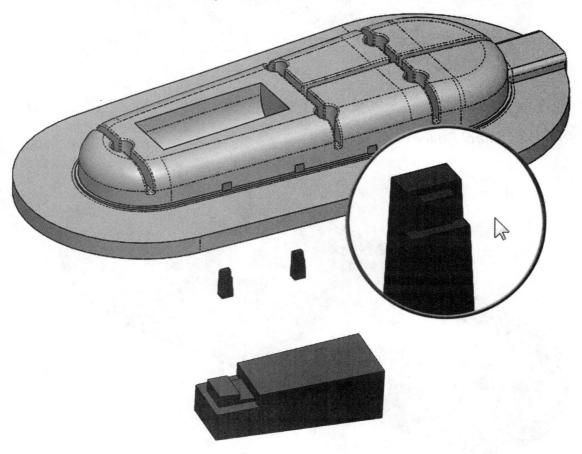

'Remote Control Top'

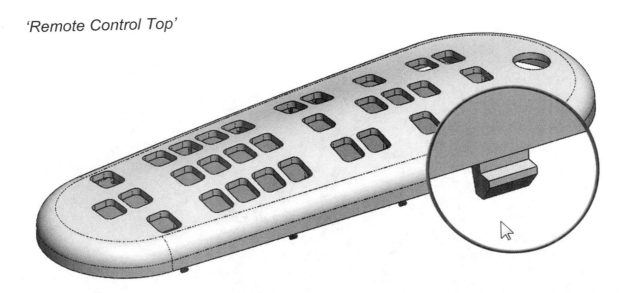

'Remote Control Top Cavity'

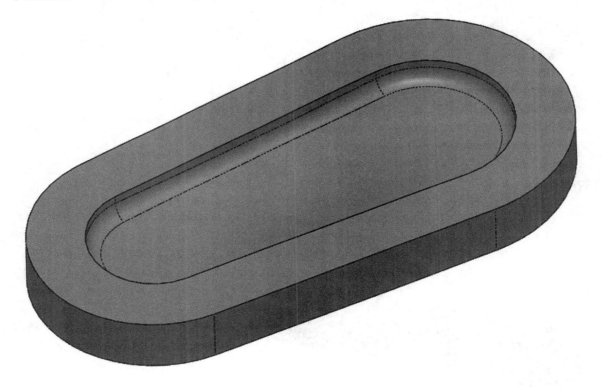

'Remote Control Top Core'

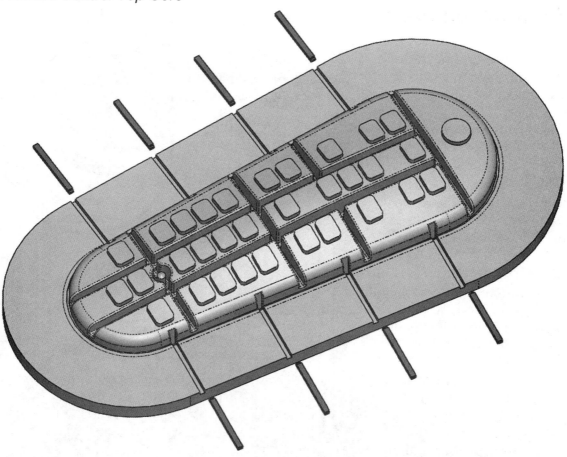

'Remote Control Top Core-Insert'

Final Comments

After going through the exercises in the book, the reader will have acquired a good understanding of several topics including sheet metal, welded structures, 3D Sketching, Top Down design, Editing tools, Multi Body parts, Surfacing, Mold tools and a handful of tips and tricks that can help a designer work faster and more efficiently.

Even though most of the topics included in this book are usually considered advanced, they are presented using simple examples that break down the concepts and make them easy to understand. Topics covered explore the most commonly used options to enable a reader to use them efficiently for the different design tasks at hand.

The Sheet Metal, Mold tools and Welded structures topics fall within the scope of the individual manufacturing trades, each of which has very specific tools, techniques and *fine details* that are (usually) learned after studying and by experience; take for example mold making: includes most of the tools available in SolidWorks, going from the very basic to advanced surfacing and more.

The amount of trade training that can be shown in a CAD oriented book is limited, since the objective of the book is to teach how to use SolidWorks and not a trade. If a reader has experience in one of these fields, he/she will quickly understand the purpose of specific tools, but we also tried our best to explain in simple terms the terminology and processes, so that the reader unfamiliar with a trade can understand the purpose of a tool and the reason for certain operations.

We welcome all of your suggestions and ideas. As this book evolves new functionality and examples will be added, but rest assured we'll always try our best to make sure the models shown maximize the functionality covered in as few pages as possible, helping you to understand and master each topic. If you have any comments about this book, by all means please send it our way, we try to reply to all email in a timely fashion, and we are always happy to hear from our readers.

Please visit our website as we have other materials available that we are sure will be of interest to you.

Sincerely,

Alejandro Reyes,
Certified SolidWorks Professional, Instructor and Support Technician.

alejandro@mechanicad.com
Mechanicad Inc.

Notes:

Index

Notes: